Wiring a House

Wiring a House

4th Edition

REX CAULDWELL

The Taunton Press, Inc., 63 South Main Street,
PO Box 5506, Newtown, CT 06470-5506
e-mail: tp@taunton.com

Copy editor: Seth Reichgott
Indexer: Jay Kreider
Technical Reviewer: Cliff Popejoy
Jacket/Cover design: Alexander Isley, Inc.
Interior design: Lori Wendin
Layout: Cathy Cassidy
Illustrator: Gayle Rolfe
Photographers: Rex Cauldwell unless otherwise noted

Library of Congress Cataloging-in-Publication Data
Cauldwell, Rex.
Wiring a house / Rex Cauldwell. -- 4th ed.
p. cm.
Includes index.
ISBN 978-1-60085-261-9
1. Electric wiring, Interior. 2. Dwellings--Electric equipment. I. Title.
TK3285.C38 2009
621.319'24--dc22
2009009136

Printed in the United States of America
10 9 8 7 6 5 4 3 2 1

About Your Safety: Working with electricity is inherently dangerous. Using hand or power tools improperly or ignoring safety practices can also lead to permanent injury or even death. What is safe for one person under certain circumstances may not be safe for you under different circumstances. Don't try anything you learn about here (or elsewhere) unless you're certain it is safe for you. Please keep safety foremost in your mind.

Acknowledgments

No one person creates a book. It is the accumulation of many hands and minds. I would like to thank Julie Trelstad, who guided me through the conception of the original edition many years ago, and all the people at Taunton who have worked so hard to make this book the great success that it has been over the years.

It should be noted that a work of this kind is very technical, and it is very easy for errors to slip by editors even though a great deal of effort has been taken to find such. If you note any problems, have questions, or would like something added or changed, you can email me at ltmtnele@yahoo.com. It should also be noted that the way houses are wired on the East Coast may not be the same as on the West Coast and that all areas do not adopt the same exact code. Even under the NEC, I have to wire differently in each county I work in.

Many companies helped with this book by providing photographic material and technical information about their products. I would like to give heartfelt thanks for their help to Jim Gregorec of Ideal Industries, Inc.; Vivian Beaulieu of Marchant and Field; Ben Bird of Certified Insulated Products; Doug Kirk of King Technology, Siemens; Constance Malenfont of TayMac Corp.; Jean Miskimon of DeWALT public relations; M. F. Reed of Tytewadd Power Filters; and Jay Thomas of AFC Cable Systems. I would also like to thank the good people at Carlon®; Bussman; Greenlee Textron, Inc.; Hunter® Fan Company; Hubbell® Incorporated; Lutron®; Magnus® Industries, Inc.; and Technology Research Corp.

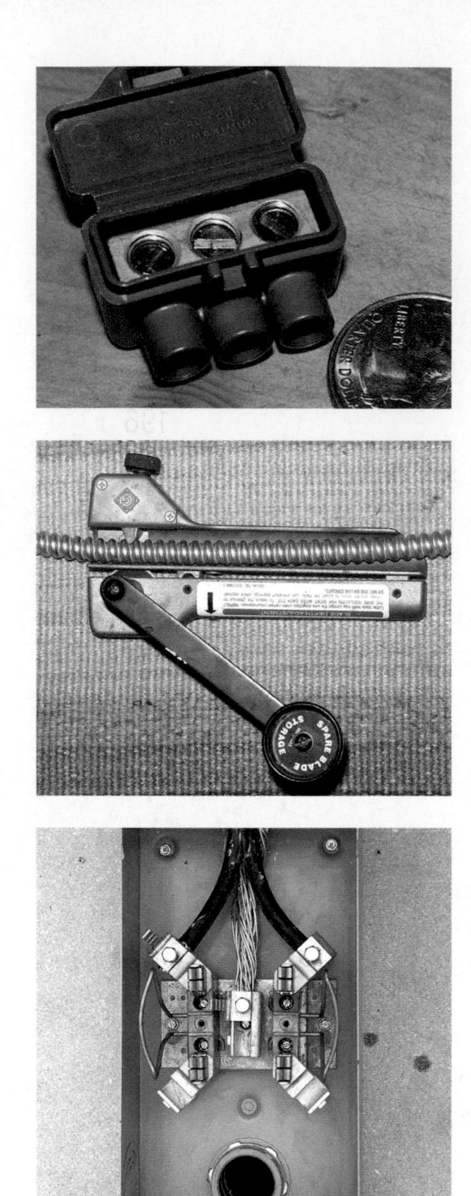

Contents

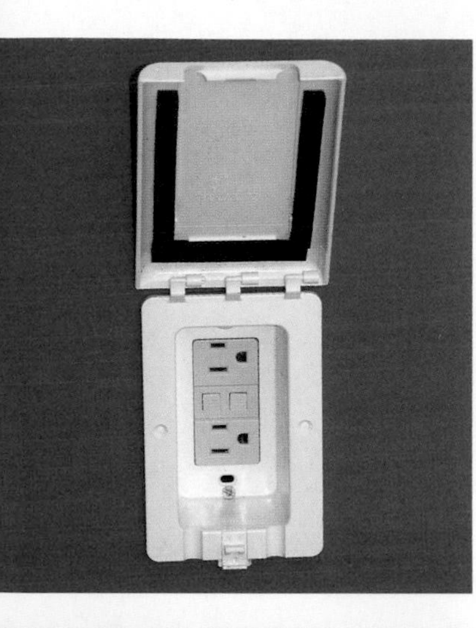

PRIME
TRANSFORMER
POWER/SURGE/GROUND
ALARM TEST
TRANSFORMER
TRANSFORMER
TELEPHONE LINE
OUT IN OUT

Preface to the Revised Edition

This new edition takes into consideration all of the latest code changes. If you haven't wired a house in a decade or two, there are a few code changes you should be aware of. (Note that states can adopt codes at different times and may exempt sections they don't agree with.)

One change was in marking the white insulated wire in switch legs. At the time *Wiring a House* was originally written, code did not require the installer to indicate whether a white insulated wire, such as a traveler in a 3- or 4-way switching circuit, was not a neutral, or whether a common switch was using a white wire for its incoming power. Code now requires the white insulated wire to be taped (any tape color except white, gray, or green) to make sure you know that it is not a neutral.

There is the new rule on bathroom receptacles. Now you can put a whirlpool tub, a bath heater, a heated towel rack, or bath lights on a single bathroom receptacle circuit as long as that circuit is not overloaded and stays in that one bath.

Common 125-volt 15- and 20-amp circuits in outbuildings, like workshops or sheds, need GFCI protection. The wiring to the outbuilding (multicircuit) should be done with four conductors with an added ground rod at the panel.

To keep children from grabbing appliance cords and pulling the appliance down onto themselves, all countertop receptacles are now required to be above the countertop and cannot be installed on a horizontal surface. Exceptions are made for peninsulas and islands designed for handicapped persons.

Installers are now allowed to use nonmetallic (NM) cable in residential installations at any grade level, including basements and attics. Also, common 15- and 20-amp 125-volt receptacles are now required to be tamper resistant (shutters built into the receptacle). AFCI requirements have now evolved to encompass the entire house except in GFCI areas, such as garages, kitchens, baths, and laundries. And when you install these same receptacles in wet and damp locations (such as outdoors) they must now be rated weather resistant (special receptacles)—even GFCIs. You might want to verify that your locale has adopted these changes before you wire.

All 15-amp and 20-amp 125-volt receptacles installed outside (you are required to have one in the front and one in the back) must utilize waterproof covers that keep the receptacle waterproof even when a cord is inserted.

Above Code

Above Code is my way of ensuring that the homeowner has a safe, high-quality, trouble-free, long-lived electrical installation. An example of using Above Code is in wiring bathroom receptacles. If you were to simply "meet code," you could load up a bath receptacle circuit with the lights, fan, heater, and so on. A single hair dryer on a low setting would wind up kicking the breaker and leaving you in the dark. In fact, you could even put the bath lights on the GFCI circuit so that the lights would go out when the GFCI tripped and you would literally be walking on a wet bath floor in the dark—and all by code. My Above Code system shows you how a bath should be wired so that doesn't happen. In addition, my system tells you how to wire better, what products to buy, and which ones to avoid.

The following list is dedicated to all those who have requested a singular list of the Above Code sections of the book so you wouldn't have to thumb through and write them all down—as well as some common-sense code requirements. I didn't previously do this because I really didn't expect my method of wiring to become so popular—assuming most of the populace would prefer to wire the cheapest way possible no matter what I suggested. So to all those who have calluses on your thumbs from turning pages, I dedicate this list to you.

- Splices outside must use silicone-filled wire connectors. Buy them prefilled, or fill them yourselves.
- A subpanel is to have its own main breaker.
- An outside 125-volt receptacle, if not a GFCI receptacle, must be of the heavy-duty type (around $3 to $5), preferably nylon. This is to prevent it from becoming cracked or broken, as happens when heavy-duty extension cords are wiggled in them.
- Do not put the overhead bath light on the load side of a GFCI.
- Smoke alarms—For new installations, remember that an AFCI will cut off power to the bedroom alarms if the AFCI trips due to a fire, arc, or overload. Verify that all your smoke alarms have battery backup and that the batteries are fresh.
- Wire gauge—Use nothing but 12-gauge cable for all the 125-volt receptacle and switch wiring with the following exceptions. If you are on a budget, you can use 14 gauge on the three-way lighting circuits due to the extreme cost of the cable. This means that all the 14-gauge three-way lighting is to be kept away from the 12-gauge cable and on a 15-amp circuit. If you have the smoke alarms on their own circuits, you can use 14-gauge wire.
- Each bathroom 20-amp, 125-volt receptacle will have a dedicated GFCI circuit with nothing else on that circuit.
- Service conductors should be copper. This makes for easier wiring, as they can be two gauge sizes smaller than aluminum.

- Splice single-strand aluminum conductors to single-strand copper conductors with the new covered 3-hole bus strips that are sold in home centers.
- Have a switch (for the overhead light) at *all* the entrances and exits of the kitchen.
- Watertight-while-in-use covers—Install for all the outside receptacle outlets.
- Color coding of the cable sheath (for non-colored-coded sheaths)—Either buy color-coded sheath cable or paint your own (side only).
- Receptacles to be of high quality—No Quickwire or Speedwire type of receptacles where the conductors are pushed in small holes in the back of the receptacle/switch and released by inserting a small object into a release hole.
- In damp areas orient the splice connecter downward.
- Bath/fan light or light-only fixtures need to accept incandescent bulbs with standard screw-in bases—not candelabra or other style bases.
- Be sure all switches are grounded (code).
- Do not put the kitchen lights on the kitchen receptacle circuit by accident. What this means is that it is against code but commonly done knowing the inspector won't find it.
- Avoid fluorescent lights that don't have "outside light"–type bulbs.
- Avoid putting 4-ft. fluorescent lights anywhere. Avoid any fluorescent in the garage, as they do not work well in cold temperatures.
- Use three-conductor cable on all the three-way switch circuits, even those with electronic dimming.
- Bring power for all switch circuits into the switch—not the overhead light or fan in the bedroom—and bring a three-conductor cable from switch to overhead light or fan light.
- Always bring the power cable into a switch or receptacle box on the upper left side.
- All splices are twisted together before the wire nut goes on unless you are using push-in splice connectors.
- Minimum service is 200-amp 40/40 panel.
- Circuit panel is Siemens® or other high-quality unit with the same specs.
- When possible, install receptacles with the hole for the ground pin on top.
- Wire strippers are to be used to strip wire, not knives or pliers.
- Dimmers on incandescent lights.
- Consider light switches adjacent to bedside.
- Consider spot or track lights over bed.
- Cannot put the bath light on the bath circuit.
- No multiwire circuits.
- Use insulated staples.
- No handy boxes.
- Use all deep boxes within walls for receptacle and light switches.
- In lightning-prone areas, install surge protection in layers: surge breakers in panels, hardwire protection at on/off panel boxes, at wells if needed, and point-of-use surge strips at loads.

- For lightning-prone areas use my Above Code grounding system.
- Wire a common free-standing stove with #6 copper fused at 50 amps.
- Take the design of a flat appliance plug into consideration: Is the cord end going up or down? Flip the receptacle as necessary as to keep the plug cord from folding over the plug end. Typically this is at the washer and fridge.
- Even if your main house water heater is 3,500 watts, wire it with 10 AWG to allow for a future 4,500-watt upgrade.
- For overhead bedroom lights, always use metal "fan approved" boxes even if a fan is not being installed.
- For overhead paddle fans, always use metal boxes—not plastic ones.
- Always use full-size breakers.
- For 200-amp service, always use a 40/40 panel (this is considered a full-size panel). The exception would be when you add a second panel to the first 40/40 and that panel has high-amperage breakers. If your high-amperage breakers pull close to 200 amps with only a few breakers, getting a physically large panel is a waste of money.
- Always opt for overhead lighting instead of lights or lamps plugged into switched receptacles.
- If the basement floor is bare concrete, have all general-purpose receptacles ground faulted, even if the area is considered finished.
- Use screws, not nails, to attach the service panel.
- Leave just enough sheath on the cable as it enters the service panel to identify what it goes to. If you are grouping cables as to function (for example, lighting circuits), use colored tape to identify the group.
- On the back of each switch and receptacle plate, write the breaker number that controls it.
- Use only 2-in. or larger galvanized pipe as masts—smaller-diameter pipes bend too easy.
- Jump from 200 amp to 300 amp to 400 amp to 500 amp to 600 amp. Do not use amperages ending in 25, 50, or 75.
- For 300-amp service, use one 200-amp panel and one 100-amp panel. For 400-amp service, use two 200-amp panels. For 500-amp service, use two 200-amp panels and one 100-amp panel. For 600-amp service, use three 200-amp panels.
- Do not use cheap electrical tape. Use Scotch® Super 33+ (7 mil thick) or Super 88 (8.5 mil thick).

I am pleased that *Wiring a House* has become a standard reference for engineers, pro electricians, and DIYers. I believe this new edition will prove to be a helpful resource for answering those head-scratching questions that often plague wiring jobs. This new edition contains a large number of new photographs and drawings, in addition to a completely new chapter on AFCIs. New buyers, I hope, will find this edition to be their "real-life" wiring reference—the one that develops dog-eared pages. For those who already own a copy of the earlier edition, I hope you will find this book even more useful and will add it to your library as well.

—*Rex Cauldwell*

Safety First

Safety is the responsibility of anyone who is working with electricity. Current flows most easily along the path of least resistance, but it will follow any path that's available, and that includes you.

Think first. If you're not sure that what you're doing is safe—or if you're not sure what you're doing—stop.

Never work on an energized circuit. Before doing any electrical work, shut off the power and use a voltage tester to make sure the power is off. It's important that you test it yourself—don't depend on circuit breakers, fuses, or someone else.

Wear appropriate gear. Always wear safety gear, including rubber-soled shoes, gloves (if working with cables, wires, or metal boxes), safety glasses, and a dust mask (when sawing or working overhead).

When possible, work with one hand. Electric shock can be fatal when the current path is from one hand to the other. This is because the current passes through the heart. You can likely survive a severe shock between a hand and foot that would cause death if the current path was from one hand to the other.

When possible, don't work alone. When working in a hazardous location, it is best to have someone with you who can turn off the power or give CPR if needed.

Learn first aid. Because of the inherent dangers in using power tools and in working on electrical equipment, make an effort to learn first aid, particularly CPR. At the very least, have a first aid kit and mobile phone close by.

Check your community's requirements. While this book is based on the requirements of the National Electric Code, many jurisdictions adopt different codes from different years and can add or delete specifications. Before undertaking any electrical work, be sure to check with your local building department for its most current requirements.

The Basics

Not knowing the basics of electricity can get you into a lot of trouble—painful trouble. I remember when my Uncle Bud started teaching me about electricity back in the 1950s. I accidentally tripped on some black wires in the attic and got shocked as I grabbed onto a funny-looking porcelain knob to stop my fall. This was my first lesson in basic wiring: Don't touch this and don't trip over that.

Today, the rules are a little more complex than those by which my Uncle Bud and my Grandfather Bunt worked. (My Grandfather Bunt didn't even have a meter to test for voltage. He used an old-time method that I can't print here for fear that somebody might try it and get electrocuted.) Although they didn't have the modern test equipment that we do, they talked the same electrical language, using terms such as "current flow," "voltage," "power," and "resistance." But I'm sure they didn't understand their meanings as well as we do today. In this chapter, I'll help you make sense of the basics of electricity so you understand what you'll be doing and why.

How Electricity Flows

Wire works much like a garden hose, but instead of conveying water, it conveys electricity from one location to another. When you turn on a hose faucet, water entering from the spigot pushes on water already in the hose, which pushes water out the other end. Electricity flows in much the same way. An electron flows in one end of the wire, which knocks an electron, which in turn knocks another electron, until an electron eventually comes out the other end.

All electrical current originates from and comes back to the three side terminals of the utility transformer.

The water analogy can be used to describe the other elements of electricity. To get water to flow, we need water pressure. To get electricity to flow, we need electrical pressure. Electrical pressure, or voltage, can be provided from either an electrical utility or a battery. And just as greater water pressure means more water flow, higher voltages provide greater electrical flow. This flow is called "current." With both water and electricity, the diameter of the hose or wire limits what you get out of it in a given amount of time. This flow restriction is referred to as "resistance."

Elements of electricity

Electricity is the organized flow of electrons along a conductor. It is generated through heat, pressure, friction, light, chemical action, or magnetism. Electricity is a complicated phenomenon, but for an "everyday" understanding, it's helpful to know the meaning of these four terms: voltage, current, resistance, and power (watts) and their relationship to each other. Ohm's Law, a basic principle of electricity, states that voltage is directly proportional to the current and resistance. You can calculate any one value if the other two are known ($E = IR$, where E is the voltage, I is the current, and R is the resistance).

Voltage is the electrical pressure provided by a battery or other power source. The higher the voltage, the more current the source can produce. Voltage is measured in volts; a voltage quantity is always followed by a capital V or the word "volts" (120V or 120 volts). A multimeter measures the voltage between two points on a circuit, or if the current and resistance are known, it can be calculated using Ohm's Law.

The Main Wiring Runs in a Residence

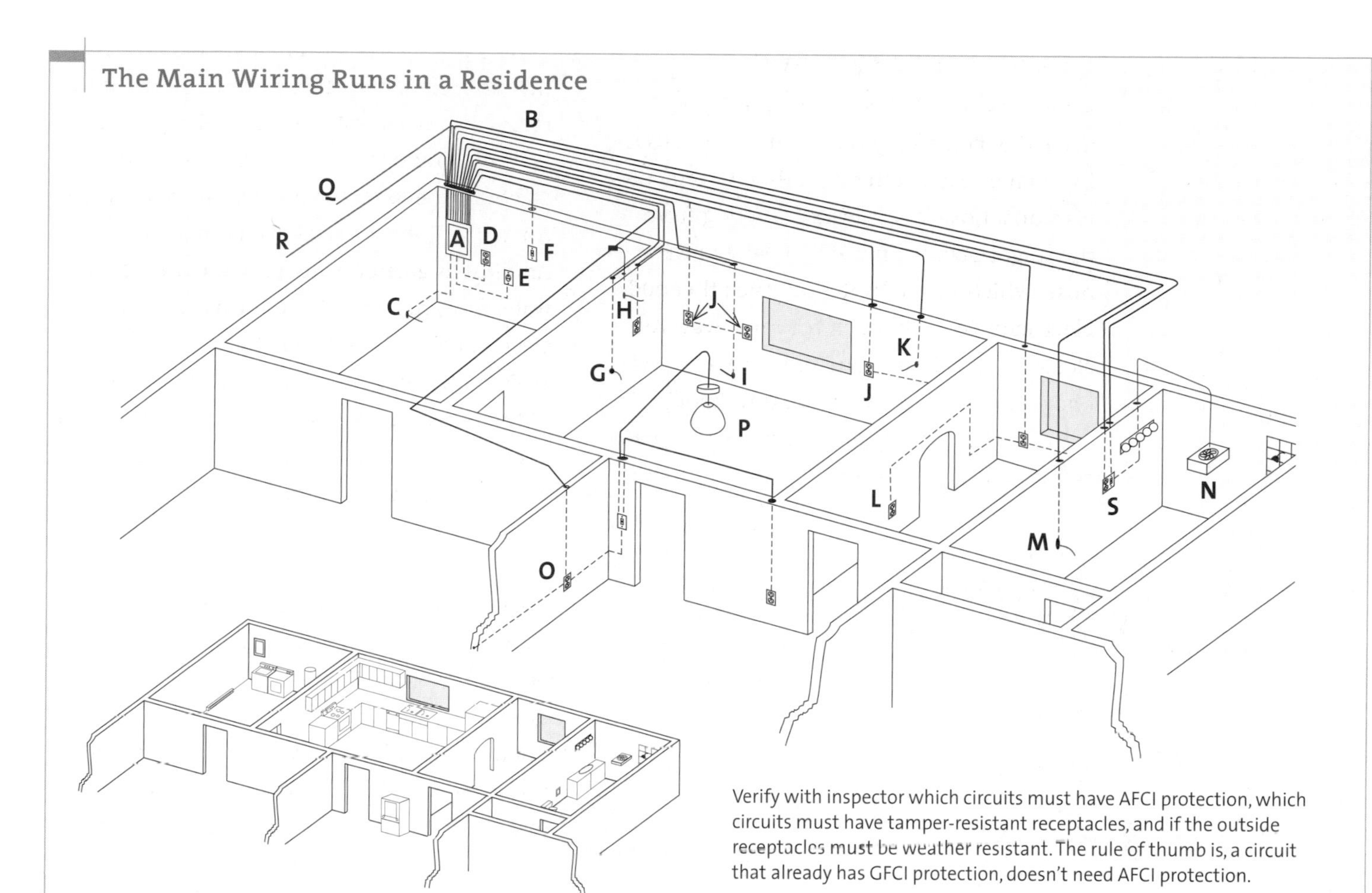

Verify with inspector which circuits must have AFCI protection, which circuits must have tamper-resistant receptacles, and if the outside receptacles must be weather resistant. The rule of thumb is, a circuit that already has GFCI protection, doesn't need AFCI protection.

A. Main service panel

B. "Home-run" cables from main panel to loads (run through attic or ceiling or through basement)

C. Cable for baseboard heater (dedicated circuit)

D. Utility-room receptacle (dedicated circuit)

E. Dedicated circuit for dryer receptacle

F. Cutoff switch (optional when in sight of panel) for water heater (dedicated circuit)

G. Range cable (dedicated circuit)

H. Range fan fed off living-room circuit

I. Dishwasher cable (dedicated circuit)

J. GFCI-protected receptacles on countertop (2 circuits)

K. Refrigerator cable (dedicated circuit optional)

L. Cable for dining-room receptacles

M. Cable for in-wall heater (dedicated circuit)

N. Bath light and fan

O. Living-room circuit

P. Kitchen overhead light powered off living room circuit

Q. Cable for smoke alarms and 3-way light switching

R. Circuit for outside GFCI protected receptacles if they don't tap into living room, garage, bedrooms, or other such rooms. Never tap into the kitchen/dining, bath, or laundry room circuits.

S. GFCI bath receptacle (dedicated circuit)

Inside the Utility-Pole Transformer

A utility-pole transformer has an insulated terminal on top, which connects a high-voltage line from the utility company to a primary coil (see the drawing at right).

The primary coil is connected to ground/neutral via a center tap, which then transfers its energy into a secondary coil to step down the voltage from 7,500 volts to 240 volts. Grounding the middle of the secondary coil splits 240 volts in half, resulting in two 120-volt sources from one 240-volt source.

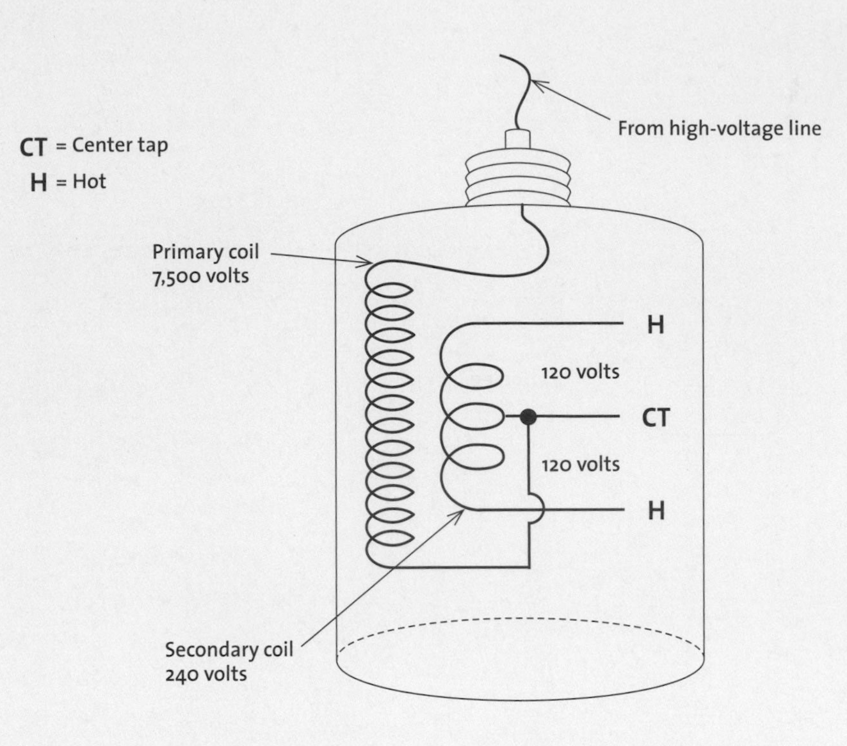

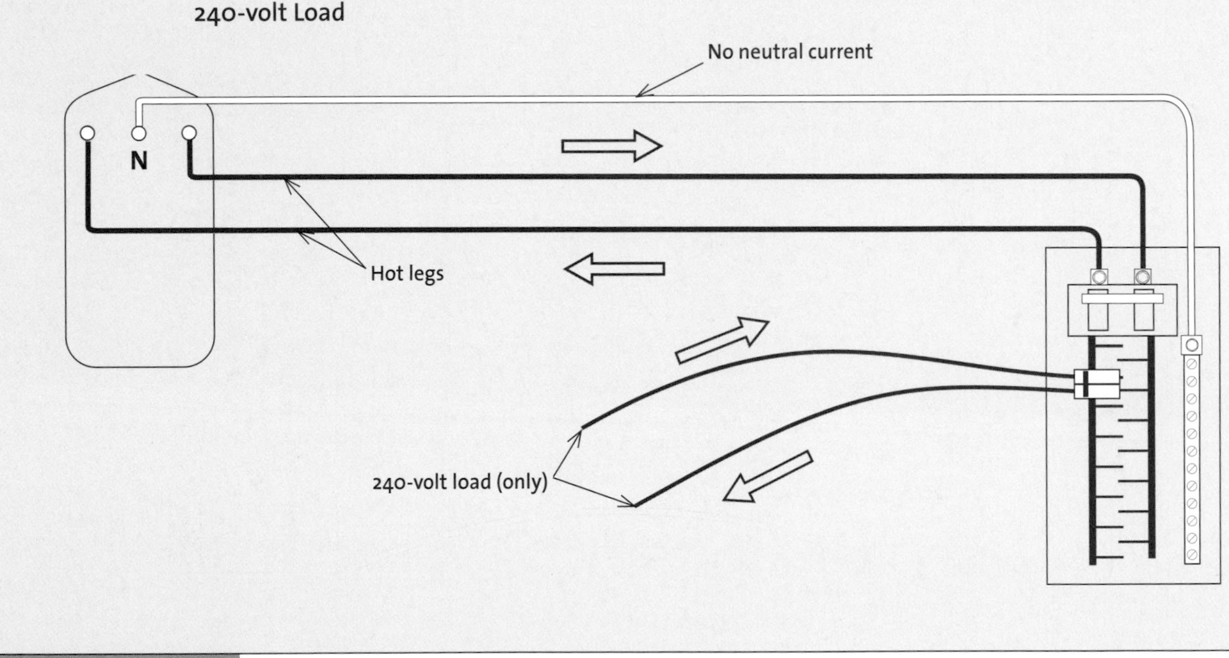

Current is a measure of the quantity of electrons flowing in a conductor. Electrons orbit around the nucleus of atoms just as planets orbit the sun. When an electron is knocked out of orbit and starts traveling down the circuit, it knocks other electrons out of their orbits, creating current. There are two types of current: alternating and direct.

Current can be increased by raising the voltage or by lowering the resistance. Current

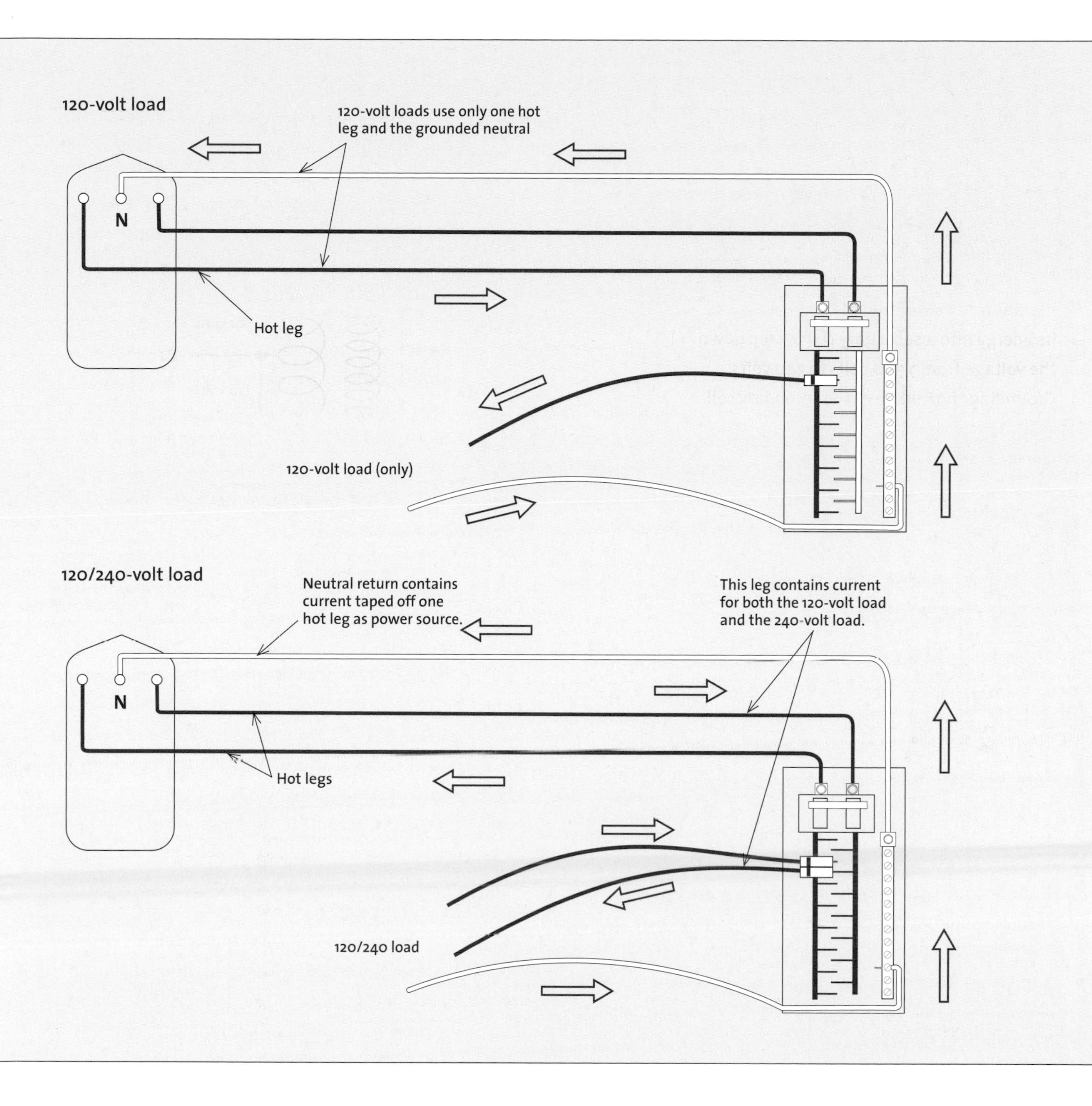

is measured in amperes, or amps (e.g., 20 amps). Electrical current can be measured by inserting an ammeter in series with the wiring (cutting the wire and attaching a meter lead to each wire end), but it is much safer to use a clamp-on meter. This type of meter clamps around the wire and measures the electromagnetic waves emitting from the wire to determine the current. If the voltage and resistance are known, current can be calculated using Ohm's Law: I = E/R.

Resistance is the inherent physical opposition to current flow (like a dam trying to hold a river back). The opposition is created by electrons refusing to be stripped from their atoms and sent down the wire. The higher the material's resistance, the more energy it will take to strip off the electrons. The higher the resistance, the less current flow. The only way to increase the current in a circuit with a given resistance is to raise the voltage, which increases the electrical pressure.

What "Neutral" Means

The neutral conductor is the conductor that leads back to the center tap of the utility transformer. It is connected to earth at two locations: the utility transformer and the main panel. That is why it is sometimes referred to as the grounded neutral conductor. Its designation is white or a white tape on a black cable.

Neutral current is commonly referred to as return current, because once the current passes through the load, the white insulated neutral conductor makes the complete circuit back to the utility transformer. All electrical current must make a complete loop. The current starts at the utility transformer and must therefore return to the transformer; the neutral is simply the return part of that loop.

The neutral cable or wire can kill you just as readily as a "hot" cable. The same current that flows in a hot conductor flows out the neutral. If you place yourself in series within that loop, even on the neutral side, you will be electrocuted.

Resistance is measured in ohms; a number quantity that is always followed by the Greek letter omega (Ω) or the word "ohms" (30 Ω or 30 ohms). Resistance is measured by a volt-ohm meter (VOM) or a multimeter, or it can be calculated if the voltage and current are known: R = E/I.

Resistance can also be the property of an appliance that converts electrical energy to heat energy. For example, let's take a common 240-volt, 4,500-watt water heater that's pulling 18.75 amps. As the 18.75 amps flow into the water-heater element, the element produces heat as it opposes the electron flow (the element has a resistance of 13 ohms). This opposition produces heat—exactly 4,500 watts worth—which turns the cold water hot.

Power is the product of current and voltage. This is listed in watts or volt amps on all appliances. Volt amps come from the formula P = volts × amps, or power equals voltage times current. Practically, we see power as the dissipation of heat, as in the water-heater example. All appliances dissipate heat or provide energy; thus, you will always see the wattage listed on the appliance.

Common VOMs and multimeters cannot measure wattage (although commercial units can), but it can be calculated if two of the following are known: voltage, current, or amps.

1. If voltage and current are known, use P (watts) = EI. For example, if E equals 120 volts and I equals 10 amps, the power is 1,200 watts (120 × 10).

2. If current and resistance are known, use $P = I^2R$. For example, if I equals 10 amps and R equals 12 ohms, the power is 1,200 watts (100 × 12).

3. If voltage and resistance are known, use $P = E^2/R$. For example, if E equals 120 volts and R equals 12 ohms, the power is 1,200 watts ($120^2/12$).

DC and AC

Electricity that flows in one direction is called direct current (DC). Naturally occurring examples of DC are static electricity and lightning. Although DC generators exist, DC is normally chemically generated, such as from a battery. The current flows from its negative terminal, through a load (a light or other appliance), and then to the positive terminal. DC drives our cordless tools and appliances.

Alternating current (AC) is the workhorse of the modern electrical industry. The voltage alternates back and forth in what is called a sine wave, and unlike DC, it can be easily produced and distributed by generators. Thus, AC is what we use throughout our houses to run our plug-in appliances.

Although AC originates at power plants and is distributed around the country via high-voltage transmission lines, as far as the house is concerned, electrical power starts and stops at the utility transformer—specifically the three terminals on its side. The two end terminals are the ends of the secondary winding (the windings step down the utility voltage from several thousand volts to the 120/240 volts used in your home). The center terminal, or center tap, is connected into the winding halfway between the two.

The two outside terminals produce 240 volts. Current leaves one terminal, goes to the house (meter base, service panel, and load), and comes back to the other terminal (then reverses). Unlike 120-volt circuits, 240-volt circuits don't use the neutral (center) tap. No neutral is required to operate 240-volt-only loads.

For 120-volt circuits, current leaves one end terminal and comes back via the cable that connects to the center tap. The center tap is the grounded neutral. The voltage from one end terminal to the center tap is called one phase, or leg, of the incoming power.

The opposite end transformer terminal also sends current into the house, and that current also comes back via the neutral to the center tap. This is the second phase, or leg, of the house power. Each phase represents a hot bus in the service panel.

For loads such as electric stoves and electric dryers that require both 120 and 240 volts, we have different load currents simultaneously flowing on both hot conductors and the neutral. Here, if the 120-volt current is "in phase" (that is, going in the same direction along the neutral as the 240-volt current), then the two currents in the neutral are added together. If the 120-volt current is "out of phase" (going in the opposite direction) with the 240-volt current, then the two currents are subtracted. This is why the neutral cable can be smaller than the hot cables.

A Residential Electrical System

The house wiring begins at the service entrance (SE)(see the drawings on pp. 10–11). In this section, I want to show you a typical house wiring system. I'll go into more detail about all these circuits in later chapters. It's also important to remember that different houses require different setups. I've already given you some background on where the power comes from. Now let's see how it all gets hooked up inside the house.

Let's go back to the service drop, or triplex (this example is an aerial feed). The utility wires splice onto the service-entrance wires next to the house at a drip loop (the section of power line that hangs down slightly before it goes into the masthead). The utility's responsibility stops at the drip loop; from there, the owner's begins. The homeowner pays for problems on the house side of the drip loop. The utility pays for problems on the utility side. The masthead, which looks like a metal hood on top of the mast and points down, and the drip loop, which circles up, keep water from following the wires, flowing into and down the mast, and entering the meter base.

Once the service wires enter the masthead (also called a weatherhead), they travel down the steel mast (or riser) into the meter base. From there, the service-entrance cable goes directly into the main service panel. The main panel is where all wiring begins. From there, wires travel to various receptacles, lights, and appliances throughout the house. In addition, the main panel is where the house electrical system obtains its earth ground via the ground rods.

The drawing on p. 9 shows some of the main wiring runs: the utility room, the kitchen, the living and dining rooms, and the bathroom. These rooms are illustrated because each one has its own special requirements. Cables going through walls perform several different functions, but if needed, all of them can be routed together through the same holes in the studs. The cables in the utility room contain the 12-gauge utility-room cable (for the receptacle only), the four-conductor 10-gauge dryer cable, the 10-gauge cable for the water heater, and the 12-gauge cable for the baseboard heater. Armored cable (see p. 24) is used for the exposed run from the water-heater cutoff box to the heater. In the utility room, there is only one 120-volt receptacle—for the washing machine. Most utility rooms require more than one receptacle. Lighting for the utility room (not shown) is pulled off of the living room circuit.

The bathroom receptacle is powered off its own 12-gauge circuit. Ground-fault circuit interrupters (GFCIs) protect all of the outlets; the in-wall heater has its own dedicated circuit. All habitable rooms must have switched lighting at the point of entry. Receptacles and switches in the living and dining rooms are powered off of individual circuits for each room. The cable enters through the ceiling and connects to a receptacle, which feeds other receptacles and switches in the room.

The kitchen circuits include two 12-gauge small-appliance circuits, the lighting circuit, the 6-gauge (four-conductor) stove cable, the 12-gauge dishwasher cable, and the 15-amp refrigerator circuit, as required by code. Receptacles along the countertop are GFCI protected.

Not shown on the drawing are the three-way switching circuits and the smoke-alarm circuits that I would put on the 14-gauge circuit.

Wire Gauges

We use insulated wire or cable to deliver electricity to where we need it. It didn't take long to learn that the larger the wire's diameter, the more electricity we could deliver to the load and the less we would waste along the wire. The American Wire Gauge (AWG) provides numbers for the different wire diameters. The larger numbers are used to represent the smaller wire diameters. For example, a cable that has a big number, such as 18, has a smaller diameter than a cable with a small number, such as 1. Zero is called "ought"; a cable with one zero is written as 1/0, or one ought. And the system doesn't stop at zero for the largest diameter cable, either. No, that would be too simple. Numbers less than zero identify bigger cables. What's less than zero? More zeros. From 1/0 it goes to 00 (2/0), 000 (3/0), and 0000 (4/0). Each descending number indicates a larger diameter cable. To make things even more complicated, the system changes in midstream. As soon as you get past 4/0 (four ought), the diameter of the conductor is referenced by its cross-sectional area: 250 kcmil, 300 kcmil, and so on.

One of the biggest problems you will have in the field will be differentiating cable and gauges. It's sometimes hard to see a difference between cables with gauges that are only one or two sizes apart. Theoretically, the gauge number is supposed to be written on the cable's jacket, or sheath. Unfortunately, many cable manufacturers don't use paint; instead, they emboss the gauge number on the jacket. Embossed gauge numbers are almost impossible to see; consequently, many people wind up installing the wrong cable for the job. In my hometown, our new library had all of its 12-gauge and 14-gauge THHN (Thermoplastic

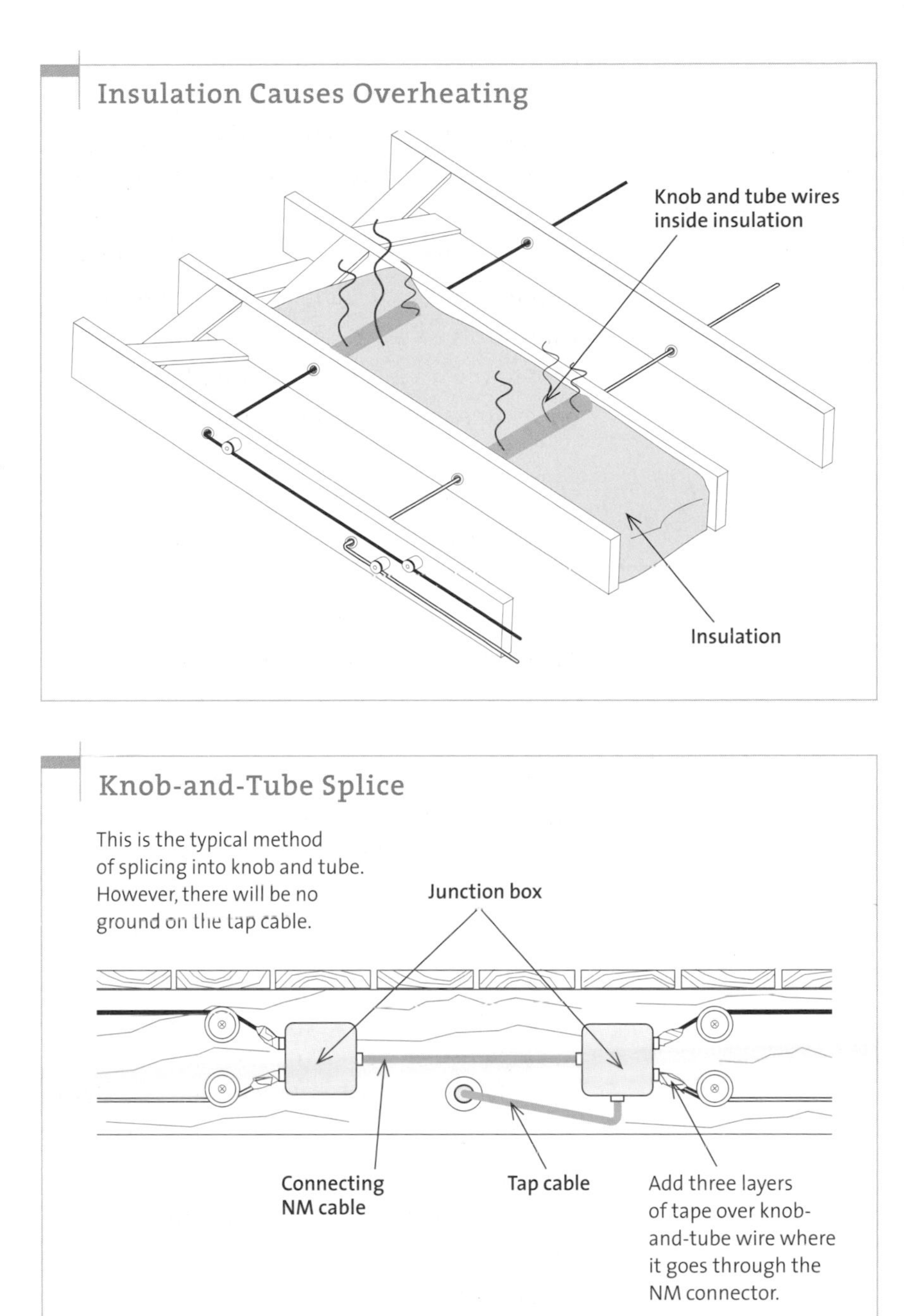

Deciphering Cable and Wire Codes

The only place you'll find more codes than on electrical conductors is in a spy novel. The gauge and manufacturer are not the only things stamped on the outside of a conductor.

A multidigit code indicates at what temperature range the conductor can physically be installed. You don't want to put a conductor that's approved only for dry locations in a wet area (for example, underground), and you don't want to put a conductor approved only for low temperatures in an area with high temperatures (such as an attic). Not only do you have to figure out the multidigit code, but you also must watch for preempt codes, which change or alter the previous code letters, depending on the installation.

The letter codes that appear after the gauge number and manufacturer's data refer to the insulation of the conductors. The letters tell you the type of insulation material, which is normally thermoplastic; the maximum temperature the insulation can be subjected to without damage—60°C, 75°C, or 90°C—and whether the conductor can be installed in wet locations, dry locations, or both.

Because the manufacturer is trying to tell us only three things, you'd think it would be rather simple, but it's not. In addition to the basic coding and the preempt coding, it is also significant if something is missing (I'm not kidding).

NM and UF Cables

Here's a sampling of the coding that appears on cables and other conductors commonly used in a residence:

- H: If present, the maximum allowable temperature is 75°C. If not present, the maximum temperature allowed is 60°C. NM cable uses the letter B instead of H, but it works the same way: with B, 90°C; without B, 60°C.
- HH: The maximum temperature is 90°C.
- N: Nylon sheath around thermoplastic insulation.
- T: Thermoplastic insulation around wire.
- W: If present, the conductor can be used in wet and dry locations. If not present, it can only be used in dry locations (sometimes damp). If the letter W appears after HH—W is a preempt coding—the insulation's maximum temperature has been reduced to 75°C for wet locations (it is still 90°C for dry locations).
- X: Cross-linked polyethylene insulation (normally used for SE cable).
- –2: A preempt suffix that means the conductor can be used at temperatures up to 90°C, wet or dry.

SE Cable

Though NM cable is fairly simple to decipher, SE cable is another story. For example, if SE cable has XHHW–2 stamped on it, the insulation is rated up to 90°C. However, the W preempts the HH and indicates that the conductor can be used in wet locations down to 75°C. The –2 preempts the W and gives a 90°C rating to both wet and dry locations. (Confused yet?)

THHN and THWN Wires

Using the codes, let's decipher THHN and THWN conductors. THHN is sometimes used in a home whenever conduit is required. It has thermoplastic-insulated wire with an outer nylon sheath and can be used in dry areas where the temperature could reach 90°C. THWN, also common in homes and run in conduit, has thermoplastic-insulated wires with a nylon sheath and a temperature rating of 75°C in both wet and dry locations.

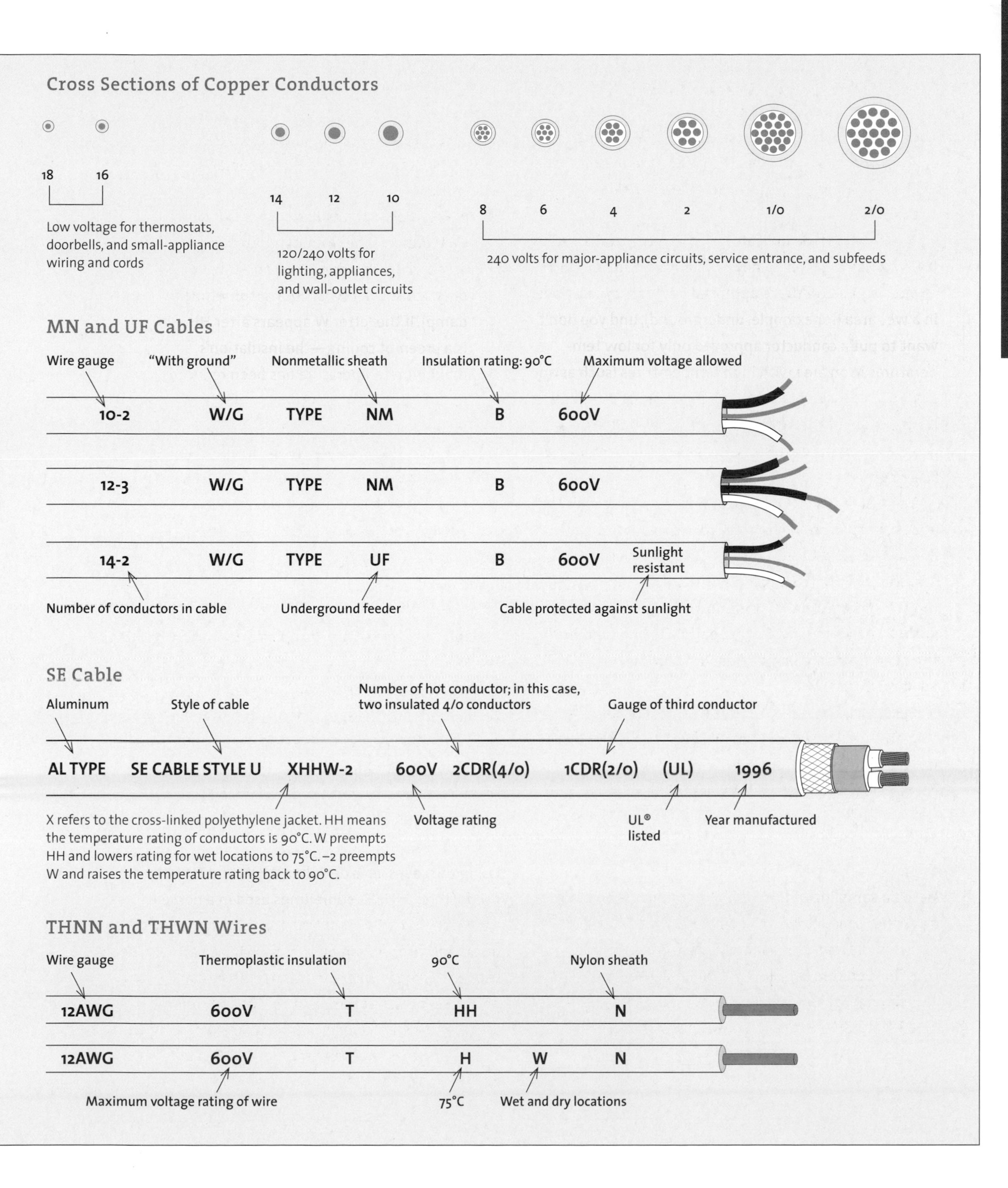
Cross Sections of Copper Conductors
18
16
Low voltage for thermostats, doorbells, and small-appliance wiring and cords
14
12
10
120/240 volts for lighting, appliances, and wall-outlet circuits
8
6
4
2
1/0
2/0
240 volts for major-appliance circuits, service entrance, and subfeeds
MN and UF Cables
Wire gauge
"With ground"
Nonmetallic sheath
Insulation rating: 90°C
Maximum voltage allowed
10-2 W/G TYPE NM B 600V
12-3 W/G TYPE NM B 600V
14-2 W/G TYPE UF B 600V
Sunlight resistant
Number of conductors in cable
Underground feeder
Cable protected against sunlight
SE Cable
Aluminum
Style of cable
Number of hot conductor; in this case, two insulated 4/0 conductors
Gauge of third conductor
AL TYPE SE CABLE STYLE U XHHW-2 600V 2CDR(4/0) 1CDR(2/0) (UL) 1996
X refers to the cross-linked polyethylene jacket. HH means the temperature rating of conductors is 90°C. W preempts HH and lowers rating for wet locations to 75°C. –2 preempts W and raises the temperature rating back to 90°C.
Voltage rating
UL® listed
Year manufactured
THNN and THWN Wires
Wire gauge
Thermoplastic insulation
90°C
Nylon sheath
12AWG 600V T HH N
12AWG 600V T H W N
Maximum voltage rating of wire
75°C
Wet and dry locations

High Heat-resistant Nylon) wires intermixed when it was built. The inspector didn't catch it because he couldn't read the wires.

What gauge goes where

Service-entrance cable for dwellings has its own reference in the NEC. The ever-popular Table 310-16 gauge finder is fine for everything except the service entrance and the subpanel feeder cables.

Above Code • With few exceptions, Table 310-16 of the NEC dictates the gauge size for the rest of the house. In practice, it breaks down to using 18 or 16 gauge for low-voltage circuits, such as doorbells and thermostats; 14 or 12 gauge for receptacles and lights; 12 gauge for most hard-wired appliances; 10 gauge for electric dryers and electric water heaters; and 6-gauge copper for electric stoves. Code makes you install 12 gauge in the kitchen/dining/pantry area, as well as in bathrooms and utility areas. All other areas are allowed to be either 12 or 14 gauge.

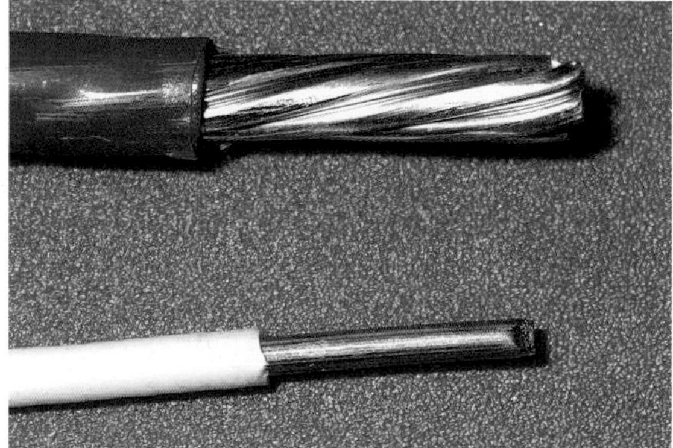

The 6-gauge stove wire can carry 2½ times more current than the smaller 12 gauge.

The problem with this allowance is that 14-gauge wiring isn't always heavy enough. It also creates the potential for intermixing 14- with 12-gauge wiring. Instead, I use 12 gauge as much as possible and 14 gauge only for the optional refrigerator circuit, three-way light circuits, and the smoke alarms. Keeping these circuits at 14 gauge saves me from having to use expensive 12-gauge three-way switch cable for loads that pull little or no current.

Copper versus aluminum wiring

When it comes to residential wiring, there are two standard wire materials: aluminum and copper. Most homes use copper wiring for receptacles and switches for good reason. Small aluminum wires have to be handled very carefully because they break easily. And they must be installed and spliced only with code-approved materials. Sometimes those materials are hard to find. For the larger-diameter wires (6 gauge and up), electricians tend to use both copper and aluminum. Aluminum has a slight cost advantage over copper, and this is more apparent in the larger-diameter cables.

When exposed to air, the surface of bare aluminum oxidizes. On one extreme is minor oxidation, which is a dull gray. On the other extreme is major oxidation, which creates a white powder that can grow like a tumor and bind everything together tighter than a thread-lock compound. The latter is a more serious problem, but even minor oxidation insulates the strands of aluminum from each other, as well as from the connector, increasing the resistance and degrading the metal. Therefore, whenever you strip insulation off aluminum, you must apply an anticorrosion coating to

Splicing Single-Strand Aluminum to Single-Strand Copper

This rectangular butt connector can splice both copper and aluminum (14 gauge through 6 gauge). To use, insert the copper wire on one end and the aluminum on the other, then tighten down the screws.

This tiny split bolt used to be my preferred method of splicing single-strand copper and aluminum together. As you can see from the box, it is approved for 16 gauge through 10 gauge—perfect for splicing 10-gauge aluminum to 12-gauge copper.

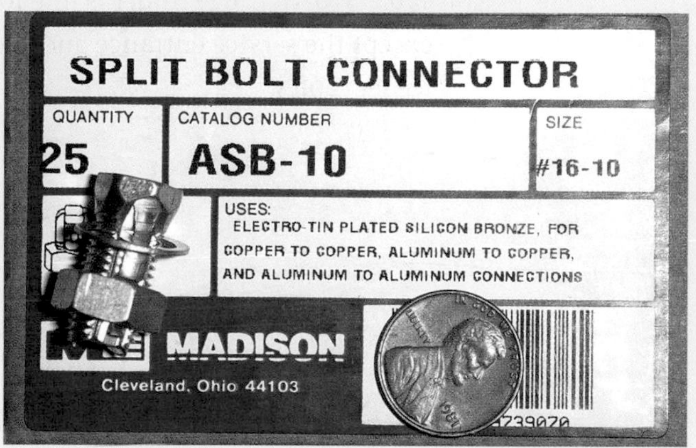

The split-bolt connector has a center tongue that separates the two wires being spliced so they don't touch each other. This is what allows the connector to splice different metal wires, such as those of aluminum and copper.

This new invention for splicing single-strand aluminum to single-strand copper follows my 20-year-old invention of using a short cut-off piece of a common bus bar. They just put it in a fancy insulated container. Now why didn't I think of that?

Round Solid Porcelain Knobs

63N3925—Old Code No. 5½ solid porcelain knob. Height, 1 9/16 inches. Diameter, 1 inch. Hole, ¼ inch. Groove, 5/16 inch. Weight, per 100, 8½ pounds. Price, each, **1c**; per 100, **92c**; per 1,000 .... **$9.00**

63N3927—New Code No. 5½ solid porcelain knob. Height, 1 9/16 inches. Diameter, 1⅛ inches. Hole, ¼ inch. Groove, 5/16 inch. Weight, per 100, 11½ pounds. Price, each, **1½c**; per 100, **$1.18**; per 1,000..**$11.20**

63N3929—No. 4 solid porcelain knob. Height, 1 11/16 inches. Diameter, 1½ inches. Hole, ⅜ inch. Groove, ⅜ inch. Weight, per 100, 20 pounds. Price, each, **2c**; per 100, **$1.57**; per 1,000....**$15.10**

63N3931—No. 4½ solid porcelain knob. Height, 1⅞ inches. Diameter, 1½ inches. Hole, ⅜ inch. Groove, 7/16 inch. Weight, per 100, 21½ pounds. Price, each, **2¼c**; per 100, **$1.76**; per 1,000....**$17.00**

Round Split Porcelain Knobs

63N3935—Old Code No. 5½ porcelain split knob. Height, 1¾ inches. Diameter, 1 inch. Hole, ¼ inch. Grooved to take two No. 12 or 14 wires. Weight, per 100, 9½ pounds. Price, each, **2c**; per 100, **$1.56**; per 1,000 **$15.00**

63N3937—New Code No. 5½ porcelain split knob. Height, 1¾ inches. Diameter, 1⅛ inches. Hole, ¼ inch. Weight, per 100, 11 pounds. Price, each, **2¼c**; per 100, **$1.75**; per 1,000.**$16.90**

Electrolet Conduit Fittings

Economical, easily installed and very satisfactory. Prices include fitting complete with two-wire porcelain cover. Three-wire covers supplied if specified.

Type "A." For inside work, meter loops, motor installations and all combination work.

63N3880—For ½-inch conduit. Each **10c**
63N3882—For ¾-inch conduit. Each **15c**
63N3884—For 1-inch conduit. Each **20c**

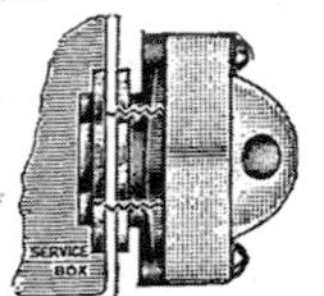

Type "O." Used on boxes having pipe knockouts where wiring changes to knob and tube construction. Complete with nipple and two washers.

63N3886—For ½-inch conduit knockouts. Each **18c**
63N3888—For ¾-inch conduit knockouts. Each **27c**

Type "B." Reversible entrance fitting for either vertical or horizontal pipe.

63N3890—For ½-inch conduit. Each **25c**
63N3893—For ¾-inch conduit. Each **33c**
63N3896—For 1-inch conduit. Each **40c**

Porcelain Tubes

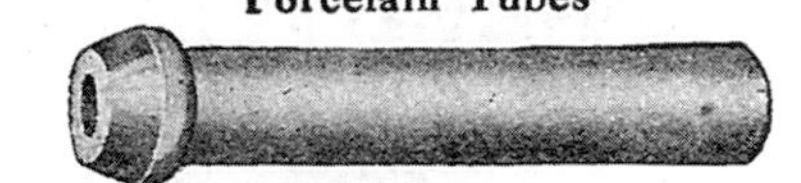

Unglazed Porcelain Tubes, 5/16 inside; 9/16 outside. Take either 14, 12 or 10 single braid rubber-covered and weatherproof wire. Required wherever a wire is drawn through a partition or joist of any kind. Length given is from underhead to end.

Article Number	Length, Inches	Each	Per 100	Per 1,000	Weight per 100 lbs.
63N3902	3	1c	$0.78	$ 7.15	7½
63N3904	4	1½c	1.25	11.95	9
63N3906	6	2c	1.73	16.80	13
63N3908	8	4c	3.30	31.35	15½

Porcelain Cleats

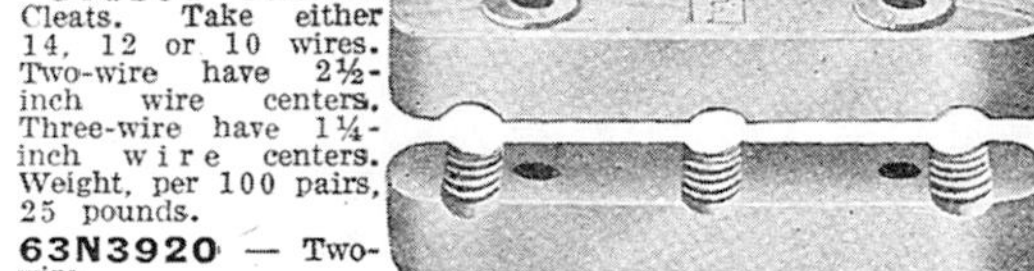

Glazed Porcelain Cleats. Take either 14, 12 or 10 wires. Two-wire have 2½-inch wire centers. Three-wire have 1¼-inch wire centers. Weight, per 100 pairs, 25 pounds.

63N3920 — Two-wire.
Price, per pair **$0.02¼**
Per 100 pairs **1.95**

63N3922—Three-wire. Price, per pair .... **.02½**
Per 100 pairs **2.20**

New Code, Rubber-Covered Copper Wire, Single Braid

Solid conductor wire, insulated with rubber compound over which is one cotton saturated braid. Recommended for any open wiring on cleats, tubes or knobs, and for loom and moulding wiring.

Each unbroken coil bears Underwriters' Inspection tag. Full coils contain 100 feet or 500 feet, or we will cut wire to any desired length.

Article Number	Size	Weight per 100 ft Pounds	Per Foot	Per 100 Feet	Per 1000 Feet
63N3015	14	3	1½c	$1.40	$10.25
63N3020	12	4½	2½c	2.20	19.75
63N3025	10	6	3¼c	3.00	27.40

New Code, Rubber-Covered Wire, Double Braid

Offers better protection than single braid wire. Should be used in all metallic conduits, flexible or rigid. Has one solid conductor, insulated with rubber compound over which are two saturated cotton braids.

Article Number	Size	Weight per 100 ft. lbs.	Per Foot	Per 100 Feet	Per 1000 Feet
63N3040	14	4	2c	$1.70	$16.20
63N3045	12	5½	2¾c	2.35	22.25
63N3050	10	7	3½c	3.20	29.90
63N3055	8	10	5c	4.55	43.95
63N3060	6	13	7½c	7.05	67.50
63N3065	4	18½	10½c	9.90	96.25

Duplex New Code Rubber-Covered Wire

Consists of two solid conductors, each insulated with rubber compound, over which is one saturated cotton braid. Conductors so insulated are laid parallel and covered over all with saturated cotton braid. Convenient for wiring in metallic and non-metallic conduits.

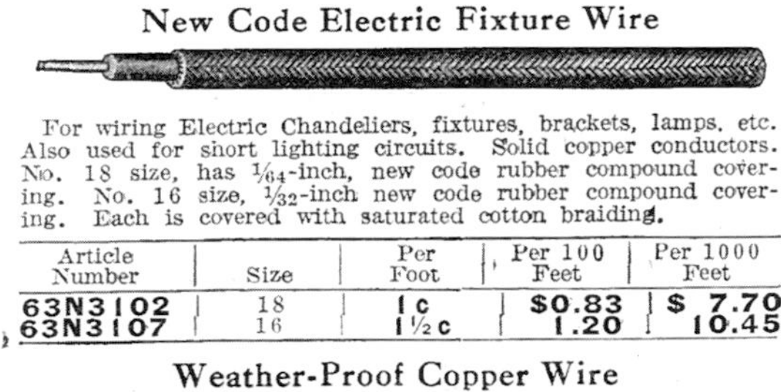

New Code Electric Fixture Wire

For wiring Electric Chandeliers, fixtures, brackets, lamps, etc. Also used for short lighting circuits. Solid copper conductors. No. 18 size, has 1/64-inch, new code rubber compound covering. No. 16 size, 1/32-inch new code rubber compound covering. Each is covered with saturated cotton braiding.

Article Number	Size	Per Foot	Per 100 Feet	Per 1000 Feet
63N3102	18	1c	$0.83	$ 7.70
63N3107	16	1½c	1.20	10.45

Weather-Proof Copper Wire

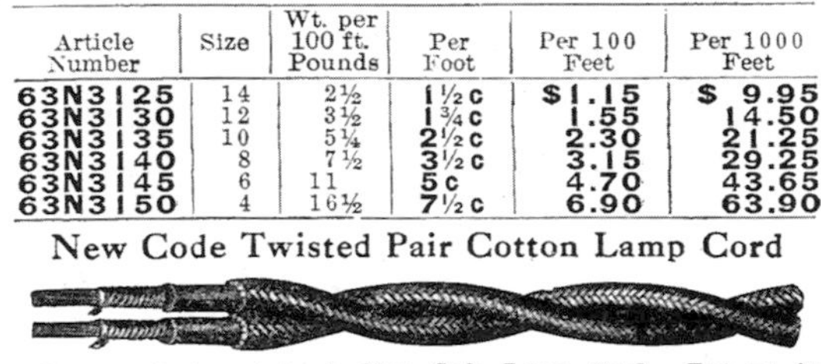

For use on outside work only. Not to be used indoors. Withstands weather better than rubber-covered wire. Conductor is of solid copper wire, covered with a triple braid of weather-proof insulating material.

Article Number	Size	Wt. per 100 ft. Pounds	Per Foot	Per 100 Feet	Per 1000 Feet
63N3125	14	2½	1½c	$1.15	$ 9.95
63N3130	12	3½	1¾c	1.55	14.50
63N3135	10	5¼	2½c	2.30	21.25
63N3140	8	7½	3½c	3.15	29.25
63N3145	6	11	5c	4.70	43.65
63N3150	4	16½	7½c	6.90	63.90

New Code Twisted Pair Cotton Lamp Cord

Two conductor, twisted, New Code Lamp Cord. Put up in 100-foot or 250-foot coils, or will cut to exact length of any required number of feet. Full coils tagged with Fire Underwriters' Inspection stamp. Conductor consists of fine copper

These photos, one from an early Montgomery Ward catalog and one from a Sears, Roebuck and Co. catalog, show wire and cable from 1900 to 1920. The knobs, tubes, and cleats were used for knob-and-tube wiring systems. Note that duplex cable was around even then. (Photos by Scott Phillips)

the bare wire—not just on the part connecting to the terminal—to prevent oxidation. I also recommend wire brushing the aluminum to remove any oxide on the surface before applying the antioxidation compound.

Splicing aluminum wire

It's easy to splice large-diameter aluminum wire with one of the dozens of devices sold everywhere for that purpose. I crimp wire together with specialized, expensive, long-handled crimpers. You can even splice service-entrance conductor (SEC) with this tool because it is a nonreversible device. For non-SEC large-diameter splicing jobs, you can use crimpers or large lugs and large split-bolt clamps. Bimetal tongue split-bolt clamps can splice large-diameter copper cables to large-diameter aluminum cables. There is even a special toothed-type fitting that tightens with a large socket and ratchet and cuts through the insulation to make a splice.

These turn-of-the-century electrical components provided electricity to us around 1920. Many, including the light bulb, still work.

Preparing aluminum cable (left to right). Remove the insulation to expose the conductor. Next, wire brush the conductor until it's shiny. Finally, use your fingers or an old toothbrush to cover the bare conductor with an anticorrosion compound.

Above Code • The best way to connect a single-strand aluminum wire to a single-strand copper wire is with the new insulated mini three-screw bus bars (sold at most large electrical stores). In the past we used a simple split-bolt clamp with tongue. The Madison #13-10 is approved for copper-to-copper, copper-to-aluminum, and aluminum-to-aluminum connections. Once you've made the splice, however, you must insulate any bare metal with three layers of high-quality electrical tape. Amp Electric makes a crimp splice but it is entirely too expensive—around $50 to $100. Ideal Industries, Inc. makes a purple wire nut that will make the splice, but allegedly there have been some installation-related problems.

What about small-diameter, single-strand 10- and 12-gauge aluminum that was once used on 15-amp and 20-amp branch circuits? That experiment failed many years ago; however, most manufacturers have largely ignored the problem. I developed several ways to deal with splicing aluminum circuits. One was cutting up neutral buses. The bus is approved for both aluminum and copper wires, so it seemed like a logical adaptation; I modified it to fit into a smaller outlet box. Today, Ideal®, Burndy®, and several other companies have such devices on the market. The biggest problem with these devices is that they take up too much room in an outlet box.

The Wire We Use

New houses use dozens of types of wire and cable. To make things even more confusing, wiring systems have evolved over the years. Some older homes may have a combination of several different wiring systems. The following is a brief overview of what you might find in your home. With renovations, you need to make doubly sure that the wiring has been done correctly. I once inspected a house that had all its old wiring replaced. The man selling the place even bragged about it. Upon testing the receptacles, I discovered that not one indicated the presence of a ground wire. To find out what the problem was, I inspected several receptacles and, sure enough, there was new NM wire. But the ground wire on each cable was cut back to the jacket. Apparently, the seller hired a couple of old-timers to do the work, and they didn't know what to do with those "confounded bare wire things."

Knob and tube

The oldest wiring that you might see in a house is knob and tube. This wiring system gets its name from the porcelain knobs and tubes that the wire runs in and around. The wiring is still legal to use and can even be installed for new extensions off an existing knob-and-tube system, assuming you can find the materials. Knob and tube consists of two parallel running wires (no ground wire). The copper wires have a rubber coating and a single-cloth braid—black for the hot wire and white for the neutral or return (most wire is so old and dirty that both colors look the same).

Knob and tube doesn't need to be replaced just because it's old. As long as the insulation is intact and you don't need a ground connection on the receptacles it feeds, there is no reason to replace the wires. But you do need to check a couple of things. Look for any sections of bare wire—spots where the insulation has hardened, cracked, and broken away. Sometimes this happens from old age, other times from overfusing and overheating the wire. If you move into a house that has overfused 14- and 12-gauge circuits (normally at 30 amps), be especially wary about the wiring. To be safe, keep the old wiring fused at 15 amps even if it is 12 gauge. If you find any bare spots where the insulation has broken away, have the house rewired. Some knob-and-tube systems had a two-fuse protection system: one for the hot wire and one for the neutral. You are no longer allowed to fuse the neutral, so if you have this system, you'll need to wire around the neutral fuses or replace the wiring entirely.

Another problem with knob-and-tube wiring is that it was designed for open-air heat dispersion. That is, it was never designed to be covered with insulation. The problem is exac-

Knob-and-tube wiring got its name from these components. Tubes on the left went through bored holes. Cleats (bottom right) and knobs (top right) held cables on runs along flat sides of joists and walls.

erbated if the house has blown-in insulation in the walls, because the knob and tube in the walls can no longer dissipate the heat. So, what do you do? You can continue to use the wire for light-duty circuits, but fuse it at 15 amps, as previously mentioned. If the gauge is already 14, I would not advise using it at all, except for very light-duty loads.

Grounding is another problem. Kitchen, bath, and utility room circuits need a ground, as do specific appliances, such as a computer. Thus you'll need new circuits for those loads. The NEC does not require splice boxes to be used when splicing knob and tube. It does state that you should solder the splice as they did back in the old days or use an approved connector. Practically speaking, soldering isn't done anymore, so a noncode splice is almost guaranteed. If you see such a splice with modern plastic electrical tape, the NM wire probably isn't soldered. If you see the old rubber tape, it is probably soldered per code.

The best way to splice into knob and tube is a moot point, because you cannot get a ground from it and the old circuit shouldn't be loaded any more than it already is. If you need to add on to an old knob-and-tube circuit, run a new cable back to the service panel instead. For more information on knob-and-tube wiring, read Article 394 of the NEC.

Armored cable terminates in an outlet box in a 1930s home.

Even close up you can just barely differentiate the black (hot) from the white (neutral) conductor. The insulation on both wires is severely deteriorated and may break away if the wires are moved.

Armored cable (AC)

Perhaps just as old, and sometimes running alongside knob-and-tube wiring, is a cable system we generally call armored cable or Greenfield (from Harry Greenfield, who co-invented it). This cable came in many strange versions, but evolved into the now-familiar spiraled metal (normally steel). You can buy the armor separate or with cable already inside. It was extremely popular from 1910 to 1930. BX is the trademarked colloquial term that comes from the very popular GE product.

Aluminum AC came out in 1959, as did the requirement of having a bonding strip, or wire, run the complete length of the armor but electrically connected to nothing. The bonding strip's purpose was to lower the overall impedance of the armor. The basic problem with AC

Cutting MC and AC Cable

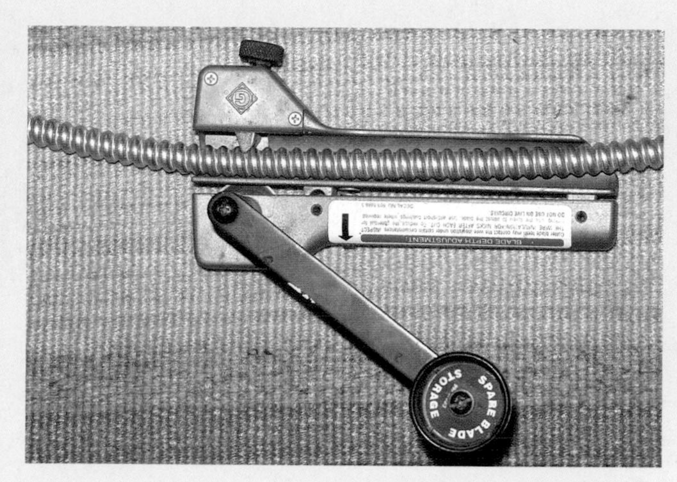

Insert cable in the long slot down the center of the tool. Position the cut line over the cutting wheel. Tighten the knurled knob on top of the tool until the cable is wedged tightly in the slot.

Turn the handle to engage the cutting wheel. The wheel will slice through the metal but not the wires.

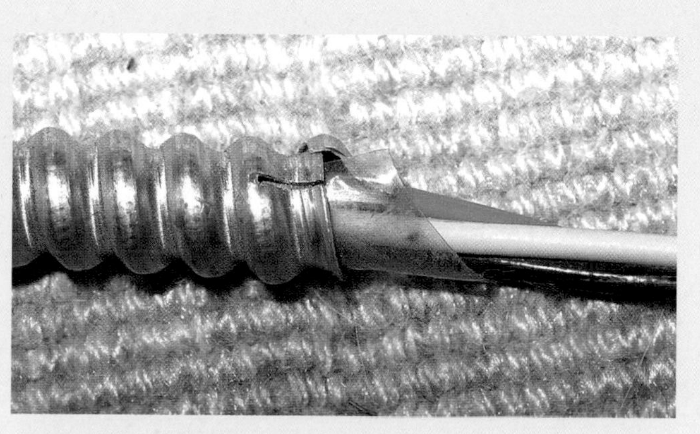

Remove the cable from the tool, then trim back and remove any remaining cut spirals.

Insert plastic anti-short bushing between the metal and the plastic sheath (the bushing here is orange, but the material comes in different colors for different sizes).

Take Note • Everyone asks where the term BX came from. The answer was always "Bronx." I never agreed with the answer because the old texts list it as B.X., with a period after each letter. Each initial must have represented a different word. But what were the words?

Perhaps this is the answer: Flexible metallic conduit originally was wound around a string, which was used to pull the coils tight. The manufacturers realized that the string could be replaced with conductors. This epiphany resulted in a version "B, eXperimental," or B-X, which was the first armored cable. Besides, the cable originated in Brooklyn, not the Bronx. Right or wrong, this answer makes more sense than Bronx.

was that the long, continuous spiral of metal produced tremendous inductive impedance, a special type of resistance associated with coiled wire. Therefore, when working with old AC, do not remove the bonding strip (assuming it had one—some versions didn't) or tie it into the grounding system. Bend it over the end of the AC to hold the anti-short bushing in place. More often than not—especially in short runs—the strip will be missing.

AC was first put in homes that had ungrounded receptacles, because grounding wasn't thought to be that important. When that assumption was proven wrong, the steel spiral was used as the grounding circuit in three-prong grounding receptacles. My county seat used ungrounded AC until the introduction of computers. Using the steel AC as the grounding circuit has kept things working fine to this day. If you have this ungrounded system in your home and the armor is intact, you can do the same. If you see that the conductor's rubber insulation is cracking and breaking apart within the receptacle and switch boxes, it would be a good idea to rewire the entire house. For repairing short runs, pull the old wire out of the AC and slide in new wire—normally THHN/THWN or NMB. To cut into the armor, you need a special tool. You can use a hacksaw, but it is almost impossible to cut into the cable without damaging the conductors.

General requirements for armored cable (see article 320 of the NEC):

- Secure the cable within 12 in. of the box and do not exceed 4½ ft. on runs.
- Must be secured, not just draped over things. However, running it through a framing member is considered secure, assuming members are less than 54 in. apart.
- Can use the armor as a ground but you must leave in the bonding strip, even though the strip can be cut off in the box.
- Must use insulating bushing. Most installers use the bonding strip to hold in the bushing.

Metal-clad cable (MC)

Metal-clad cable has taken the place of AC. MC cable looks a lot like AC, but within its metal jacket it has a green insulated ground wire (the metal jacket cannot be used for a ground). You can use this wire system in lieu of NM cable throughout the house, but because of code changes it is more commonly used in commercial work. However, if you have a multi-story house you may come across it on your upper floors. (At one time, code limited the number of floors on which a residence could use NM cable.) Code no longer prohibits NM cable in a residence, regardless of height.

MC cable must be kept dry, but it can be embedded in plaster and run through block. It also must be supported within 12 in. of the outlet box and no more than 6 ft. apart after that. Its advantage over NM cable is that rats and mice cannot gnaw on it. Its biggest disadvantage is that it is very labor-intensive to install, and you need to use special bushings to prevent the metal's sharp edges from cutting into the wires. (If you are planning to run rigid conduit, MC cable can be a lot faster to install than NM cable.) You also need a special tool to cut into the jacket, special metal boxes to terminate the cable, and special connectors.

My advice? Don't use MC cable in a residence unless you have a specific reason to do so. Once I had a customer insist on MC cable because she owned an old house and thought that mice would chew on the wires. When she saw how problematic (and costly) MC cable was to install, she eventually changed her mind. As for the mice, her cats took care of that problem.

General requirements for metal-clad cable (see article 330 of the NEC):

- Cannot use the metal clad as a ground; it must have an integral grounding conductor.
- It must have insulating bushings; the cutting of the metal shield leaves it razor sharp.
- The minimum bending radius is 12 times the outer diameter.
- Must be secured within 12 in. of the outlet box and at least every 6 ft.

Nonmetallic sheath (NM) cable

NM cable consists of two or more insulated conductors in a single nonmetallic jacket (two insulated conductors in an NM jacket is called a duplex). It is not a modern invention. Duplex cable has been around since before

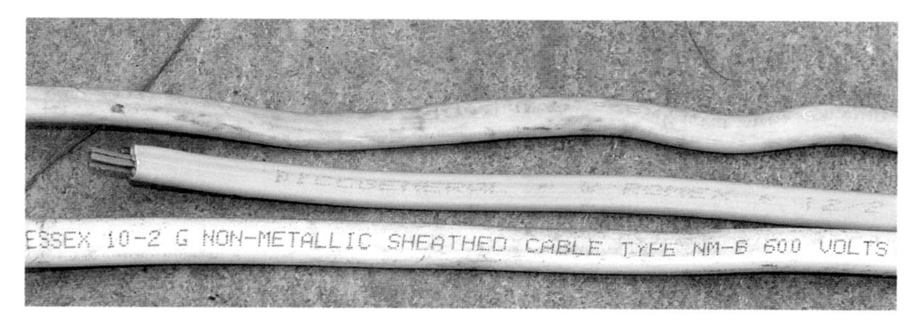

Avoid buying NMB cable if you cannot read the writing on the sheath. The printing on the embossed cable (top) and the painted cable (middle) are both illegible. Newer NM cable is color coded to indicate cable gauge.

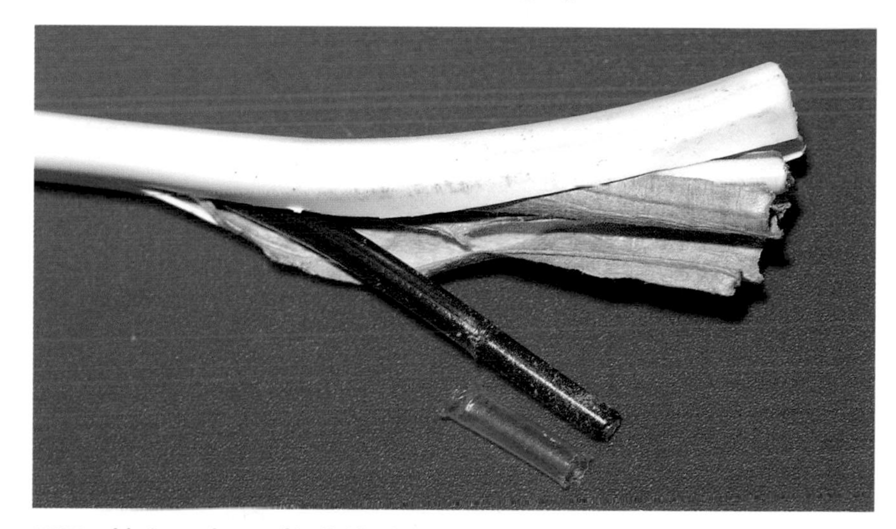

NMB cable is made up of individual thermoplastic-insulated conductors, each slipped into a clear nylon sleeve, then wrapped in paper insulation and a thermoplastic jacket. The clear sleeve gives NMB cable a higher temperature rating than the older-style NM cable.

Old NM cable is made up of individually insulated conductors, lots of paper insulation, and a hidden zip cord to rip the cable, all contained inside a woven sheath.

1918. Early duplex NM cable consisted of two insulated conductors in a silver-colored jacket or a braided, brown-jacketed cable. The silver-colored jacket cable was used to wire houses in the early 1950s; the brown-jacketed cable was used to wire houses in my section of Virginia in the 1920s and 1930s.

Modern NM cable (also called Romex™, a trademark of the Southwire Company) came into use around 1965. There are four basic types of NM cable: old-style NM, NMC, NMB, and UF. Each one has a specific job. Old-style NM was the most common cable used in home wiring until a few years ago. It has been replaced by NMB cable because of the need for a higher insulation temperature. The new wiring is similar to THHN; it even has the same clear protective sheath over the insulated conductor. Old-style NM is rated at 60°C; NMB is rated at 90°C. This increase in temperature rating was long overdue because of problems of overheating within fixtures, resulting in blackened and hardened conductor insulation.

Take Note • Unless you have a specific reason not to, use nonmetallic sheath (NM) cable. Code requires that nonmetallic sheath be supported every 4½ ft., though I like to support it more frequently: every 3 ft. to 4 ft.

Generally, NM-style cable has two or more thermoplastic insulated wires, one bare grounding wire, paper insulation surrounding the conductors, and a thermoplastic jacket or sheath. NMB cable is the NM style that is currently used around the home for receptacles, lighting, and small-appliance circuits. Modern NMB cable can be used for both exposed work (as long as it is not in physical danger) and concealed work in dry locations, but it cannot be embedded in masonry, concrete, adobe, dirt, or plaster. NMC cable is the same as NMB cable, but its outer jacket is corrosion resistant.

Work Safe • The most common error made with PVC and EMT conduit is using too small a diameter when trying to slip in NM cable. Small (1/2-in.) conduit allows you only enough room to slip in one cable. Comparatively, 3/4-in. and 1-in. conduits allow for a lot more room and freedom of design. Just make sure you pick boxes that are prepunched for those two sizes.

General requirements for nonmetallic sheath cable (see article 334 of the NEC):

- No restriction on number of stories except by architect.
- To be supported every 4½ ft. and within 8 in. of nonmetallic outlet box without an internal clamp (12 in. other boxes).
- NMB can no longer be in conduit outside the house. Use UF instead.
- Cannot be subject to excessive moisture or dampness (UF can).
- NMB cannot be embedded in concrete, masonry, adobe, dirt, or plaster.

UF (underground feeder, article 340) cable looks like NM cable, but UF has the grounding wire, the neutral wire, and the hot wire embedded in solid thermoplastic. UF cable is used in wet locations, such as underground circuits to outdoor lights, where NM cable would not be allowed by code. All NM cable, even UF, is normally sold in 250-ft. rolls. Some stores that cater to do-it-yourselfers sell the cable in shorter lengths; however, the markup may be considerable.

Use UF cable for wet and buried locations.

Type-U cable with XHHW insulated conductors is perhaps the most common service-entrance cable used. It has thermoplastic insulation, glass-reinforced tape, a braided neutral, and two insulated conductors.

Service-entrance cable

Type SE cable brings power into the house from an overhead splice at the utility cables. SE cable comes in different styles: U, R, and

Forget using a hacksaw to cut large-diameter cable (XHHW); instead, use a reciprocating saw with a fine-tooth blade.

A common scissors-type cutter works well for cutting cable and individual conductors.

USE. Style U is flat and approved only for above-ground use. This is the cable you see attached to the outside of most houses. It has two black insulated wires (sometimes one has a stripe for phase identification) surrounded by many strands of braided bare wire for the neutral, and is covered by a layer of glass-reinforced tape and protected by a PVC jacket. The jacket is flame-retardant, moisture-resistant, and almost entirely UV-resistant.

Style R is round and has three insulated conductors with one bare grounding wire covered with a layer of glass-reinforced tape and protected by a PVC jacket. Like style U, R is approved only for above-ground use.

Individual insulated conductors rated USE (underground service entrance) are normally used when the service must be located underground or within conduit. USE conductors are available in gauges of 6 and higher.

THHN/THWN is the most common SE cable used in conduit. Smaller gauges can be used as branch circuit wires in the house, as long as they stay in conduit.

SE cable can also be used to provide power to some appliances, such as 240-volt-only heat pumps and electric furnaces. However, it cannot be used where an insulated neutral is required. USE cable cannot be used to feed 120/240 appliances, such as electric dryers and stoves. Such appliances need three insulated conductors plus ground (two insulated conductors for the 240-volt load and a return insulated neutral for the 120-volt load), and standard SE cable has only two. However, SER cable can be used to feed 120/240-volt loads (subpanels and heavy-current loads), because it has three insulated conductors with an additional bare grounding conductor. SE cable is covered in detail in article 338 of the NEC.

THHN/THWN is also used for service-entrance cable, but it has to be in conduit. If you want to run the cable out of the weatherhead for an aerial drip loop, contact your inspector to verify that it is allowed, as it must be approved for outside.

Conduit Systems

Conduit systems provide physical protection to cable or wires. The rule is that if the conductors have a jacket around them as a group, then that is their physical protection. If the conductors do not, and if they are individual conductors, such as THHN/THWN, then they have to be run in conduit. In addition, if the cable is open to significant abuse, then the cable or wire must be in conduit.

The two most common conduit systems for branch circuit wiring are plastic (PVC) and electrical metallic tubing (EMT). Both types of conduit can be used indoors or outdoors. PVC is glued together and has the advantage of being watertight and not needing to be grounded. A common application of both

The two most common types of rigid conduit are PVC (plastic) and EMT (electrical metallic tubing).

types is in a basement or garage that has finished block walls. You would normally attach a piece of treated wood onto the wall and mount the conduit and boxes to it. This is much easier than trying to mount it on a block wall, unless you have a hammer drill that can drill into the cement easily. The wiring inside the conduit can be NM or THHN/THWN. If you have a complete conduit system, use the latter. THHN/THWN is cheaper and faster to install and it takes up less room in the conduit. However, most people just use the vertical sections. They'll run NM cable overhead and then slip it into the vertical conduit as it goes down to the receptacle and switch circuits. This gives you the advantage of running the flexible NM cable overhead and having the physical protection where it is needed.

A newer-style conduit system that is gaining in popularity is called "liquidtight." Commonly available at most supply houses, it is very easy to install. This nonmetallic, nonconductive, noncorrosive conduit is extremely flexible and can be used to protect wire/cable going to heat pumps, air conditioner hookups, spas, and anything else that has outside wiring. PVC can be used in its place, but it is harder to install and cannot take vibration. (Because of this, inspectors sometimes fail heat pump installations where PVC conduit is used.) Liquidtight conduit installs like a garden hose; all you need is an adjustable wrench.

Running groups of cables

Running a large number of cables through joists or across the tops of attic beams can get

Liquidtight flexible conduit is very popular for running wire and cable outside. It can be bought in rolls or by the foot and is available in various diameters.

you into trouble. On one job (which failed inspection) a man had all the cables bundled tightly together. It looked beautiful—15 to 20 cables all wrapped together going through the joists as neat as a pin. But such beauty has a major engineering drawback. The cables on the inside of the bundle will overheat because they cannot get any moving air. The heat will break down the insulation, which will eventually lead to a short. To avoid this problem, always spread the cables through several holes when running through joists or studs. When running the cables on top, spread them flat over a large area. Locate attic cables to one side—choose a side where there is little headroom and fewer chances of someone tripping over them.

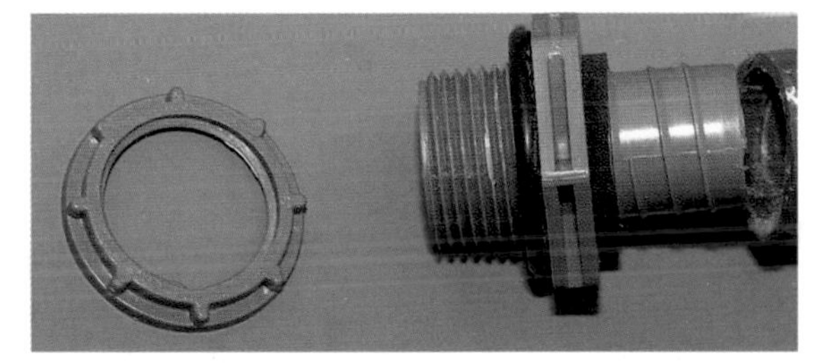

Liquidtight fittings seal on both sides. The threaded end fits into a box knockout and is sealed by an O-ring. The opposite end screws into the conduit.

The finished product. The fitting is attached to the conduit and the O-ring is attached to the male threads.

CHAPTER 2

Tools of the Trade

Electrical tools haven't changed that much since I wrote the first edition of this book. My everyday tools are pretty much the same. I still use the same broad-nosed electrician's pliers to cut cable and pull wires. I still use a needle-nose pliers to bend stripped wire ends into half-circles to put under screws. However, old dogs can learn new tricks, especially when it makes work more comfortable and efficient. These changes are included along with the basics.

If you are a first-time reader, pay close attention to my favorite tools, such as my Klein® electrician's pliers and my Gardner Bender® automatic wire strippers. Other tools, such as the common cable ripper, are notably absent. The problem with rippers is that as they dull they tend to bunch the sheath up around the cable, forcing me to use my utility knife. Rippers are also hard to use in tight locations, like a crowded outlet box. And even if you get the sheath ripped, you'll still have to use a knife to cut it away from the cable. I've found that using a utility knife is equally safe and easy.

Buying quality tools is important, and so is maintaining them. You can judge how serious people are about their work by the quality and maintenance of their tools. It's equally important to use the right tool when working with electricity. One of the first lessons an electrician learns is that the right tools can make a hard job easy and safe—and the wrong tools can make an easy job hard and dangerous. (To borrow from an old phrase: There are old electricians and there are foolish electricians, but there are no old, foolish electricians.)

In this chapter, I will tell you what I've learned about electrical tools from my many years of using (and abusing) them. And I will throw in a few tricks of the trade as well. I'll show you the tools an electrician uses every day, including how to use them safely and how to maintain them to ensure a long life span—for both tool and user. And here's an important note: When on the job site, many tools look the same, and someone may claim yours. To help prevent this, put your name on all your tools with paint, stickers, or other means. My favorite method is paint. If you paint your tool a solid color, you will be able to recognize it from across the room—especially if it's in someone else's hands.

I now use a tool pouch instead of a belt. A free-standing pouch takes the weight off my hip and keeps me from losing the tools every time I bend over or climb into a crawl space.

General Tools

General tools are the tools you'll be using most often. Although I can't cover every tool, I will discuss the ones that are most important to me, how to use them, and their advantages and disadvantages. Stay away from dime-store specials. Quality tools may be expensive, but they are worth it; if you don't buy quality electrical tools, you should buy a good, prepaid insurance policy.

My tools may look worn and torn, but I make sure that they all work well. Whenever a tool becomes worn or broken to the point that it can't do its job safely, fix it or replace it. In addition, use a tool only for its intended purpose. Most hinged tools need a drop or two of lubricant about once a month. Some need a lubricant spray to clean out the dust and debris. The lubricant should be nonsticky so that it won't attract dust, but it should leave a slick film to allow for smooth operation.

Tool carrier

When I wrote the first edition of this book, this section was called "Tool Belt." However, I've since changed the way that I carry my tools—now I use a pouch. I'm not sure if I did it to create a more efficient system or because I grew tired of carrying around the extra weight, but the new system works better. With my old belt,

Insulated and Grounded Tools

Accidents happen. For that reason, every tool you use needs to be double insulated or grounded. The old ungrounded all-metal antique you may be using needs to be retired, not used. Both types of protection will save your life if you accidentally cut a hot wire.

Most hand tools can't be grounded (although some electronic hand tools do have ground leads). Instead, look for good-quality insulated handles. If the plastic starts slipping off or appears cracked, repair or replace the tool immediately.

Double insulated means all live electrical parts are double insulated from the user. If an internal short occurs, the tool will direct the current back to the panel and trip the breaker. A grounded tool has all metal surfaces connected to the ground wire in the cord. In this case, you must use a three-prong grounded receptacle to use the tool safely.

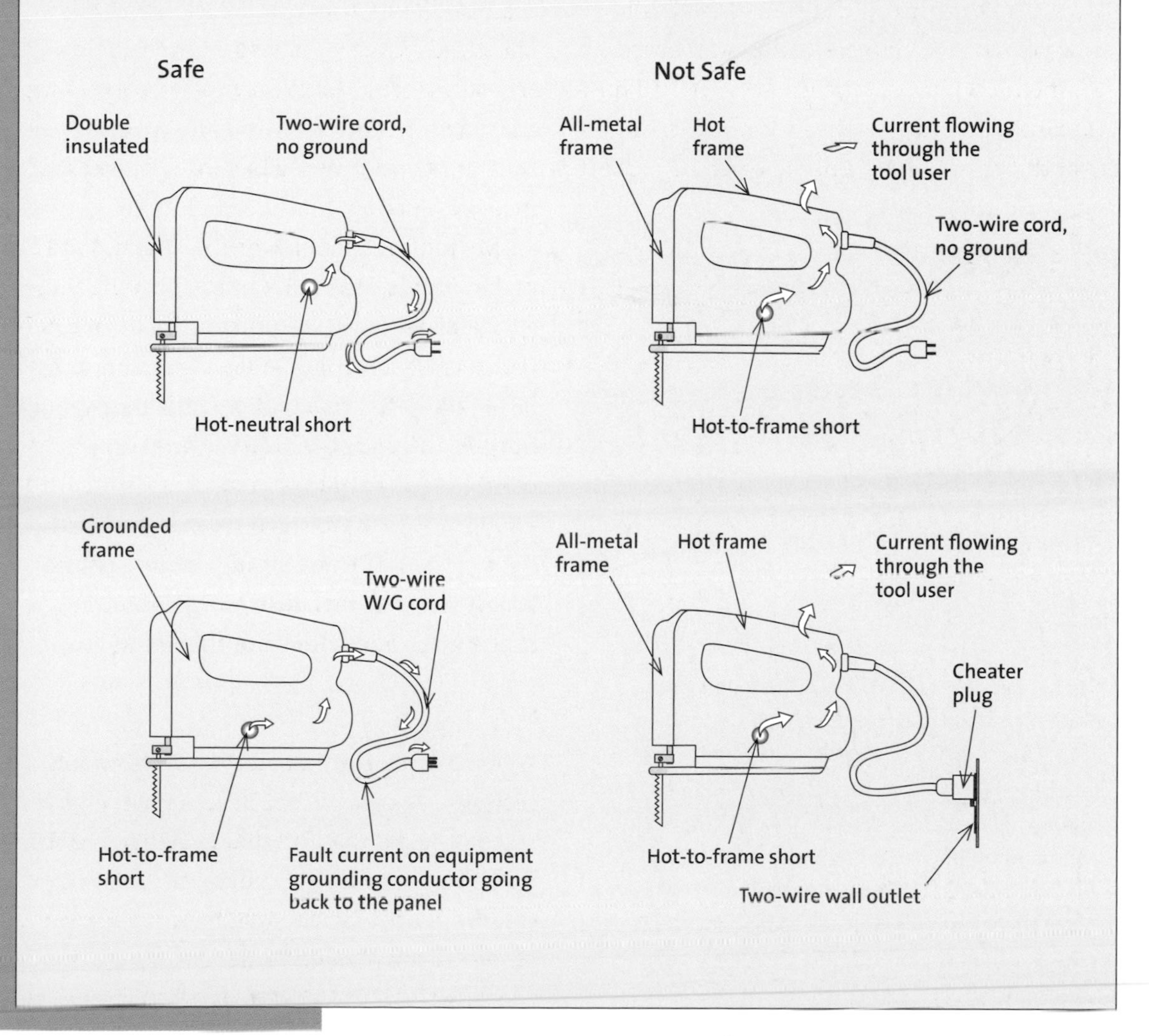

My most-used tools: electrician's pliers (Klein), needle-nose pliers, diagonal cutters (called diagonals), and end cutters (staple pullers). Don't start wiring until you have these tools.

Until I got organized, I think I left enough tools behind me on job sites to supply half the county. Finally, I got a little red wagon, which served as a home base for my tools. Later, I upgraded to a green garden wagon, which I still use. It holds three large plastic containers: one for boxes, one for AC and DC tools, and another for small hand tools. If I lay the tools in the wagon (as opposed to in the window or on the floor), they won't be left behind.

it seemed that my tools used to fall out every time I bent over or climbed into a crawl space. It also seemed that I was always taking off the tool belt to do something. The stand-alone pouch has solved all my problems.

The tools in an electrician's tool belt or pouch will vary slightly depending upon his or her specialty. In any case, each tool should have its own place so you can withdraw it without looking. Stock your pouch with only your most-used tools; otherwise, you'll wind up looking like Rambo ready to go to war. I keep less-used tools in a secondary bucket or in my newest play-tool: a little red or green wagon. The wagon has been a godsend.

I now have a home base for all my tools. Rather than put the tool down on a counter or on the floor, I put it in the wagon, eliminating the time I used to waste hunting for "lost" tools. When I move to a different location, I just pull the wagon and all my tools come along for the ride.

Work Safe • Even though we all occasionally do it, one should never use a knife to remove insulation from solid wire. The knife blade notches—and weakens—the wire, which may cause it to break in the future. For the same reason, you shouldn't use bladed pliers to cut and pull insulation off a solid wire.

Side cutters

Side cutters, also called lineman's or electrician's pliers, are to an electrician what a sidearm was to a gunslinger of the Old West. Although used mostly for cutting and pulling wire, side cutters can be used for other jobs, too. The blunt, wide nose can twist wires together before a wire nut or any other mechanical splicing device is applied.

Good side cutters are among the most misused of the lineman's tools. Being heavy-duty, they are occasionally used for a hammer and other things I am ashamed to mention. The only way I stopped abusing my side cutters was by keeping a hammer and other miscellaneous tools by my side (normally in the bucket mentioned earlier).

Long-nose pliers

Next to my side cutters, long-nose pliers are my most-used tool. They are great for bending wires into a loop for insertion around receptacle and switch screws and for pulling wires into narrow locations. Never use them to twist nuts off screws; the twisting action will spring the pliers' jaws off to one side and eventually ruin them. Try to find long-nose pliers with an integral wire cutter so that you won't have to change pliers just to cut the wire ends.

Diagonal cutters

Sometimes wire needs to be cut extremely close to the surface of an object so that there is minimal wire protrusion. The thin head of a diagonal cutter allows it to fit into narrow places where other pliers can't go. I often use mine when I have to cut one wire in a bird's nest of wires.

End-cutting pliers

Also called "nippers," end-cutting pliers are used primarily for pulling staples. But they also make good straight-in close-cutting pliers. Sometimes a wire needs to be cut, but access to it is only straight in from the front. Side cutters and needle-nose pliers all cut from the side of the jaw, meaning that the entire head of the pliers must be at a right angle to the wire. End-cutting pliers cut dead ahead.

Wire strippers

There are basically two types of strippers: cut-and-pull and automatic. Until recently, automatic strippers stripped the insulation

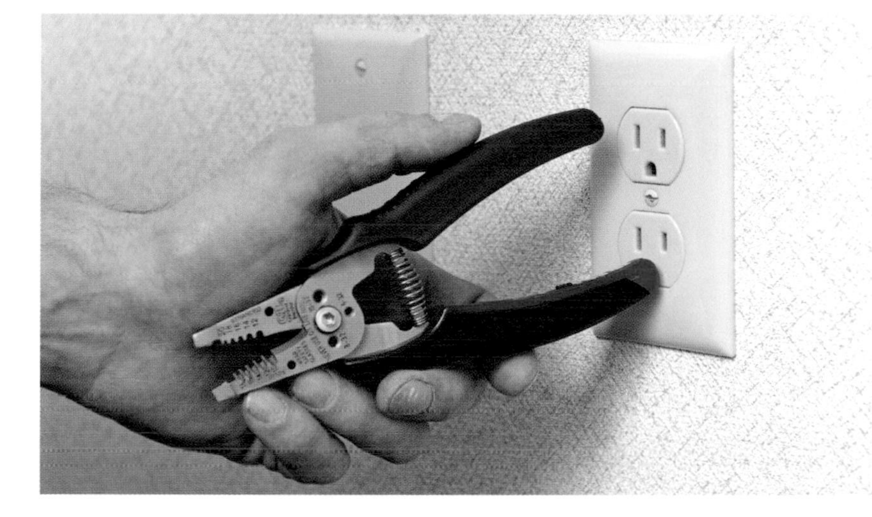

This wire stripper rises above the pack by employing a noncontact voltage-presence indicator in its handle.

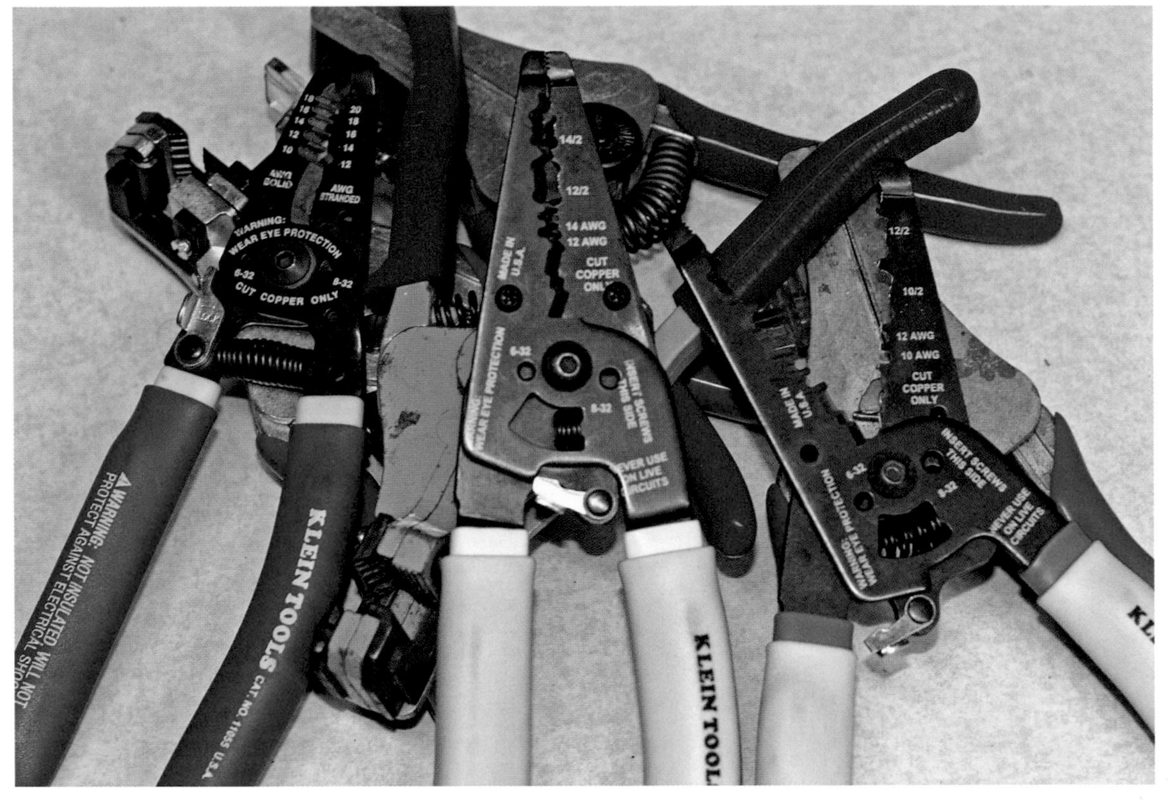

When choosing strippers, you have many to pick from. Forget the large automatic strippers (the ones on the bottom) because they haven't worked properly since conductor size went from NM to NMB. Find a good cut-and-pull tool that strips wire in the gauges you use most often.

off a wire with a pull of the handle. These old standbys got it right first time every time. When cable switched from NM to NMB, the diameter of the insulation got a bit smaller, but the tool didn't change accordingly, which means they rarely work properly. However, cut-and-pull strippers still work well. Some manufacturers have added cable-sheath stripping to these, so no more endangering your thumb by stripping off sheathing with a knife.

Bottom line, whichever stripper you buy, get one that strips the most common wires: 16, 14, 12, and 10 gauge. The 16-gauge wires are used in light fixtures; the others are used in common house wiring.

Screwdrivers

An electrician needs many kinds of screwdrivers, and they all can't be kept in a tool belt. I keep one high-quality multipurpose screwdriver (a driver with interchangeable heads) in my pouch and the rest in the tool bucket.

There are many types of screwdrivers: flat head, Phillips head, Torx®, square drive, and offset are the most common ones used by electricians. Offsets are used in tight places where a standard screwdriver is too long. I use mine to remove screws from an appliance when there's only about 2 in. between the screw head and the wall.

Stripping Cable Safely

Stripping the sheath from a cable is an art. You feel just like a surgeon cutting the skin but nothing underneath. To do this operation, keep the cable flat, not twisted, and support it on a solid surface. Never, under any circumstances, support the cable with your knee, as I have done, while cutting the sheath. As soon as you think everything is going fine, the blade will be buried deep in your leg (I have a scar to prove it). And don't try to cut into the sheath while holding it in mid-air. Once the cable is properly supported on a table or bench, use a sharp utility knife and gently slice down the center of the cable. The cable center contains only the bare ground wire, so there will be no harm if the blade occasionally goes in too far. Once cut, peel back the sheath and slice it away.

Sheath strippers are new on the market. They work just like cut-and-pull wire strippers but are faster at stripping off sheathing than the old way of using a knife—and they pose less risk of cutting the insulation. Look for a designation on the flat of the blade that corresponds to the conductor size and number of wires in the cable.

As an alternative stripping method, I prefer cutting into the jacket, then tilting the blade under the jacket where it's still attached. The blade slides over the top of the insulated conductors as it cuts the jacket.

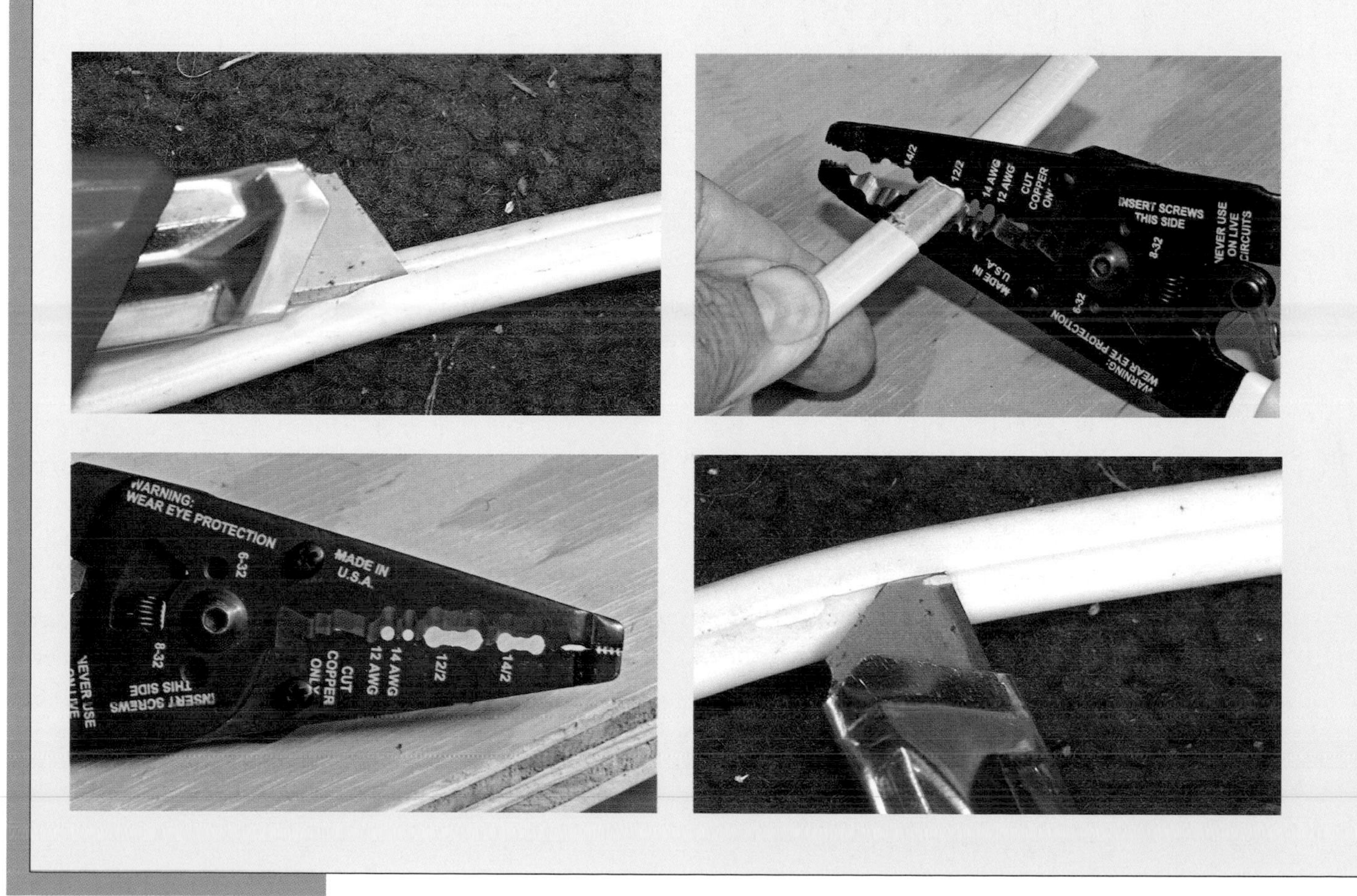

Specialty screwdrivers, such as these speed drivers, are nice to have around because they have handles that spin. As you twirl the handle, the rotation of the L-shaped shaft drives the screw. Speed drivers only work on loose screws; hard-turning screws still need a common rigid-shaft screwdriver.

Don't use the wrong screwdriver for the wrong job. For instance, never use a straight blade in lieu of a Phillips head. And remember that even straight blades have different blade thicknesses—you'll need both wide and narrow.

Eventually, you may want to invest in a few specialty screwdrivers. One screwdriver, available in both flat blade and Phillips head, has a clip at the bottom that holds the screw on the blade. When you have to extend the screw into an area where your hand can't hold it in place, this driver is worth its weight in gold. A tiny flat-blade screwdriver is handy for releasing wires in push-in receptacles. And don't forget the thread-starter driver—I couldn't do my job without one.

Work Safe • Screwdrivers, like side cutters, are often misused. We use them as chisels, prybars, and punches. To reduce the temptation of misusing your screwdrivers, carry the tools that you really need—the ones for which the screwdriver is doubling—in your tool bucket. If you find you're constantly using a screwdriver for a specific purpose other than driving screws—for instance, as a chisel—modify the screwdriver to make it work better for that job and then use it only for that purpose.

Utility knives

Most utility knives use the same cheap retraction mechanism that jams when it gets full of dust and dirt. However, the new low-cost breakaway-blade utility knives work great. I prefer the utility knife to other types of knives because the blade is sharp and thin, and dull blades can be replaced in a snap.

You may prefer to use a standard knife instead. But on the whole, a general-purpose pocketknife makes a poor substitute for a utility knife. The point of a pocketknife is rarely razor sharp.

Electrical tape

The electrician will need both black electrical tape and several different colored electrical tapes. White tape is used to identify any wire that is a neutral (grounded conductor) and doesn't already have white coloring on its insulation. Gray may be legally used for neutral marking, but I have never seen it done. Green tape is used to identify grounding wires, the wires that ultimately connect to earth ground via a ground rod. Any color tape other than white (gray) or green can be used to identify different hot wires if you want to keep track of them. Colored tape should be used for color-coding only. Once you've used colored tape, you will realize that it may not be the same quality as good black electrical tape. Black tape is used for general taping purposes—splicing, insulating a metal box, holding things together—as well as identifying a hot wire from the panel or utility.

Above Code • Do not use the cheap electrical tape found in grocery stores and discount houses. I prefer a high quality tape, such as Scotch Super 33+ (7 mil thick) or Super 88 (8.5 mil thick). Both are good tapes that work from 0°F to 260°F and resist ultraviolet rays.

For splicing in extremely hot applications above the temperature range of standard tape, I recommend using 3M 69 Glass Cloth electrical tape. For covering splices with a solid rubber or silicone rubber tape, use 3M® 130C or Scotch Tape 70. All tapes are available through professional electrical suppliers.

Work Safe • Some extension cords have ground-fault circuit interrupters. If yours doesn't, always be sure to plug the cord into a GFCI-protected outlet. Contractors are now required to use GFCI protection, whether the job is new construction or a renovation.

Extension cords

Extension cords are valuable tools for anyone working in the trades. Cords come round or flat; I prefer the round ones because they are much easier to loop for storage. I really like the ones with the lighted ends, because you never have to guess whether or not the cord has power.

Typical extension-cord gauges (from thinnest to heaviest) are 18, 16, 14, 12, and 10. A cord with a smaller gauge number, say 10, provides more power than a cord with a large number, say 14. Always remember to use a GFCI or equivalent with extension cords. You can buy a 3-footer (extension cord) with GFCI to add onto the end of your existing cord.

Ladders

Ladders, like extension cords, are necessary for most trades. I have a 4 ft., a 6 ft., an 8 ft., and a long extension ladder—all fiberglass. I chose fiberglass because it doesn't absorb water, it doesn't warp, it is strong, it is corrosion resistant, and it is nonconductive. Electricians should never use aluminum ladders unless they want to have a short life. Wood is fine, but it cannot take abuse and weathering the way fiberglass can. My wooden ladders generally lasted less than five years.

Ladder Safety

Ladders may appear tame, but they are one of the most dangerous tools in this chapter. My left ankle is all steel from a ladder mishap. One minute everything was fine. A second later, I hit the ground and heard the sound of breaking bones. If you value your life or property, take the following precautions:

Use a ladder with swivel feet, which prevent the ladder from kicking out when you are working on it. However, the feet won't work if they're just sitting on the ground (top photo, left). Stomp on the first rung to sink the claw into the ground (top photo, right).

When working over a wood deck, clamp or nail a board behind the ladder's feet to keep them from kicking out (bottom photo).

Work Safe • With heavy-duty tools, I never use any cord less than 12 gauge; however, lighter-duty tools could use 14 gauge. The longer the cord, the heavier the gauge you need. I have two 50-ft. 12-gauge cords, one 50-ft. 10-gauge cord, and one 100-ft. 10-gauge cord. To keep the longest from getting tangled, I wind it around a 10-ft. board with Vs notched on both ends.

Make sure the cord you buy has enough outlets. You should buy either a cord with multiple outlets or a separate 1-ft. to 3-ft. cord with multiple outlets. Never use the cheap multiple-tap fixtures that plug into an outlet—they can't take the current flow.

Ladders come in different grades, which may be very confusing. All you have to remember is to get a type 1A fiberglass ladder of whatever length you want. (The 1A designation means that the ladder is extra-heavy-duty.) Sometimes you can go by color to determine the ladder grade. Louisville® ladders, for example, are color-coded: orange for class 1A and yellow for class 1.

There are all kinds of accessories available for ladders, and you will eventually buy a few of them as the need arises. For instance, because ladders leave scratch marks where they touch a wall (indoors or out), you may want to cap the ends. A low-cost method is to wrap a towel around the ladder ends that make contact with siding or walls. Another option is to buy special end caps.

AC-Powered Tools

Many types of AC-powered tools are required on a job site. The extra torque provided by drills, saws, and other corded tools—as opposed to cordless, battery-operated tools—is a nice thing to have. Here, I will tell you

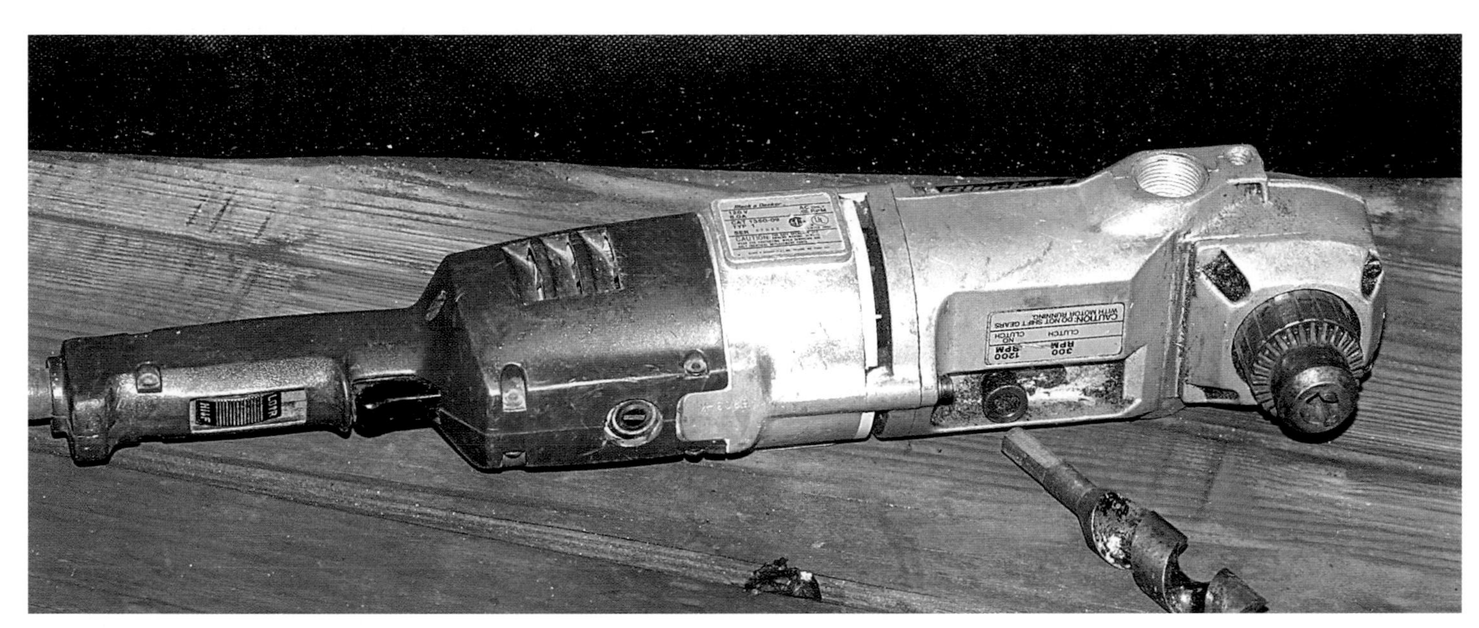

My right-angle drill excels at hole cutting and deep drilling. Its right-angle design, along with a short auger bit, allows it to fit between studs. Its tremendous torque makes it extremely dangerous to those who don't know how to use it.

my experiences with such tools, both good and bad.

Safety is paramount at home or on the job site, and tool maintenance is important to your safety. Follow the manufacturer's instructions for tool use and maintenance, and pay attention to the power cord on the tool. The cords on AC-powered tools get worn and damaged (I've lost count of the number of cords I've seen with damaged sheaths or bare wire showing). These should be fixed or replaced as soon as you notice any signs of wear. And although I shouldn't have to say this, never break the ground lug off a three-wire cord to fit a two-wire outlet—this is compromising safety for convenience. When the time comes to replace the cord's plug, install a high-quality plug as opposed to a dime-store plug. Never use a power tool around water—or even excessive moisture—and always plug into an outlet protected with a GFCI. Finally, never use a power tool that you don't feel comfortable with or that intimidates you.

Drills

Roughing-in wiring in a residence requires heavy-duty drilling. Although most drilling can be done with a ½-in. drill and a sharp spade bit, a professional has a wide spectrum of drills and bits. I use two AC-powered drills: a ½-in. pistol grip and a heavy-duty ½-in. right-angle drill. A ⅜-in. drill just doesn't have enough power. Everyday, heavy-duty use would eventually ruin it.

Heavy-duty drilling rigs, however, are very dangerous and should be used only by professionals—and with extreme care. They have so much torque that, without a clutch, they can break an arm or break a wrist. I currently use an old, outdated Black & Decker® TimberWolf (that has decided to last forever), which has a clutch, on low speed, that engages when the bit sticks. It has two speeds: 300 rpm for drilling large-diameter holes and 1,200 rpm for speed drilling. I use the low speed whenever I am using a self-feed or a hole-cutting bit and sometimes when I'm using an auger bit. This

Drill Safety Checklist

To operate a right-angle drill safely, follow these precautions:

- Maintain a good, tight grip on the tool.
- Always assume that the drill may twist out and fly around. So keep your head away from any place where it could get smashed by a runaway drill, and don't hold your arms in an awkward position. Also, do not stand in a precarious position to drill. Anchor yourself in a good, solid position.
- Wear soft, flexible leather gloves and safety glasses.
- Feel what is happening through the vibration of the drill. Many times you can pull back as you feel the drill start to work harder, become jammed up, or even hit a nail.
- Never drill with a damaged or dull bit.
- If you feel the bit jam, let go of the trigger immediately. Try not to let go of the drill. If the drill twists out of your hand, jump out of the way immediately, which leads me to the next rule.
- Always have a clear area behind you.
- Because of the last two precautions, heavy-duty drilling on a ladder should be avoided if at all possible.

drill is what you want all drills to be: heavy-duty and well designed. Right-angle drills also come in medium-duty models, which can also cause injuries. I don't recommend them for the professional because they will invariably get used in heavy-duty situations. My pistol-grip drill is a Makita® ½-in., but don't be deceived by its looks. Being ½ in. and without a clutch, it can break your wrist in a heartbeat just as a right-angle drill could.

Drill bits you'll need

For cutting through wood, I use several types of bits. Which bit you use will depend on the situation and whether you'll need to cut through a double top plate, a corner with nails, or a stud. It is important that you know what each bit can and cannot do so that you can choose the right bit for the right job.

Spade bits There are several spade designs on the market. I've found that spade bits with two protruding end cutters cut faster than the standard flat-bladed variety (see the photo below). Spade bits come in short (4-in.),

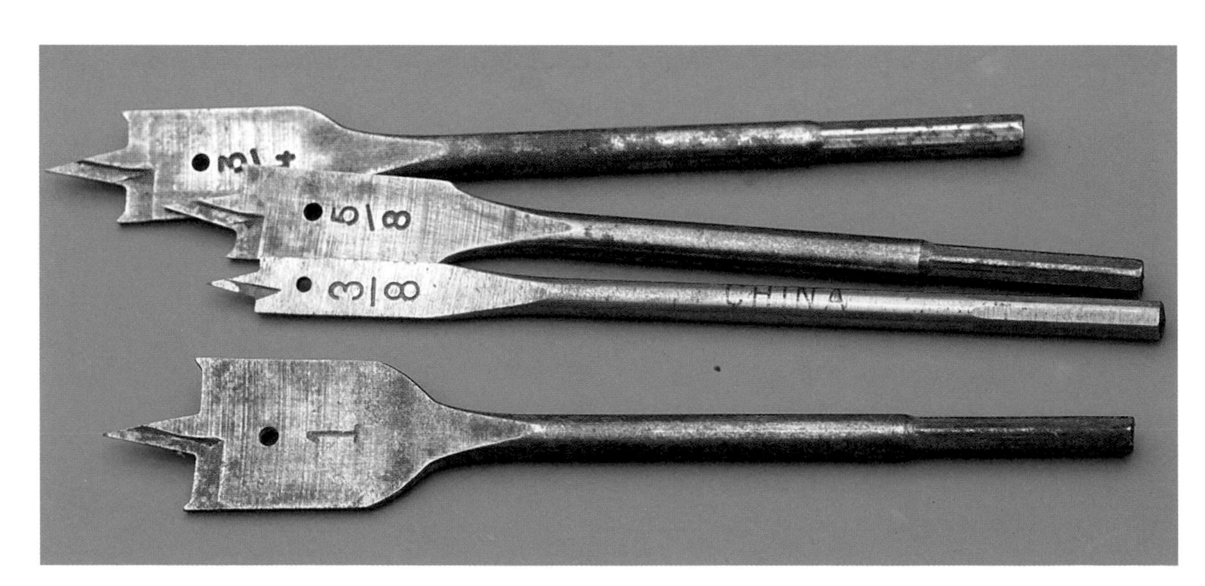

Spade bits with protruding cutters allow for fast cutting.

standard (6-in.), and long (16-in.) lengths. One problem with spade bits is that they tend to break at the shank if you don't keep the drill centered in the cutting hole. DeWALT® solved this problem by making extra-thick shanks. DeWALT also makes a ¼-in. hex shank attachment for hex drivers, which allows the blade to be installed without opening the chuck.

Work Safe • Be extremely careful when using an auger bit in a pistol-grip drill (I don't recommend it). The torque can make the drill twist with the bit, possibly breaking your wrist. Some manufacturers make a side handle that fits into the drill for extra leverage.

Auger bits Auger bits are long-bladed spiral bits used to drill long distances through wood or several studs at once. One problem with long drill bits, and especially with auger bits, is that they tend to bind and get stuck in the hole. When this happens, remove the jammed bit. First, unplug the drill and remove the drill from the bit. Next, use a small pipe wrench to turn the drill-bit shaft counterclockwise and unscrew the bit. To prevent this from happening in the first place, coat the drill bit's flat sides with wax, soap, or wire-pulling compound. The slick coating keeps the bit from binding on the sides of the hole. Another problem with auger bits is cutting into nails. You'll normally feel a repeating "thunk" as the edge of the cutter keeps cutting into the steel. This won't be a problem if you're using a bit that can cut through steel, such as the Greenlee® Nail Eater II™. If you hit a nail with this bit, the auger will cut through it, instead of jamming the bit, which could hurt the person holding the drill. (Before drilling, try to remove any nails that will be in the way or find another route.)

Auger bits are used for fast and deep hole cutting. Extension arms attach to the bit via two setscrews, allowing you to cut deeper holes.

Auger bits are available in any diameter needed. I recommend both a long bit (18 in.) and a short bit (6 in.), as well as an extension.

Self-feed wood borers The large boring bits that cut away all the wood within the perimeter of the cutting circle are called self-feed wood borers. These bits come either in one solid piece or with blades. I prefer the blades, because they cut very fast and

Self-feed wood borers cut large-diameter holes quickly. The bit on the right has replaceable cutting blades. Such bits should only be used in a heavy-duty right-angle drill.

Hole saws are safer to use than self-feed bits. They come with and without an arbor.

accurately. However, if either one cuts into a nail, look out! The bit simply stops turning. When this happens, the drill body itself will start to turn, hitting the hapless worker in the head or breaking his or her arm. This type of bit should be used only by professionals—and preferably in a drill with a clutch.

Hole saws The bits that cut only the outside perimeter of the cutting hole are called hole cutters, or hole saws. I started using hole saws in lieu of self-feed bits because I had so many close calls. I prefer carbide-tipped hole saws (special order), but the most common type is bimetal. I prefer the carbide tooth because it can cut through a nail as if the nail weren't even there. A nail will ruin a bimetal cutter.

Hole saws come with and without an arbor (the part that holds the drill bit). The arbor fastens to the cutter head, and then fits into the drill chuck to transfer the drill's rotating power to the cutter head. I prefer to buy a hole saw without an arbor and then buy the arbor separately. That way I don't have to throw the arbor away every time I buy a new hole saw. Hole saws that come with an arbor are more expensive than those without an arbor. One arbor can be used with a number of hole saws, which keeps the cost of the cutter heads to a minimum.

An alternative to the two-part cutter-head-and-arbor system is the one-piece arbored hole saw. These hole saws tend to cost more because you have to buy the arbor each time you buy the hole saw. I don't use them much because they are normally only bimetal tipped.

Work Safe • Never cut with any saw, especially ones with long blades, unless you know what is behind the wall. It could be plumbing lines or electric cable—even the main service-entrance cable. If you cut into the latter, you and the saw will be in big trouble, because there may not be an overcurrent device (breaker) in the system, except what the utility has on the pole.

Either system, whether with an integral arbor or with a separate arbor, will require a pilot bit, which centers the hole saw in the hole to be cut. Pilot bits have a habit of breaking, so have several spares on hand. One company recommends using a carbide-tipped pilot bit; I don't. A standard twist-drill bit cuts into the wood faster and is cheaper to replace if the bit breaks.

Twist drills Standard steel-drilling bits (called twist drills) are needed for drilling wood and steel. A common application is for drilling pilot holes through a floor and into the crawl space to act as a position locator.

Saws

I use corded saws on the job for heavy-duty cutting. A circular saw does the common work of cutting straight edges when I'm trying to cut through plywood or heavy wood. My primary AC tool for cutting everything is the old standby, the reciprocating saw. With the proper blade it cuts anything.

It is with the reciprocating saw that you'll make most of your accidental cuts through hot wires. That's why it's very important to have a double-insulated tool or a grounded tool. I've cut many hot wires without even a minor shock. I've lost only the blade—and the wire.

Besides the standard blades for reciprocating saws, extra thick and bimetal types are available in varying lengths. The reciprocating saw makes only rough cuts, because the blade tends to move around in the cut. Some blades move more than others do; extra thick blades seem to track better. I normally use the bimetal type because it has some bend in it; it gives without immediately breaking. That blade

Work Safe • Work Smart • Think Ahead

Never drill metal—or anything, for that matter—if the workpiece is not stationary or firmly supported (clamped, for instance). One time I thought I could drill through a small piece of metal while I was holding it. The bit caught the metal, the metal turned in a circle with the bit, and it gouged out a very large and deep hole through my thumbnail. Learn from my mistakes; don't repeat them.

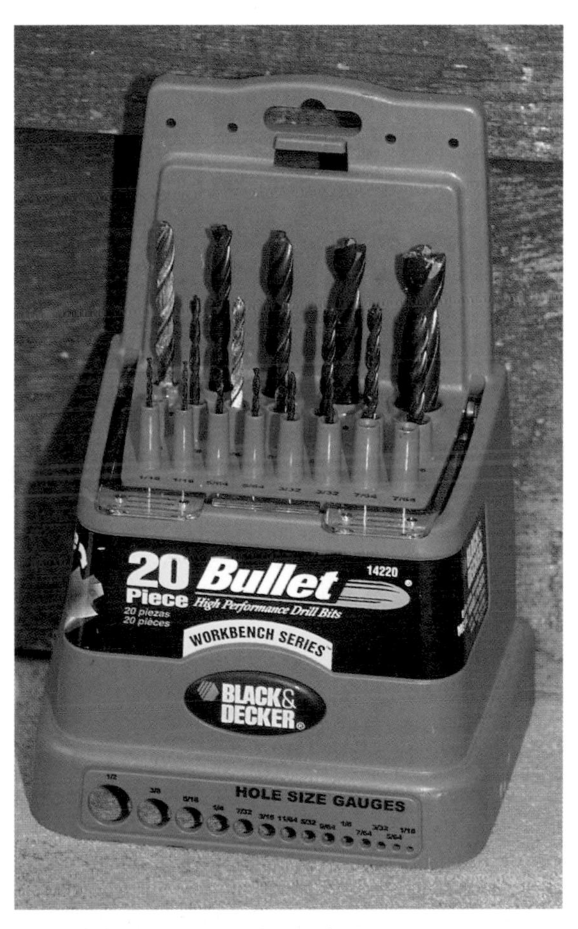

Twist drills are commonly used for drilling wood and steel.

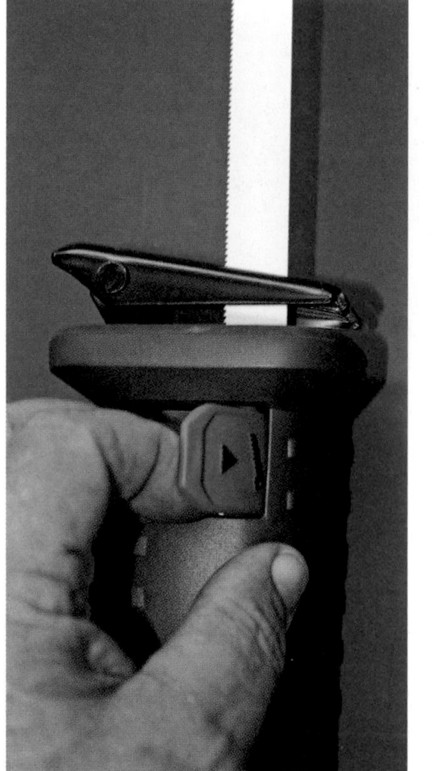
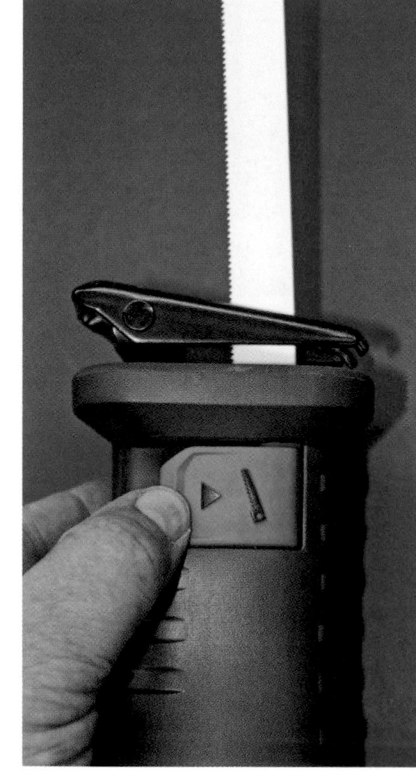

Do not purchase a recip saw unless it has a toolless blade-change mechanism. On this model, you raise the lever, insert the blade, and close the lever.

costs a little more, but its longer life more than repays the investment.

The newest design change in recip saws is the toolless blade change (long overdue). I only recommend recip saws that have this feature. Before you buy one, test it a few times to see if you like how it changes blades.

A good jigsaw is another important tool in the electrician's arsenal. This is the tool for straight cuts, as compared to the reciprocating saw and its wandering blade. I don't like saws that use a hole-in-the-blade design. They seem to break right where the hole is. I prefer tools with a bayonet design—they break, too, but not as easily. You need a variety of blades no matter which style you use. And you'll work safer if you let the jigsaw cut at its own pace.

Cordless Tools

My first experience with battery-powered tools was unpleasant. First-generation cordless tools were poorly designed and useful only for very light-duty applications. My distaste lasted for many years. Today, the market is full of high-quality cordless tools. Now I'm not without my battery-operated drill, saw, and screwdriver. Cordless tools offer a few advantages specific to each one. The most obvious, of course, is that the tools don't require an extension cord, so they are very portable and won't electrocute you.

All cordless tools need battery chargers. In the past, the battery needed about an hour to charge. Today, some chargers require only 15 minutes to do the job. For continuous work, you'll need several batteries on hand. I'd stay away from the charger design that requires you to push a button to start the charging cycle. You may come back later to find that the battery isn't charged because you forgot to push the button, or that you pushed it and it didn't work. A good charger indicates when it's charging and when it's done. It also analyzes the battery to see if it has any problems. For example, it will tell you if the battery is weak and cannot supply full power. Some of the better ones charge two batteries at once.

Drills

Cordless drills, when used with a sharp bit, can drill most of the holes in the house and won't break your arm doing it, as a heavy-duty AC model may. And they can't electrocute you, even if they cut into a hot wire, because the case is made of plastic. I do at least 90 percent of my drilling with cordless drills. They offer me speed, versatility, and safety. I can use a

These are the most common cordless tools used. Most manufacturers sell them as a kit with a carry-all bag. The advantage of this kit is that it comes with two batteries and a dual battery charger.

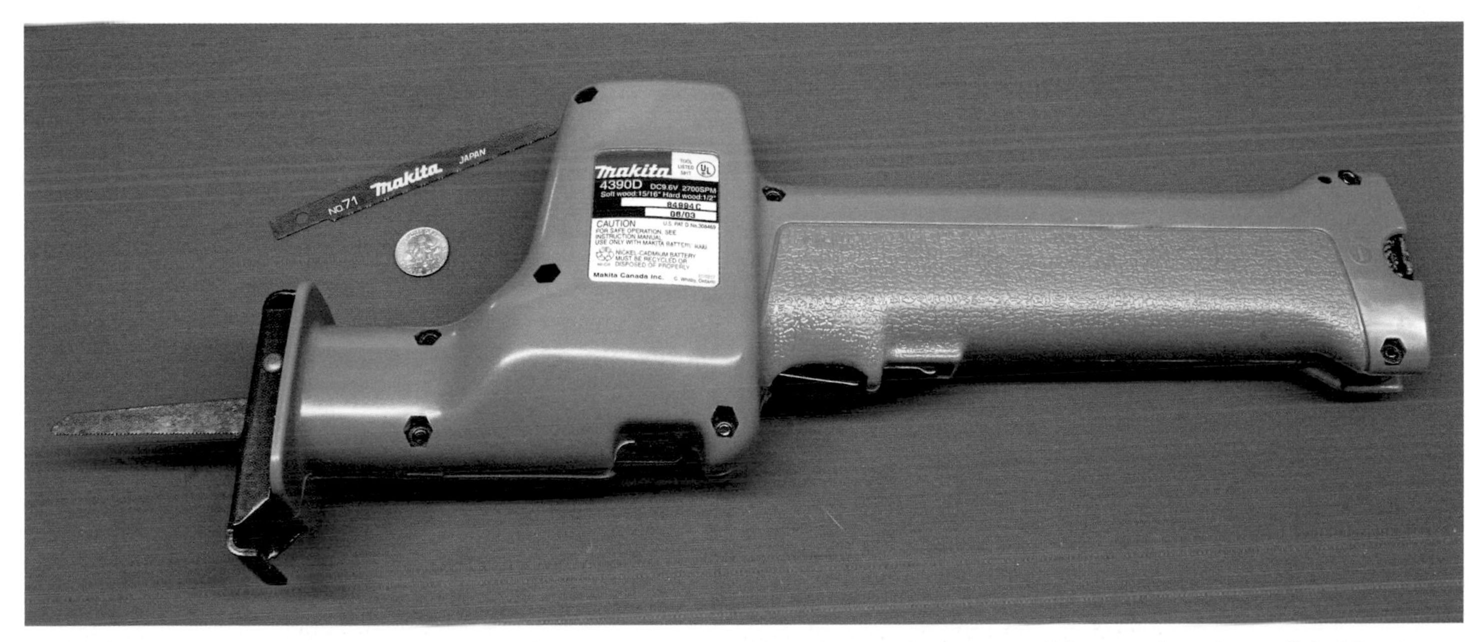

This Makita mini recip saw has a tiny thin blade that allows me to cut things that are almost impossible to reach, such as nails holding boxes onto a stud.

cordless drill on a ladder, around water, and in crawl spaces and other awkward places where I would normally have to worry about pulling, dragging, or lifting an extension cord.

Cordless drills are available in several different voltages, ranging from 9.6 volts to 24 volts and higher. The 9.6-volt through 18-volt units are the current favorite for the average do-it-yourselfer because they are lightweight and have a decent amount of power. The more serious do-it-yourselfer, or the professional, may need a higher-voltage, larger-chuck unit. However, lately I have been regressing. That is, I have forgone the one-upmanship of always getting the larger-voltage drills (those batteries are massive and heavy) and have gone back to getting a lighter drill with less voltage (and a smaller and lighter battery). I do seem to like them better.

A small cordless screwdriver can't be beat when changing out a broken switch plate.

Today's cordless drills are available with various size chucks. I only use a ½ in. (even in the smaller, lighter drills) because it can accept larger-diameter bits, which makes it the most versatile. Most cordless drills now come with keyless chucks. At first I couldn't wait to get them, and now I wish I hadn't. Keyless chucks are great for changing bits quickly, but beware of their minor problems. Without using a chuck key, you can get the bit only so tight, and if you're drilling deep the bit may stick in the wood and start spinning in the chuck. In addition, to tighten the bit in the drill you have to place your hand on the drill chuck head and pull the trigger. The chuck tightens until the motor stalls. Once the motor stalls, you have to assume the bit is tight in the chuck. This process wastes valuable battery power.

Circular saw

Another battery-operated tool that I use often is a cordless circular saw. This is not a toy—it cuts with the best of them. It works like a champ to open up areas and to make fast trims and cuts. Carpenters love this saw for overhead cutting. It's nice just to pick up the saw and cut rather than look for a grounded outlet and drag out the extension cord. If you're working in areas where no electricity is available, this is a good saw to have around.

Screwdriver

Cordless screwdrivers are getting stronger and smaller. The bigger ones have more power, but I have moved toward the smaller ones lately due to their size advantage. I use them primarily for getting into multigang switch and receptacle boxes. A four-gang switch box has 16 screws to pull before you can get into the box.

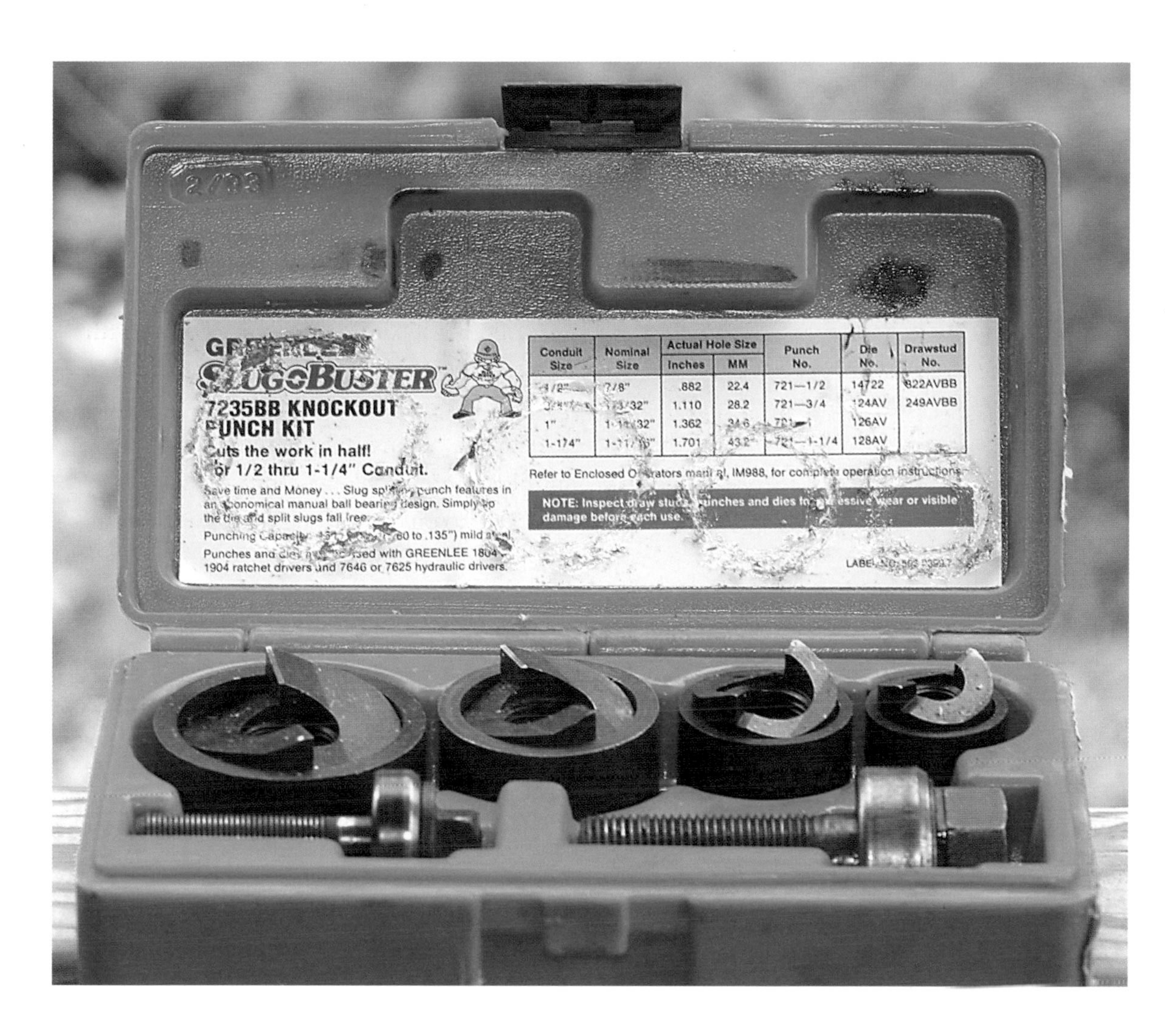

The Greenlee hole puncher kit. **Consider buying 1½-in. and 2-in. hole punchers in addition to this basic kit. To use the puncher, you'll need to drill a starter hole for the bolt.**

Specialized Tools

The tools in this section are very specialized. I don't recommend that people doing minor wiring jobs buy them, because many of them are expensive and you may only use them once or twice. Instead, these tools are for professional electricians and serious do-it-yourselfers (those folks who plan on doing some big-time wiring work around the house).

Hole punchers and stepped drill bits

All electricians know that the prepunched holes in meter bases and control panels are rarely where you want them. A lot of labor and material can be saved if you place the conduit and the cables where you need them. There are many different ways to punch holes in metal boxes. The most obvious is to use a common hole saw. After that would be special hole-saw bits sold to cut metal fast. In a pinch I've seen an electrician drill many small holes in a circle, and punch the circle out. The best method, of course, is to use a hole puncher (but you have to drill a pilot hole first).

For drilling small-diameter holes quickly, such as the pilot hole for a hole puncher or a knockout, nothing beats a stepped drill bit. This is a special metal-cutting bit that cuts various diameter holes depending on how far you push the bit into the metal (the farther you drill, the bigger the hole). Stepped drill bits are available in two designs: a blunt-end variety

Nothing beats a stepped drill for drilling holes through steel. It's commonly used to drill holes in service panels for NM connectors.

used to increase the diameter of existing holes and a tapered-point variety used to open new holes. I use either one to increase the diameter of the knockout holes in metal boxes. I also keep carbide-tipped hole cutters around for very thick steel.

Conduit benders

Professionals bend conduit, whereas do-it-yourselfers use premade angle fittings. Mastering the art of bending conduit takes a tremendous amount of time, patience, and money (for wasted conduit to practice with). If you want to install the conduit fast, forego the conduit benders and use preformed angles.

Fish tapes

Pulling and fishing wire can be a major problem if the wrong equipment is used. The purpose of a fish tape is to "fish" wire through walls. Here's how: Put the thin metal tape into a hole in a wall, maneuver it through the wall, and bring it out a different hole. Then attach a wire to the end of the fish tape and pull the wire into the wall as you pull the fish tape out.

However, standard metal-coiled fish tape is, in my opinion, the wrong tool to use. It is conductive, has an end that catches on everything in the wall so you can't get it back out (many times I have had to cut the metal fish tape and abandon it in the wall because it had caught on something), and is hard to get through a wall or ceiling cavity because it always tries to go in a circle. To counter this, I cut several feet of metal fish tape off my reel, bend it straight (I store it straight as well), and cut off the looped end. Now it easily slides inside wall cavities and through insulation. I no longer recommend using fish tape made of metal, and the nylon tapes I've seen stay circular just like the metal ones do, and they catch stuff in the wall cavity. However, the fiberglass tapes work very well.

What makes a good fish tape? It should be nonconductive and flexible but just rigid enough to stay straight in the wall. A 20-ft. straight-length polybutylene plumbing pipe or a semiflexible equivalent, such as PEX pipe, makes a perfect fish tape. It is rigid enough to go long distances within walls and ceilings and stays flexible enough to bend into the cutout hole and pull through. Once through, you can push the wire up into the pipe and tape it off. As you pull the pipe back, out comes the wire. This is the method I use most often. To prevent the end of the pipe from catching, I tape or insert a marble at the end.

Rotary hammer

Used to break up and drill concrete and punch through cinder block and brick walls, the rotary hammer (also called a hammer drill) is an indispensable tool for the professional. Whatever brand you select, make sure you get one with a clutch that engages when the bit sticks, because this type of drill has tremendous torque. I also use the drill for installing ground rods. Makita, Bosch®, and several other manufacturers make a special attachment for their hammer drills that will drive a ground rod into the soil in just a few seconds. It sure beats using a sledgehammer.

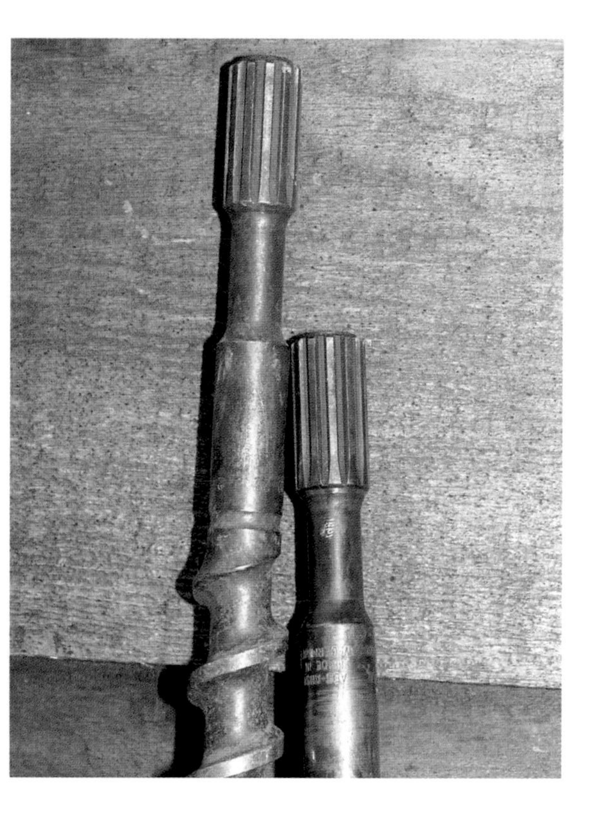

A rotary hammer is the tool to use for getting through concrete walls and driving ground rods. This type of hammer uses bits splined at their drive end. The splines keep the shaft from spinning in the chuck. Each bit costs around $20 to $100.

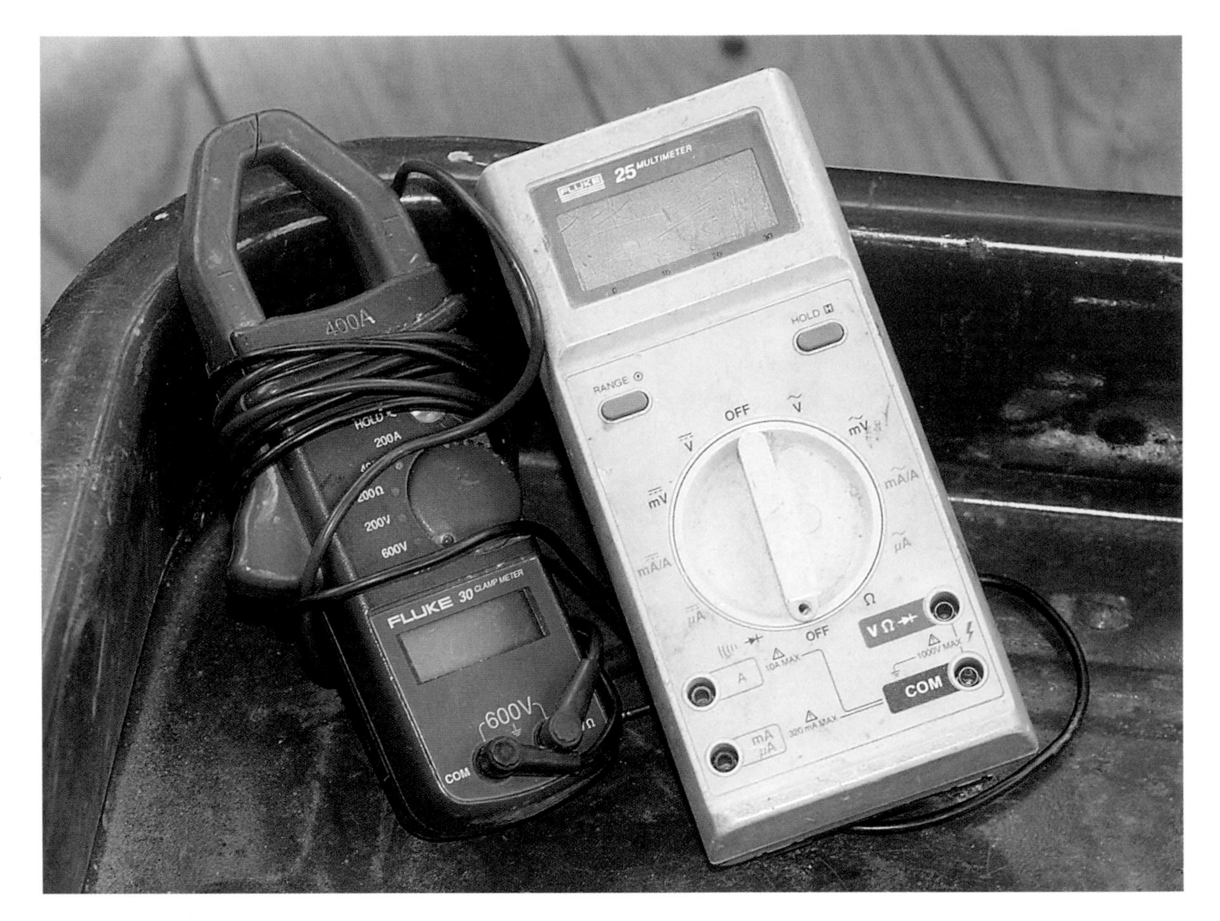

Two tools I cannot do without. The yellow meter is auto-ranging—you don't have to turn a knob to an approximate voltage—and measures both AC and DC. The black meter measures only AC. Its primary purpose, however, is to measure current. The fingers clamp around a wire and the meter gives an indication of the amount of current in that wire.

VOMs and multimeters

Volt-ohm meters are the most valuable tools in any electrician's tool chest. I still call them VOMs because that's what we used to call them in the early days. Technically, in this day and age, they're called multimeters. Having been through many of these devices—from the old analog Simpson 260® to today's digitals—I have to say that my digital Fluke® model 25 and model 30 are the best I've ever used. They do everything with extreme accuracy, and they are almost indestructible (I've dropped mine on concrete and they still work). Some digitals make you wait while they count up to get the voltage. However, both of these models give an instant voltage reading. They also work in cold weather (some models don't).

The model 25 is auto-ranging, which means that I don't have to know the voltage before I measure it and switch it to the right scale. I simply put the probes on the testing points. It has a built-in continuity tester to test light bulbs and water-heater elements, a capacitance tester to test motor-starting capacitors, and a millivolt meter, which is useful when testing thermocouples on gas water heaters.

The model 30 clamp-on meter indicates how much current the branch circuit is pulling without having to open the circuit. I can check submersible pumps, water heaters, and electric baseboard heaters to verify how much current they're pulling without getting near a bare wire. The model 30 is also a multimeter that measures voltage, resistance, and continuity (measuring continuity is a way of verifying that two points are electrically connected). However, it is not auto-ranging, meaning that you'll have to know the approximate voltage of the circuit and switch it to the proper scale. I prefer having two multimeters on hand—one as a backup.

There are many accessories available for VOMs: extralong test leads, alligator clips that allow one or both hands to be free, and line splitters. Line splitters allow current to be checked in any corded appliance.

Plug-in analyzers

I use plug-in circuit analyzers and GFCI testers to test if a receptacle is wired properly and whether or not a GFCI is still providing ground-fault protection—you'd be surprised to know how many are not. But there are situations when you cannot test a GFCI using a plug-in tester. The bottom line is that whenever possible, always use an integral GFCI test button.

In the past, a GFCI tester was simply a plug-in tester with a push button on it to simulate a ground fault. No longer. State-of-the-art testers can now give you a whole world of circuit analysis from a little plug-in detector. This unit is too elaborate and expensive for the amateur but should be in the arsenal of every electrician and official who does a lot of testing.

Besides standard wire polarity checks and GFCI testing, an analyzer also tests for the presence of a bootleg ground. A bootleg ground occurs when someone has incorrectly installed a jumper wire on the receptacle back from the neutral over to the ground lug (a common tester will give you a false "OK" reading).

I also use this tester to check the resistance along the equipment-grounding conductor. That is, if you plug it into a receptacle, it will give you a resistance reading along the bare ground wire from the receptacle back to the breaker panel. This test is very important if you are doing any lightning suppression. A surge suppressor—for example, the one at a computer—takes the surges off the line and sends them back to the panel and earth via the equipment grounding conductor. If you have a loose screw on a receptacle or a bad splice, the suppressor won't work properly. If the test gives you a high reading, then you know there is a problem.

Safety Equipment

The most dangerous problem on the job site—besides electrocution—is debris going into your eyes, which makes approved safety glasses an electrician's best friend. They must be comfortable, or you will be tempted not to wear them. Be sure to get some that have adjustable arms. Antifog and antiscratch coatings are nice, but they fog and scratch anyway. I hang my glasses from the rearview mirror of my truck, which reminds me to put them on when I leave the truck. Get into the habit of hanging your safety glasses around your neck when they're not in use. I've found that if I don't have my glasses within arm's reach when I need them, I won't use them.

After your eyes, your toes are the most in danger. Always wear steel-toed shoes. Falling drywall, plywood, and heavy control-panel lids are always threatening to amputate your toes or injure your feet. If you're around a job site long enough, something will happen. If hard-soled steel-toed shoes are not to your liking, steel-toed running-type shoes are now available.

Rubber gloves, soft leather gloves, a rubber mat, and a hard hat are items that should also be kept with you—or at least on your work truck. Rubber gloves can save your life. Once I was wiring merrily away in a panel—nowhere near a bare hot wire—and zap! What happened? I discovered that I had brushed against a hot wire where a previous electrician had gouged off a section of insulation. From then on, I started wearing rubber dishwashing gloves in hot electrical panels.

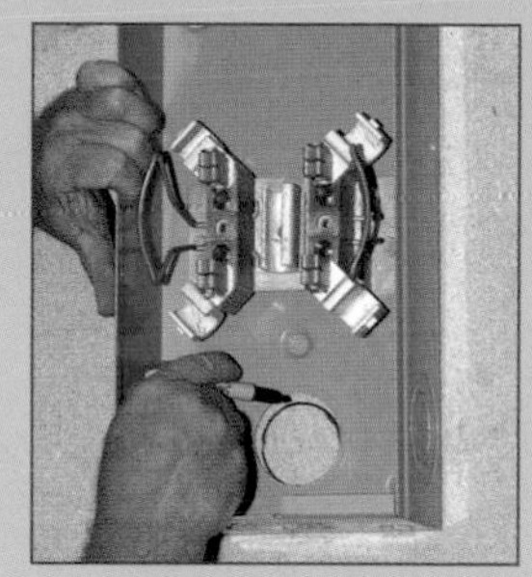

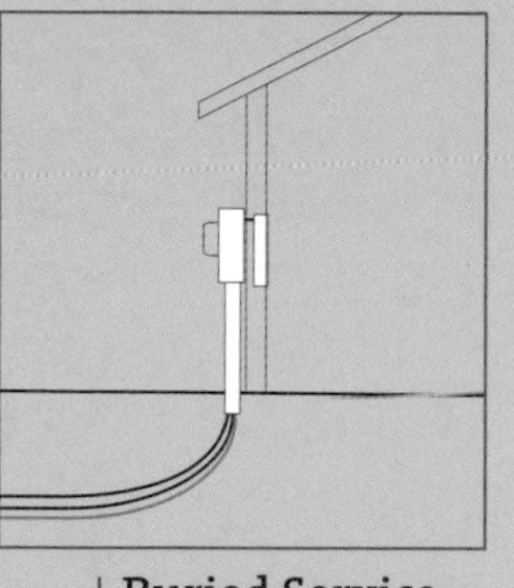

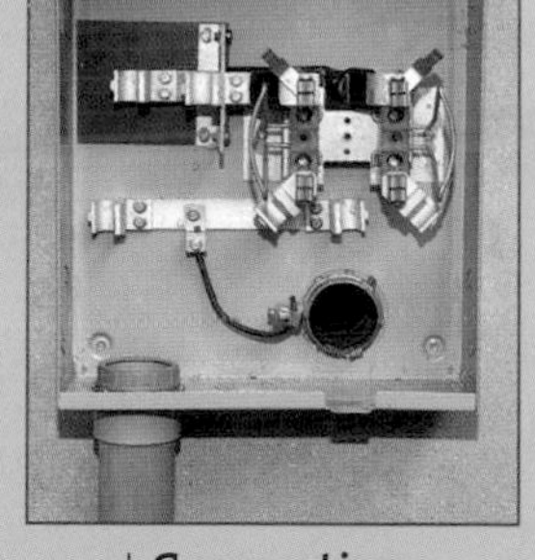

CHAPTER

3

The Service Entrance

The **service entrance (SE)** is the part of the electrical system that starts at the utility transformer and terminates at the main service panel. It includes the cable from the transformer to the residence, the meter base, and the service panel. A residence has either a buried or an aerial service entrance. In either case, part of the cable is provided and installed by the utility; the rest is your responsibility. In a buried service, the utility provides the cable and labor to install the service entrance all the way to the meter base. In an aerial service, the wiring that swings from the pole to the house is called a service drop, or triplex, and is provided and installed by the utility. The utility's responsibility stops at the point where the drip loop connects to the house's service cable. In both cases, the electrician installs the meter base.

Many initial issues in new residential construction concern the service entrance, which is a major factor in the design and construction of a home. Coordination and cooperation between the utility, the local inspector, the building and electrical contractors, and the homeowner is mandatory during this time to avoid problems and errors. This chapter will help you make intelligent decisions regarding the design and installation of the service entrance.

A buried service entrance requires a deep ditch—expect things to be quite messy.

Buried versus Aerial Service

Advantages of Buried Service

- Fewer trees to cut
- Smaller right-of-way requirements
- House looks better without exposed conduit and cable and utility poles
- SE protected from ice storms and falling trees
- No clearance problems
- Utility company provides cable to meter base
- No bent masts
- No cables crossing roof
- No damage to roofline (aerial service gets extra tricky if you have to insert a mast through a metal or tile roof)

Advantages of Aerial Service

- Won't damage your landscaping
- Easy to see potential SE problems
- No chance of cutting through buried utility
- Aerial lines can span boulders, creeks, and swamps

Buried or Aerial Service

The first decision is whether you want a buried or an aerial service entrance. (If you don't request a particular type of service, the engineer will figure the cost of both and put in the cheaper option.) If you choose the more expensive option but it is within the established company allowance, you pay nothing; however, if it is over that amount you may be expected to pay the difference. In some cases you can avoid that added expense by doing some of the work yourself. For example, with buried cable, the utility usually takes care of opening the ditch, but you could offer to open and close the ditch yourself.

Advantages of each service are listed at left. Whichever choice you make, you need to meet with the utility engineer and discuss where the utility cable, and perhaps the utility poles, will be placed on the property. In order to do that, you must know on which side of the building you want to place the meter base.

Calculating Amps, Choosing Cable

As soon as you know whether you're going with a buried or an aerial service entrance, your next task is to figure out the amperage size of the residence's service. You can guess the loading simply by adding most of the heavy-duty loads, which is what most people do. However, many times pros are required to do a load calculation. In addition, the inspector may want to see a load calculation before he or she will issue a permit. The best reason to do a load calculation is to ensure that you don't underestimate your needs and install a service that's too small.

The official electrical load of a residence is sized according to NEC guidelines. Once you

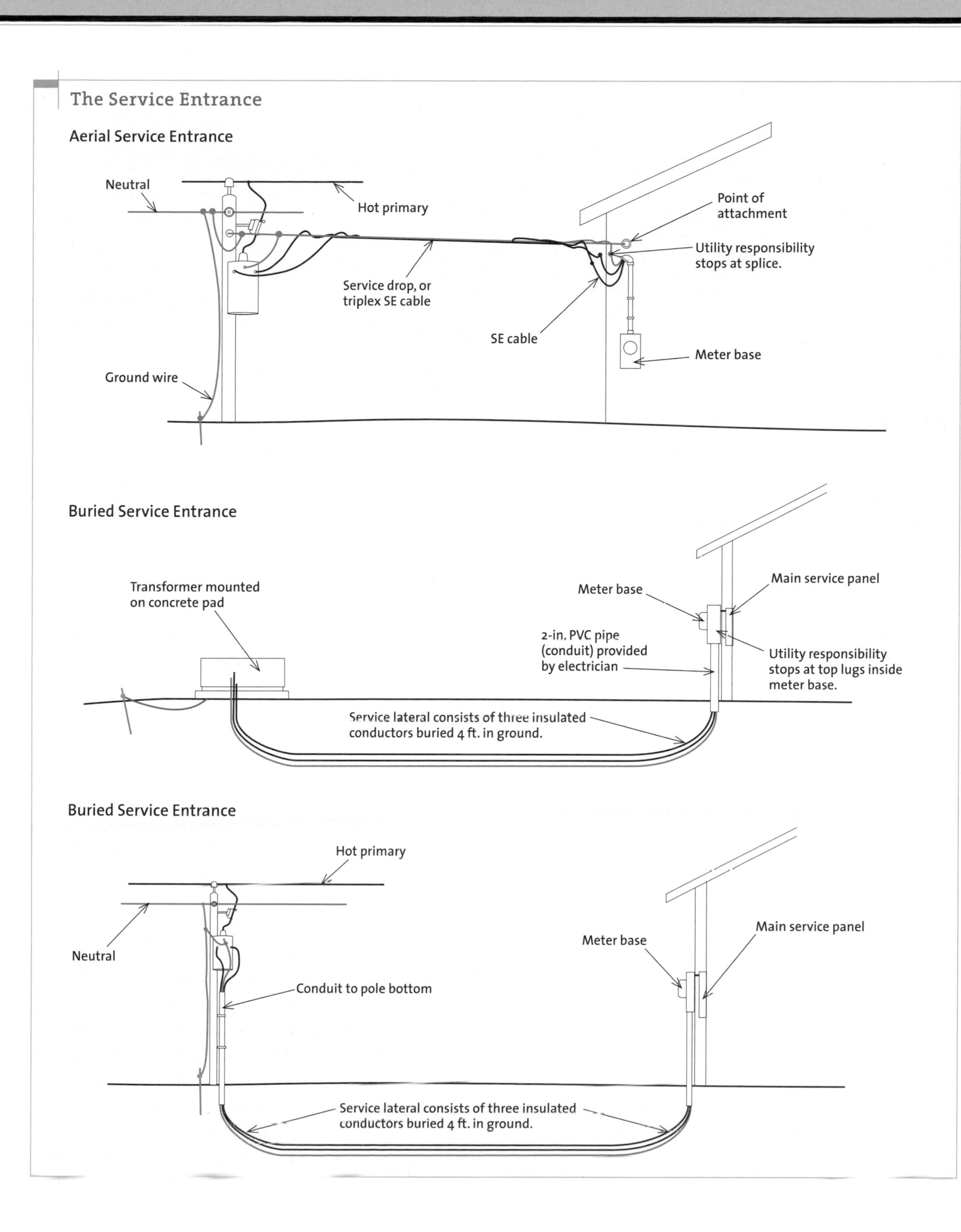
The Service Entrance
Aerial Service Entrance
Neutral
Hot primary
Point of attachment
Utility responsibility stops at splice.
Service drop, or triplex SE cable
SE cable
Meter base
Ground wire
Buried Service Entrance
Transformer mounted on concrete pad
Meter base
Main service panel
2-in. PVC pipe (conduit) provided by electrician
Utility responsibility stops at top lugs inside meter base.
Service lateral consists of three insulated conductors buried 4 ft. in ground.
Buried Service Entrance
Hot primary
Neutral
Meter base
Main service panel
Conduit to pole bottom
Service lateral consists of three insulated conductors buried 4 ft. in ground.

Drip Loop to Meter Base

Size of Service	Size of Copper Conductor	Size of Aluminum Conductor
200 amp	2/0	4/0
300 amp	250 kcmil	350 kcmil
400 amp	400 kcmil	600 kcmil

Meter Base to Service Panel

Size of Service	Size of Copper Conductor	Size of Aluminum Conductor
200 amp*	2/0	4/0
300 amp (200-amp panel and 100-amp panel)	One set of 2/0 and one set of #4, respectively	One set of 4/0 and one set of #2, respectively
400 amp (two 200-amp panels)	Two sets of 2/0	Two sets of 4/0

*Never install a singular panel above 200 amps. A typical 300-amp panel, even a 400-amp panel, has only 40 circuits. A house that needs a 400-amp service panel will need an abundance of circuits—more than what you would want to squeeze into a single panel. You'll need to split the cables into two panels in order to create enough room for all the circuits.

know the load of the house, you can choose the correct-size meter base, cable, and main panel for the amperage of the house. Here's how I figure the load for a 2,000-sq.-ft. house:

1. Calculate the square footage of the house, excluding uninhabited areas, such as an unfinished attic, garage, and carport, to obtain the general lighting requirements. Multiply the square footage by 3 to obtain the power required: 2,000 × 3 = 6,000 watts, or VA.

2. Add 1,500 VA for each small-appliance circuit in the kitchen. A minimum of two is required by code, so the number will be 3,000 VA. Then add 1,500 VA for the laundry, excluding the dryer. Total = 4,500 VA.

3. List all appliances to be included and write down their power ratings in volt amps or watts. Some common appliances include a dishwasher (1,500 VA), water heater (4,500 watts), clothes dryer (4,000 watts; do not use the standard rating of 5,000 watts for the dryer—use the nameplate rating for the load calculation; the same goes for the stove), stove (8,000 watts), water pump (1,500 VA), garbage disposal (900 VA), whirlpool tub, and so on. Don't include the air conditioner or the heating unit. Let's assume that the total in this case is 30,000 watts or VA.

4. Add the numbers up: 6,000 + 4,000 + 30,000 = 40,000 watts. Because all loads don't operate at the same time, take the first 10,000 at face value, then take 40 percent off the rest: 10,000 + [.4(30,000)] = 10,000 + 12,000 = 22,000.

Making Way for the Service Entrance

The route the utility cables take to get to the house should be decided upon before construction begins. In some locales this is no big deal—for instance, when you are adjacent to a utility line. In the country, however, it can be a very big deal. You may be quite a distance from the utility. If the cable crosses another property, a right-of-way will have to be obtained by the utility from the owner. You may also have to sign a right-of-way to get it across your property.

A right-of-way is a legal document that gives the utility permission to install and maintain equipment on your property or on someone else's; normally the utility will not install its lines without the form. It is signed and filed in the county courthouse. Once the form is signed, the utility, not you, controls the section of property specified in the right-of-way (but you still pay the taxes). If the utility has to clear trees and such from the right-of-way, they may charge you for it—therefore, it is common for the landowner to clear it.

The width of the area to be cleared depends on whether primary (high voltage) or secondary (120/240 volt) power lines will be passing through it. An aerial primary line needs at least 40 ft. of clearance. A secondary line normally requires 20 ft. or less, depending on the utility. Whether it will be a primary or a secondary line depends on the length of line and the voltage drop. The utility engineer will give you a report.

A buried service entrance requires only a path wide enough for equipment to pass through. If placed along the driveway, no additional clearing of trees may be needed.

Aerial Lines through Woods

A primary line requires a 40-ft. clearing through woods. A secondary line requires 20 ft.

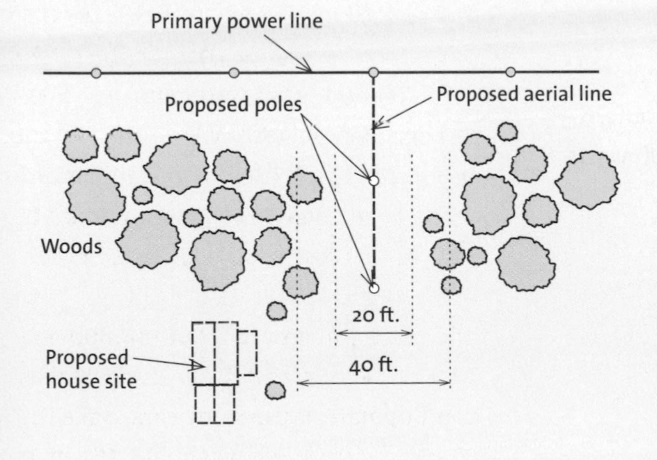

Buried Service

A clearing through the woods needs to be only as wide as a trencher or backhoe. Other equipment can go around trees or boulders.

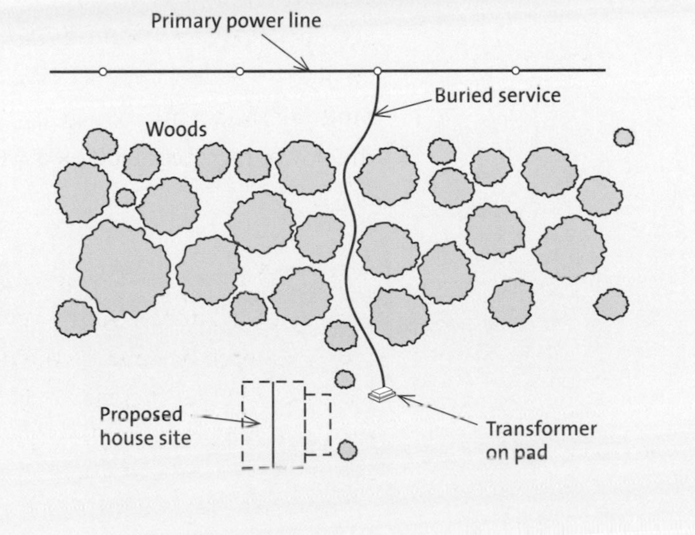

Above Code • This installation requires a 200-amp main panel and meter base. But since we are only 25 amps from maxing out the panel (the voltage of one hair dryer or window-fitted air conditioner), I recommend investing in a larger service to ensure that the homeowner has enough amperage, not only for the present load, but also for future loads. In general, I never install anything smaller than a 200-amp service (it is against code to install anything less than 100). Compared to the smaller services, the cost difference is insignificant. For loads greater than 200 amps, round up to the next 100-amp increment, such as 300 or 400 amps.

5. Add the power of only the heating unit or the air conditioner, whichever is the larger, to the total load. In this example, the heating unit is 20,000 VA. Therefore, 20,000 + 22,000 = 42,000.

6. Divide the total wattage by the service voltage to find the total electrical load of the house: 42,000 watts/240 volts = 175 amps.

Choosing SE cable

The ratings of the SE conductors and meter base must match the ampacity of the house's electrical system. This breaks down into two sections: drip loop to meter base (aerial service only) and meter base to service panel. When dealing with larger services, the big cables that are used from drip loop to meter base are very expensive and very hard to install. This alone should convince

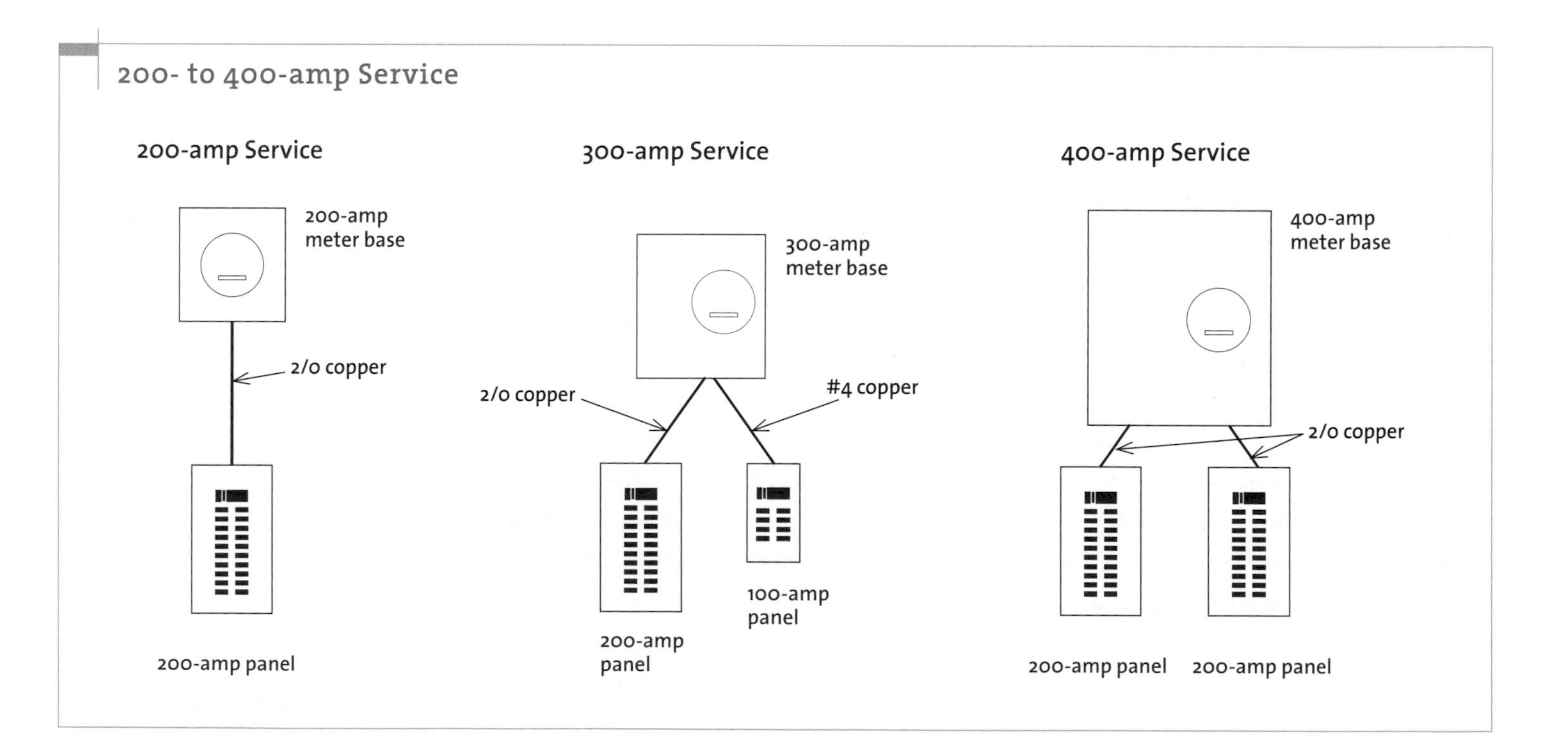

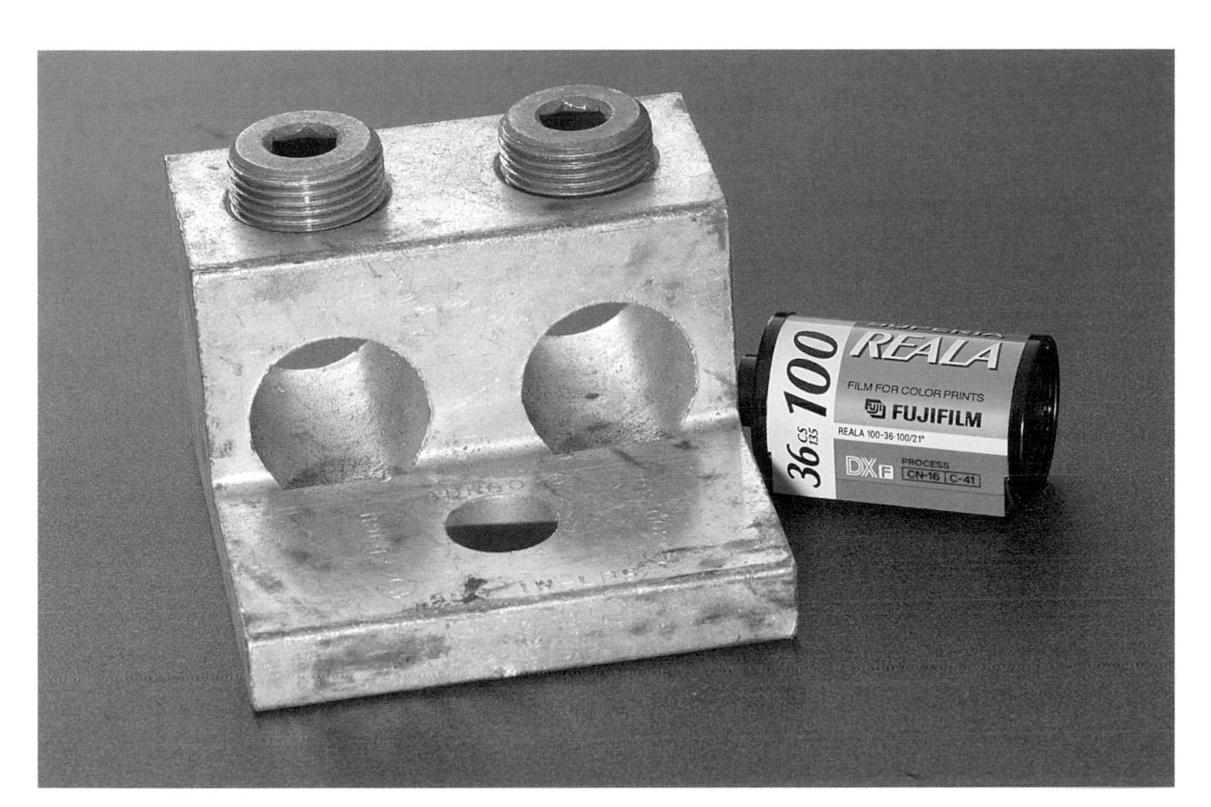

About the size of two 35mm film cartridges, this two-cable lug bolts onto a 400-amp meter base. The lug attaches onto a bus immediately below the meter itself. You'll need three of these (hot, hot, and neutral) if you are installing two 200-amp service panels.

you to choose buried. With buried, the utility provides the expensive cable and does the work. You'll also have to decide whether to use aluminum or copper SE. Copper style U (the flat cable) is out there but may be hard to find (call around to various electrical distributors). I prefer copper service entrance, because it is easier to install two-sizes-smaller gauge for the same amperage and because copper corrodes less than aluminum. Copper THHN/THWN in individual conductors is quite common and available at most electrical supply houses.

Meter bases

Meter bases normally come in 100 amp, 300 amp, and 400 amp. A 400-amp meter base is quite large and heavy and has bolt-on lugs for all the cables—even for the meter itself. The lugs, which are sometimes sold separately, may set you back an extra $5 to $10.

If you have two service panels, you will need to run two sets of SE cables from the meter base. For multiple cables leaving the meter base, you can buy a two-hole lug-bolt kit that accepts two cables side by side—even aluminum and copper. Don't forget to ask for them when you get the meter base.

Local requirements

Once you've decided where the cable is going to run on the property and what size service entrance you need, the next step is to ask

Take Note • When dealing with the utility and the inspector, be sure to ask about local procedures and codes that are not part of the NEC. (Although the NEC is a national code, local requirements may differ—and supercede—the NEC.). For example, does your area require the meter base to be adjacent to a sidewalk? Does your area require a specific way of taking the SE cable into the basement or into the house?

about the requirements of the local utility and inspector. These change from utility to utility and from locality to locality—sometimes even from inspector to inspector. Your area may have a specific system that you must follow to obtain service (an electrical hookup). For example, in my area, the first step is to call the utility and log a request in their system. They'll ask if this is a new installation or an upgrade, the amperage of service entrance for which you want to apply, whether it's buried or aerial, and other questions. Some utilities may require a deposit to defray their costs and ensure that you complete the work. The utility will then give you a service request number and issue you a proper-amperage meter base. When that's completed, your next step is to apply for a permit at the local inspection department. (Be sure to call ahead for an appointment.)

Once all the paperwork is done and you've obtained a permit, you can install the service entrance. For a buried entrance, that means from the meter base to the service panel. For an aerial entrance, that means from the drip loop, through the meter base, and to the service entrance panel. To install the service entrance, you will have to "borrow" electricity from a neighbor, use a generator, pay for a temporary power hookup from the utility ($100 to $500), or use battery-operated tools.

You are normally allowed to install one GFCI-protected 120-volt powered receptacle in addition to the service panel. This receptacle is usually installed immediately adjacent to the service panel. Once all this is done, call your local inspector for a service entrance inspection. The inspector will verify that everything is done within code. If you pass, he or she will notify the utility and they will put you on a waiting list. From that point on, it can take several days—sometimes months—before you get your utility hookup.

Once you have utility power, you are allowed to use that one GFCI receptacle for general-purpose tools. Do not heat up any other 120-volt receptacles until the final inspection. When you have run all the wire within the walls and in the outlets, you are ready for a rough inspection. When the finished walls are up and the receptacles and switches are installed and powered, you are ready for a final inspection.

Locating the Meter Base

Before the utility runs the service drop to the house, you must install the meter base. If the meter base has to be installed before the fin-

Take Note • If the inspector says that a cutoff panel is needed, then code requires that the main service panel be wired as a subpanel. That is, all the neutrals must be isolated from the grounds.

ished siding is up, attach the meter base to the rough wall and allow the finished siding to be cut around it. Another option is to shim out to the exact distance of the proposed siding and install the siding behind it later. If brick siding will be installed on the house, the meter base will have to wait until the brick is installed. Fasten the box into the mortar between individual bricks. Any type of concrete fastener will work.

Regardless of whether you choose a buried or an aerial service entrance, always try to locate the utility meter base and the main service panel back to back. This design is less expensive and will be less of a problem when it comes time to run cable to the service panel. If the panel is installed several feet away from the meter base, a cutoff panel (which is expensive and will require additional work to install) may have to be installed next to the meter base. The cutoff panel is simply a main disconnect switch, or circuit breaker, that provides protection for the SE cable on its way to the main service panel. Without a cutoff panel, if the cable is damaged—say, by a carpenter driving a nail into it—before it reaches the breaker, the only way to shut off the power is at the utility cutoff at the power pole. By the time the utility cutoff removes the power, all the service cable will probably be damaged beyond repair, and it may even start a fire. When such a fault occurs, the wires hum loudly, the utility meter power-usage disk whirls so fast you can't see it, and the utility transformer buzzes. (This is what I've heard; it hasn't happened to me. Yet.)

Mounting the meter base

The meter base is either mounted at the exact height required by the utility (around 5 ft. from final grade), or in absence of a spec,

Back-to-Back Meter Base and Panel

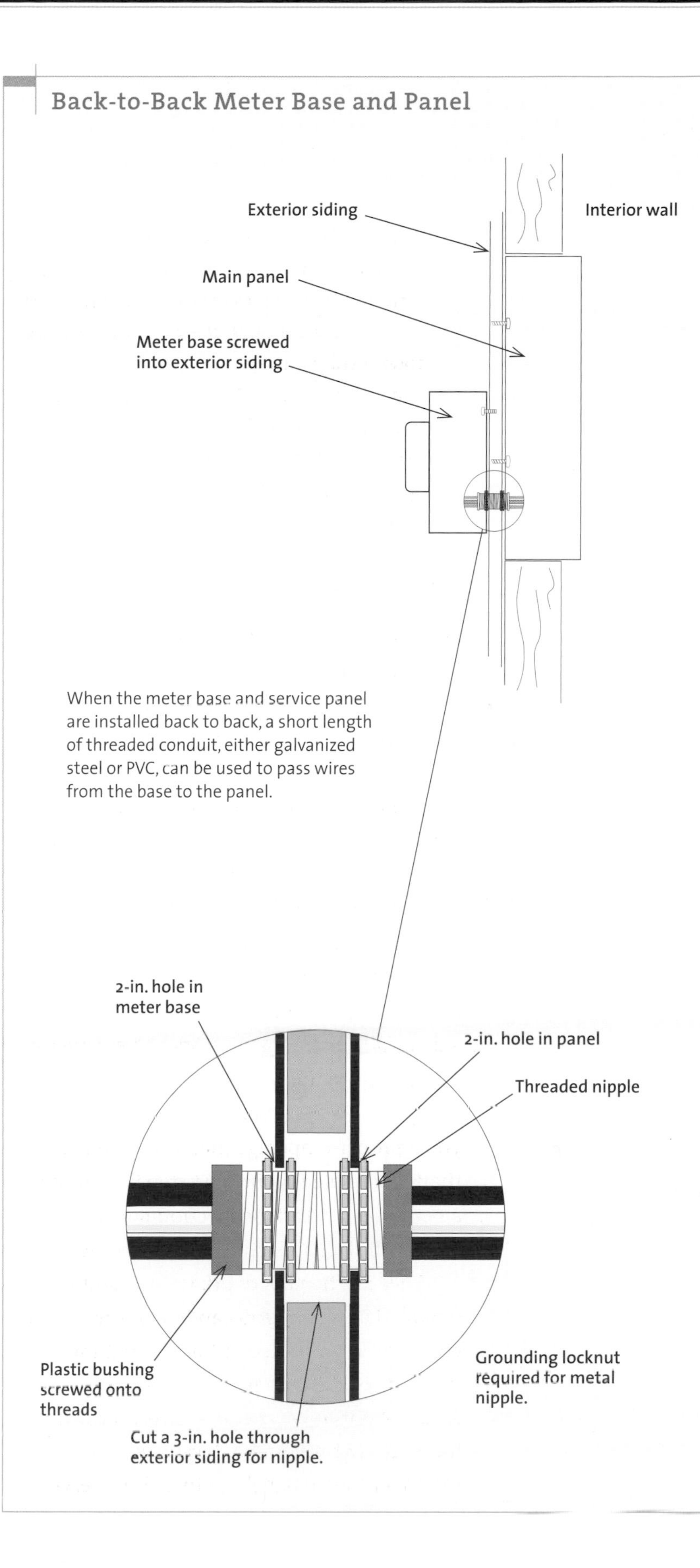

When the meter base and service panel are installed back to back, a short length of threaded conduit, either galvanized steel or PVC, can be used to pass wires from the base to the panel.

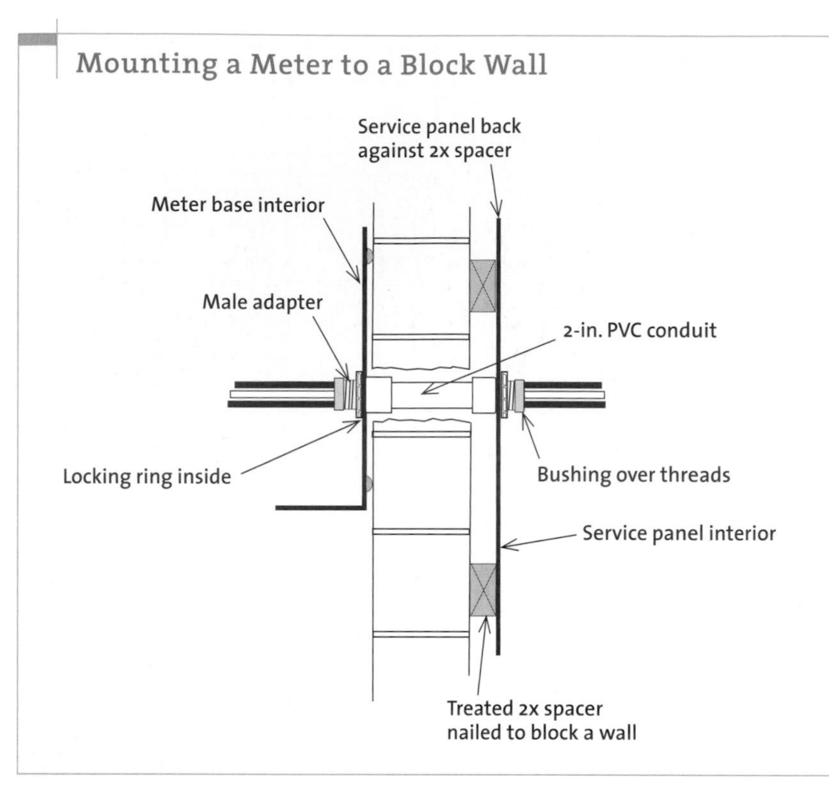

With the panel top placed 5 ft. above finished grade, place the meter base against the siding and draw a circle around the knockout hole. Once opened, this will be the entry into the house.

Work Safe • Before you mount the meter base, think about where you want the service panel to be, so that the two can be installed back to back. If the SE cable knockouts on the two panels don't align, you won't be able to connect them with a straight piece of conduit.

at head height for easy reading. Before you begin work, double check to make sure you have the right meter base—buried for buried, aerial for aerial—and the right amperage for the job. In the meter base, you'll see four miniature knockouts for four screws. Use a punch to remove the knockouts and be sure to remove all of the metal (check behind the box, because the knockout may leave a jagged metal tab). Remove a knock-out from the back of the meter base for the conduit that will be the pass-through for the cables running to the main service panel. (For the purposes of this discussion, assume the main panel is located back-to-back with the meter base.)

If you plan to exit the back of the meter base to get to the main panel, you'll want to knock or cut a hole in the exterior wall before installing it. To do this, place the panel top on the 5-ft. mark, use a magnetic torpedo level to

Punching a New Hole into a Meter Base or Panel

1 Use a stepped bit to drill a ½-in. hole in the box where you want to position the knockout. Make sure you allow extra room around the hole for the locknut to turn.

2 On each side of the new hole are the two primary parts of a punch: a cutting head (right) and a receiving head (left). Not shown is a bolt that comes with the knockout when you purchase the set.

3 Insert the bolt through the receiving head on the opposite side of the panel and the bolt end through the panel hole. The hollow side of the receiving head should face the panel.

4 Screw the cutter head onto the bolt until it's snug against the metal, then tighten the bolt with a socket wrench. As the bolt is tightened, the cutter head will slice a hole through the panel. This 2-in.-diameter hole is for a 2-in. conduit.

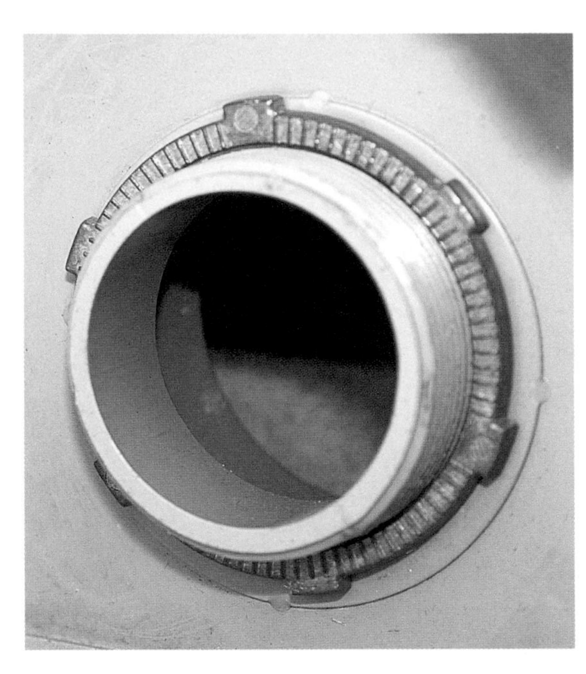
Where required, insert a male adapter into the meter base from the back so its threads extend fully into the box. Once in, screw on a locknut and tap-tighten it with screwdriver and hammer.

Screw a bushing onto the male adapter's threads. This bushing protects the cable from the thread's sharp edges.

make sure the panel is level, and then trace the knockout on the wall. The radius of this circle must be increased to allow for the conduit hub. (If the proposed hole cuts into a stud, shift the base slightly to go around it.) To cut through wood, use a drill and/or a reciprocating saw. For concrete, I use my Makita hammer drill. Once the exterior wall is opened, install any adapters on the panel that may be required.

For 200-amp service, the diameter of the conduit and fittings is normally 2 in.; a larger service will need a 3-in.- or 4-in.-diameter conduit. I prefer to have the buried utility cable enter on the left side of the meter-base bottom instead of in the center. This keeps the utility cable on the left side and leaves ample room to run the exiting SE cable out the right side.

Mount the meter base against the exterior wall with four screws or bolts (do not use nails—they can work loose). I use 5⁄16-in. hex-drive screws. Once the meter panel is secured, put a bead of clear silicone caulk along the top and sides and between the panel and the wall to prevent rainwater from running behind the panel and into the cut hole.

Buried Service Entrance

A buried service is the easiest, and sometimes the cheapest, to have installed. To obtain buried service, all you have to do is request it. The utility will normally send out an engineer to make a drawing and figure out what the price (if any) will be. Before buried service goes in, you will need to locate the water line to the well, the sewer line, the septic tank, the distribution field, and the buried telephone line so they don't get damaged.

For buried service, the utility buries from the closest aerial pole or installs a transformer close to the house. The transformer is usually inside a large green box mounted on a concrete pad. The transformer serves the same purpose as the transformer mounted on the utility pole—it changes the higher transmission voltage to a lower voltage for a residence. The utility is responsible for getting the wires, which are called the service lateral, from the transformer to the actual connection at the meter base. The homeowner, the contractor, and the

Make sure you leave a piece of conduit (plus male adapter, two locknuts, and a bushing) for the utility contractor to install when he or she brings in the buried SE cable. Simply lean it against the meter.

utility company must agree upon the location of the transformer.

The meter base and the connecting SE cable should be in place before the utility arrives to install the buried service. The electrician must provide the utility crew with a section of PVC conduit long enough to extend from the bottom of the meter base at least 2 ft. underground (approximately 5 ft. to 6 ft. total, depending on the depth of the ditch), a male adapter, two locknuts, and sometimes a long sweep at the bottom. The conduit attaches to the bottom of the meter base with a male hub, two locknuts, and, depending on the inspector, a bushing to provide a more rounded surface for the cable to bend against.

Once you cut the conduit to length, simply lean it up against the meter base and wait for the utility company to finish the installation. Do not glue the male adapter onto the long sweep, because the utility crew may need to cut the pipe to length as they install it. The conduit diameter is normally 2 in. for a 200-amp service and 3 in. to 4 in. for larger services.

Aerial Service Entrance

Aerial service can take an entire day to install, even when you know what you are doing. You must take into consideration the height of the utility cable, porches, and decks, as well as whether or not you're crossing the existing roof. Of course, there's the added safety factor—you will be doing most of the work on a ladder or on the roof.

There are several ways to interface the house SE cable with the utility cable at the drip loop, and your area may require a specific way that is not required by code. Regardless of the interface, your first concern is establishing proper clearances.

Aerial clearances

In Charlottesville, Virginia, a girl was killed when a low-hanging wire caught her by the neck and knocked her off a hay wagon. This tragedy might not have happened if the wire had been at its proper height. After that tragic death, I was hired by the utility to check for other low clearances. I found a book full. What people don't know (and the NEC doesn't emphasize) is the difference between being built to the proper clearance and maintaining a proper clearance. I'm sorry to say that even I learned this lesson the hard way. Once, when installing some cable over a city street in Mathews, North Carolina, I had my two linemen build the line as I checked the pavement-to-cable sag. I measured exactly 18 ft. at the lowest point. When I went back the next day, I remeasured the sag and it was at 17 ft. I had forgotten to compensate for the summer sun. The heat caused the wire to expand and created another 12 in. of sag. The lesson I learned that day is always to build above the minimum required height to maintain a safe clearance. For the spans I normally deal with, a good rule is to build at least 1 ft. above the minimum height.

The drip loop (installed by the utility) needs to be at least 10 ft. off the ground. To accommodate for the loop, the last clamp you install (always double-clamp the last attachment) needs to be around 13 ft. off the ground—assuming a level finished grade. The individual wires of the drip loop must be kept at least 3 ft. away from the bottom and sides of any operable window, stairs, balcony, deck, and so on. As long as the SE cable stays in its sheath, or the cable is in conduit, there is no minimum distance between it and any of the above items.

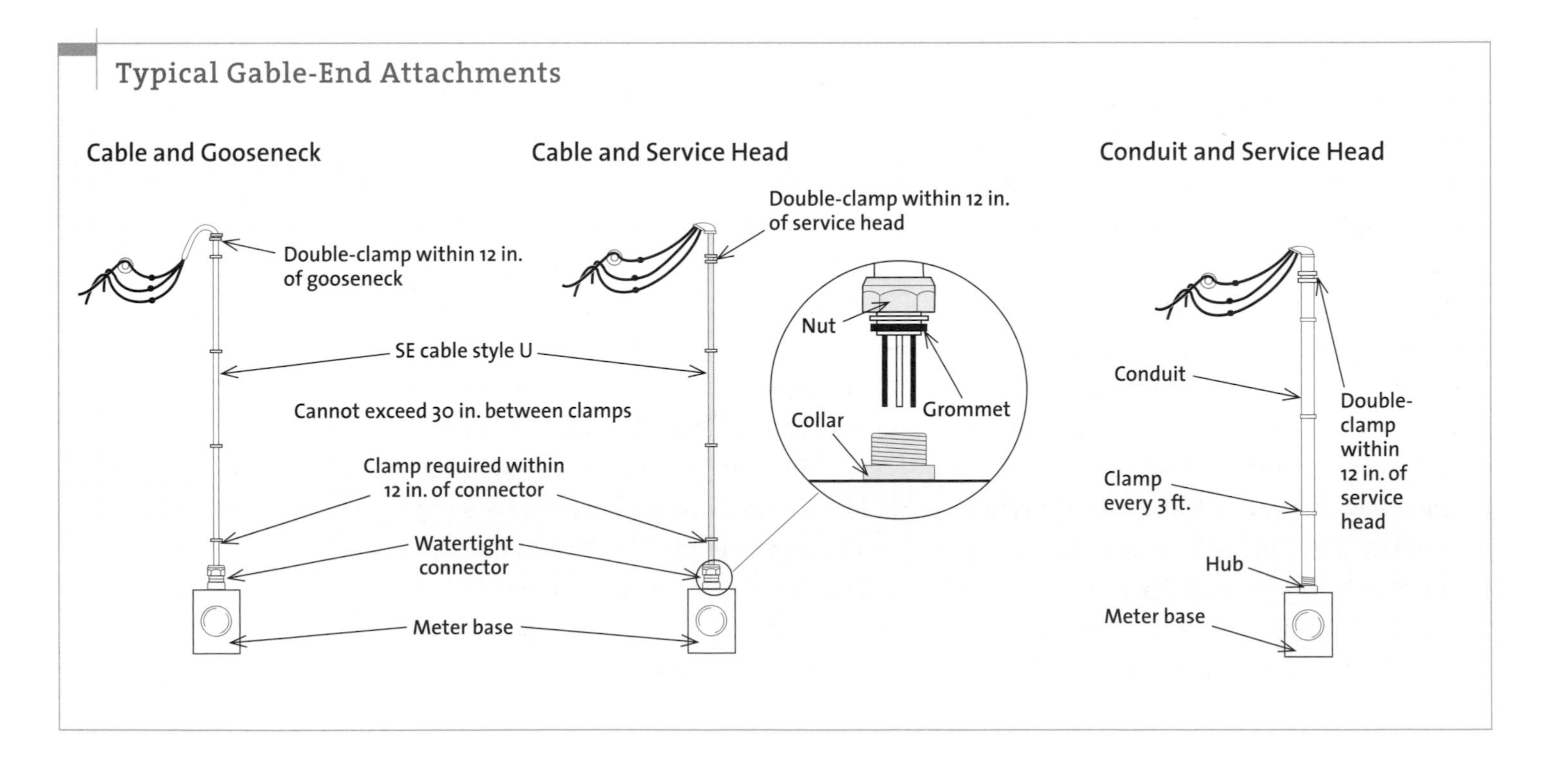

Connecting the meter base with the aerial utility

The four most common ways to connect the meter base to the aerial utility cable are as follows:

1. Run style-U cable from the meter base and clamp it to the siding (no service head). After installing the meter base, use a plumb line or level to draw a line straight up from the center of the meter base. The cable, as it leaves the meter base and clamps against the siding, must follow this line. You are allowed to arc the cable along the exterior wall or clamp it to the soffit and run it to another part of the wall. However, this looks ugly and should be avoided if at all possible.

   To attach the cable, first screw the bottom half of the watertight seal into the meter base and slide the top half onto the SEC end. Next, slide in about 18 in. of cable—enough to connect to all three terminals inside the meter base. After inserting the cable, place the first clamp 12 in. above the meter base and the others at 3-ft. intervals. (This is easier said than done: While working on a ladder, you must lift the heavy cable, place the clamp on the wall, and then screw it in place.) Double-clamp the cable where it bends down for the drip loop. (I use a cordless drill, a magnetic 5⁄16-in. hex driver, and 5⁄16-in. hex-head screws long enough to reach solid wood inside the wall. It helps if you have someone arc the cable for you as you attach the clamps.) Once the cable is up, tighten the cap on the watertight seal and add silicone where the cable enters the seal.

Three pieces of a seal: rubber grommet, anti-grab washer, and screw cap.

To splice the SE cable into the meter base, cut back the stranded neutral until it is the same length as all the strands and it fits under the neutral clamp. Cut the hot legs approximately 1 in. shorter than the neutral. Now all you have to do is bolt them down. Don't forget to smear on anticorrosion grease if you are using aluminum conductors. You do not have to ground the meter base. The neutral bus in the center of the meter base is already connected to the frame.

Minimum Clearances over Property

From Pole to House

NEC calls for minimum clearances on service drops. But ice and heat can cause aerial lines to sag below original construction height. To compensate for sag, construct above minimum requirements.

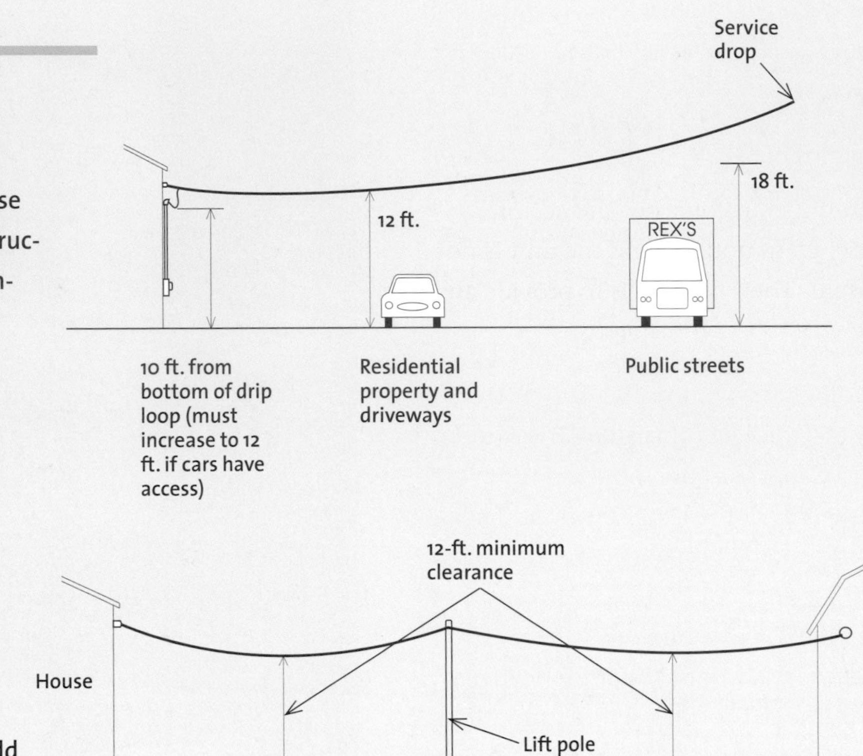

Adding a Lift Pole

When minimum clearances cannot be maintained across a span, you must add a lift pole.

Around Windows

If a window is operable, you cannot install individual conductors (a cable such as SEU, which has a jacket, is exempt from this rule) within 3 ft. of the side and the bottom of the window, but overhead is okay. If the window is not operable, you can get as close as you want. The rule also applies to doors, porches, balconies, ladders, stairs, fire escapes, or similar locations.

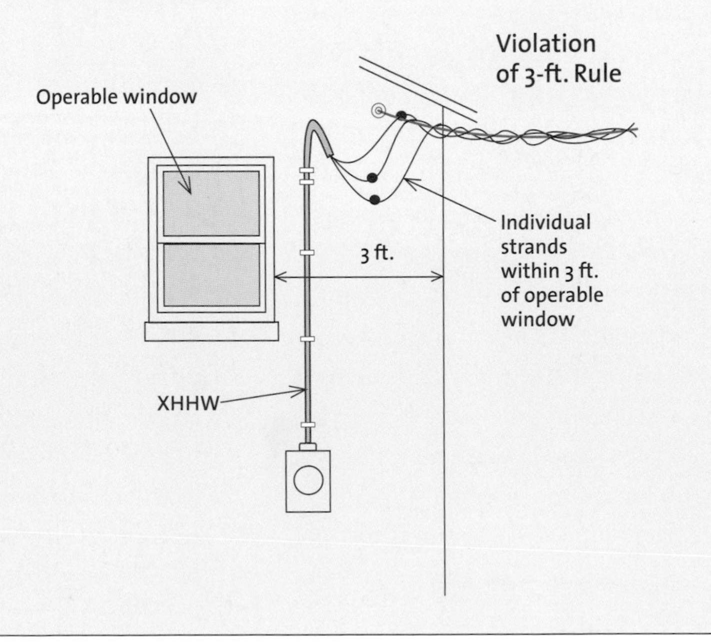

2. Run style-U flat cable from the meter base and clamp it to the siding (with service head). The installation method is basically the same, but in this case you'll also need to deal with some type of service head at the top of the cable. Inside, the individual wires are stripped out of the cable and then point straight down to come out the bottom. They attach to nothing and the unit is not watertight. Their purpose is to provide an umbrella to cover the cable where the wires break out of the sheath; this keeps water from running down into the cable. If at all possible, I avoid using such devices on

The most common meter base hookup, type-U aerial, should wind up looking like this.

Use silicone sealant where the cable enters the rubber seal. Without it, water will leak through the hub and into the meter base.

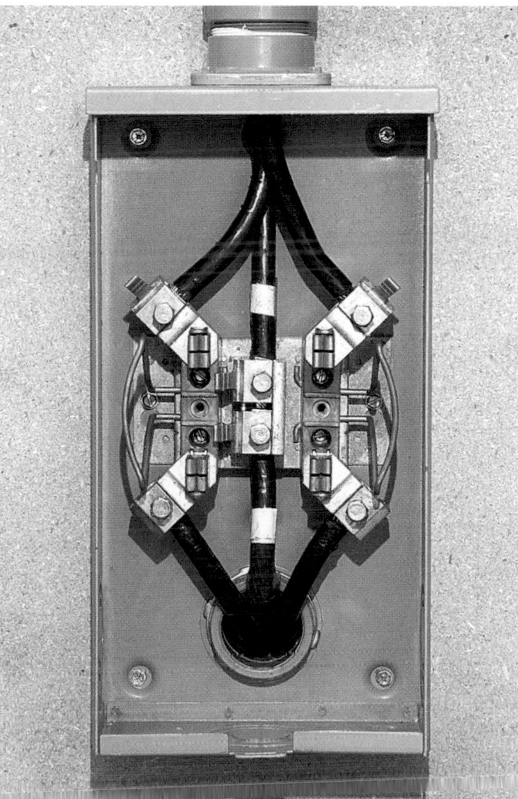

This wired aerial service meter base uses individual cables.

Above Code • For masts above the roof, I use only 2-in.-diameter or larger thick galvanized steel. Smaller pipe can't take the stress of the utility cable pulling against it and may bend. Never use PVC conduit above the roof; it can bend and snap. Always double-clamp within 12 in. of the gooseneck or service head.

Service to the Eave Side

If mast height exceeds 4 ft., bracing should be considered. Also, no coupling allowed above roof.

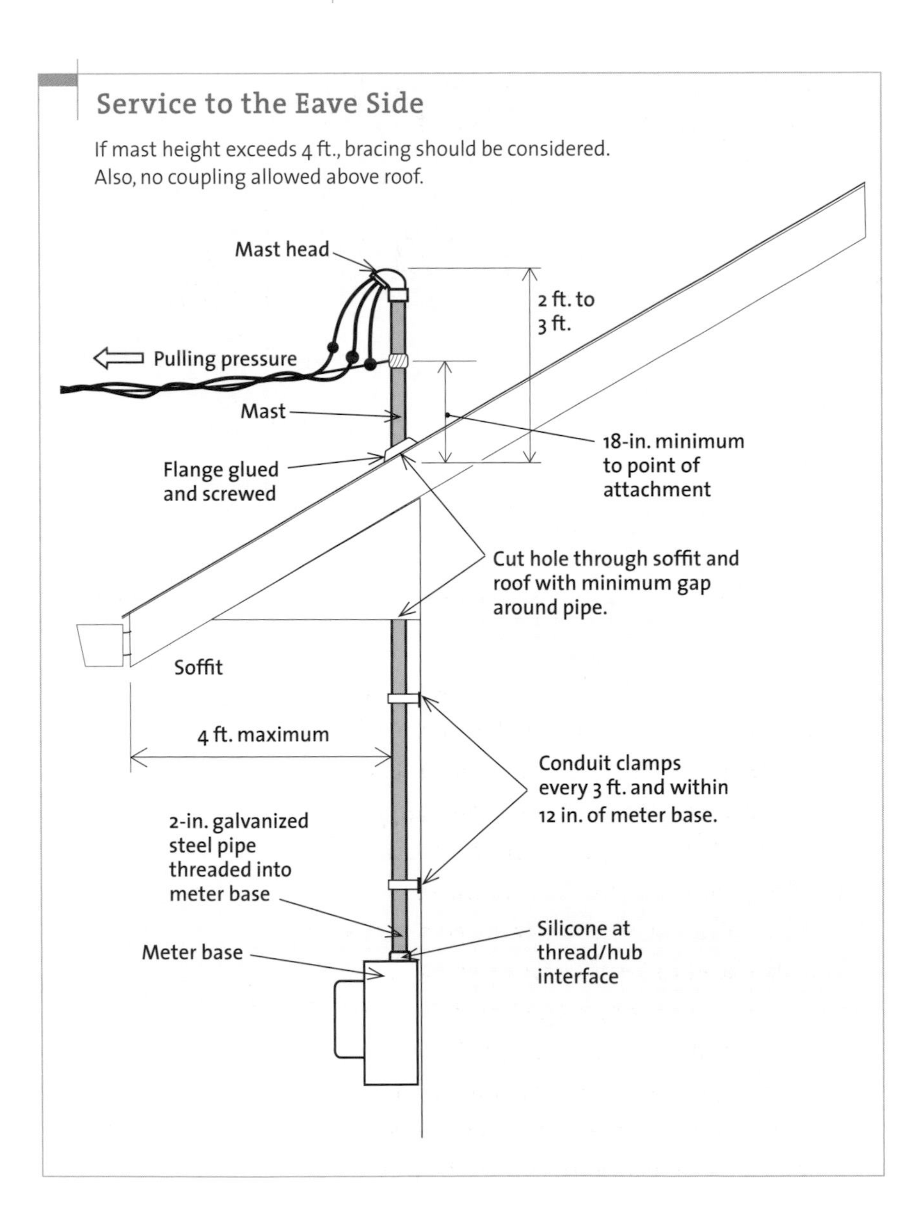

style-U SE cable because they are difficult to install. However, if your local area requires it, you have no choice.

You'll need 3 ft. to 4 ft. of cable coming out of the service head for the utility to splice into and form a drip loop. Make sure you keep the sheath of the style U intact all the way into the service head; otherwise, the service head won't be able to do its job.

3. Run steel conduit from the meter base, through the soffit, and up through the roof (service head mounted on top of a mast). This method is used mostly on single-story houses where extra height is needed for the utility clearance.

When calculating the length of steel pipe needed, allow for 3 ft. to 4 ft. above the roof. Above that, you may need to tie the upper end of the mast to keep it from being pulled over by the weight of the utility cable. The utility connects to the mast about 12 in. under the service head and places its attachment and drip loop around 2 ft. to 3 ft. above the roof (the minimum distance is 18 in.). Never have a splice (coupling) in the mast section of conduit above the soffit and roof. The threaded section of the metal is too weak to hold the stresses of a heavy utility cable.

The utility cable must not swing over any part of the roof other than the overhang. If it does, it places those who work on the roof, such as anyone cleaning gutters or repairing shingles, in serious danger. (The utility doesn't like running cables over the

400-Amp Service Entrance

When running cables larger than 4/0 aluminum, the installation gets tricky because of their thickness. Most of the time it's easier to run two cables of the same diameter, length, and material in parallel rather than run one large cable. (Never use one large-diameter cable and one small-diameter cable—all current would flow through the larger cable because of its lower resistance.) For example, for a 400-amp service, an electrician may choose to run two 4/0 cables to each leg of the meter base. In that case, it is imperative to mark each phase wire with colored tape. For example, use blue tape to mark the two wires connecting to the meter base's left terminal (mark the cable at the meter base and at the service head). The wires connecting to the meter base's right terminal will then need a different color, such as yellow, to identify them. The neutral is always white. Never use green—it is for ground. Use two conduits of the same material, one for each cable group, as opposed to putting all cables into one large conduit. In situations like this, it is better to use 2/0 copper than 4/0 aluminum because, being lighter (albeit more expensive), the copper is easier to drag up a ladder.

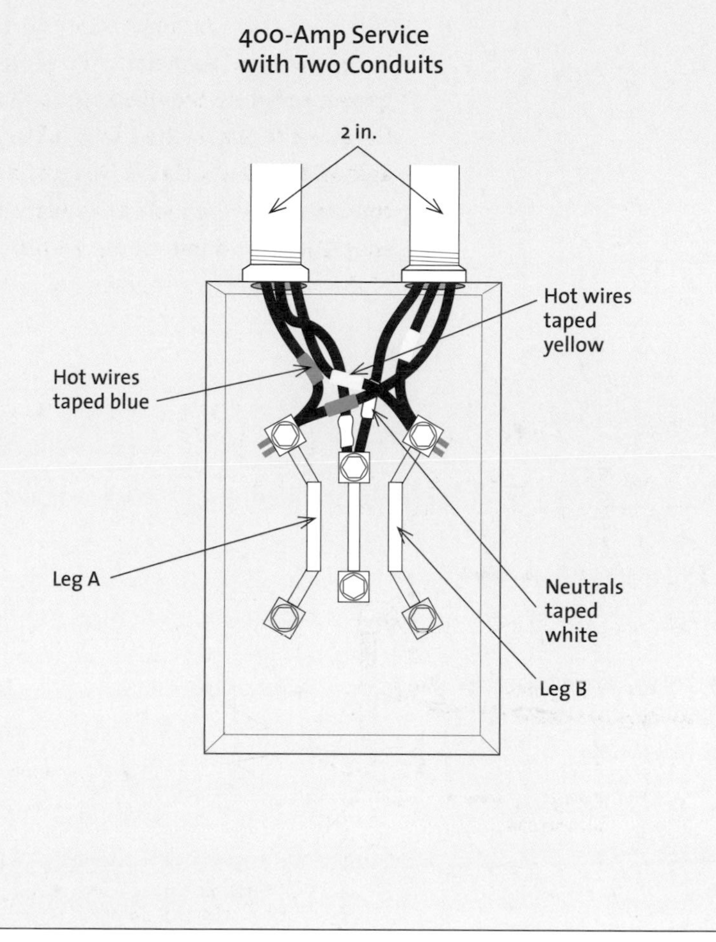

roof either, but it is worried about the cable, which would be ruined in case of a fire.) If the service entrance looks as if the cable will cross the roof, find a different route or move the location of the meter base. In some cases, the utility can place a drop pole to change the approach angle of the cable to the house.

Installing a service entrance through the soffit is not easy. To do this, drill a hole in the roof and soffit for the conduit to reach above the roof. Then carry a heavy 2-in.-diameter pipe up to the roof and slip it down through both holes. The threaded end screws into the hub of the meter base. Put some pipe dope on the threads before you tighten it with a pipe wrench. Install the first clamp within 12 in. of the base and the others at 3-ft. to 4-ft. intervals. You will also need to install a weather seal at the pipe/roof interface. The hole cut for the conduit coming through the roof should be as close as possible to a perfect fit around the conduit/mast. If there is an excessive gap between the pipe and the wood, the pipe

Work Safe • The secret to cutting a round hole on a sloped roof is to use a template. Using a scrap piece of conduit, cut one end to match the slope of the roof so that the pipe stands vertically when positioned on the roof. Trace along the base of the pipe to establish a cut line. When you make the cut, lose the line. You'll need the extra wiggle room to slide in the conduit.

may bend over as the weight of the utility cable pulls it. This will make the roof seal leak.

Once the conduit is installed, cut the SE cable to length and slip it into the conduit either from the roof or up from the meter base. Add an extra 3-4 ft. for the drip loop and 1 ft. for the meter base. Next, insert the cables through the service head on the top of the mast and mount the service head. Let the SE cable hang down from the service head. The utility will form the drip loop as they splice in their utility cable. Last, splice the cables into the meter base, making sure the neutral is noted with white tape at the drip loop and at the meter base.

4. Run schedule 80 PVC conduit up the siding and terminate it in a service head on the side of the house. Running PVC conduit up the siding is fine as long as it does not extend through the soffit and is not used as a mast. Its light weight makes it easy to work with (especially when you're working alone), but it does not have the strength to support the weight of the utility cables.

To install PVC conduit, follow the same steps as described for steel. The cable lengths of the drip loop and the splice in the meter base are also the same.

There is, perhaps, a fifth way of bringing an aerial service into a residence—one that happened to be the worst I'd ever seen. The owner had laid the style-U SE cable on the ground from the meter base pole (farms back in the old days had the meter on a pole in the center of the farmyard) all the way to the house. Dogs had been chewing on the cable but as yet hadn't bitten through to the hot conductor. The cable then went in through a window (that always stayed open), through the bedroom, and across to a panel on the other side of the room. Although this method may be a fast way to install a service entrance system, and it could double as a handy clothesline, it's not one that I recommend.

Connecting Meter Base to Panel

Once I was working on a house that was about 10 years old. The owner was complaining about dimming lights and everything acting "funny." First, I checked the voltages at the meter base that was mounted on the outside corner. All the voltages checked fine. When I checked the voltages at the service panel, which was located in the basement somewhere around the middle of the house, the voltages were funny: There was low voltage (80 volts) from one hot leg to neutral and high voltage (160 volts) on the opposite hot leg to neutral. And the voltages were fluctuating. Upon closer examination, I discovered that the neutral was

The Art of Stripping Large Cable

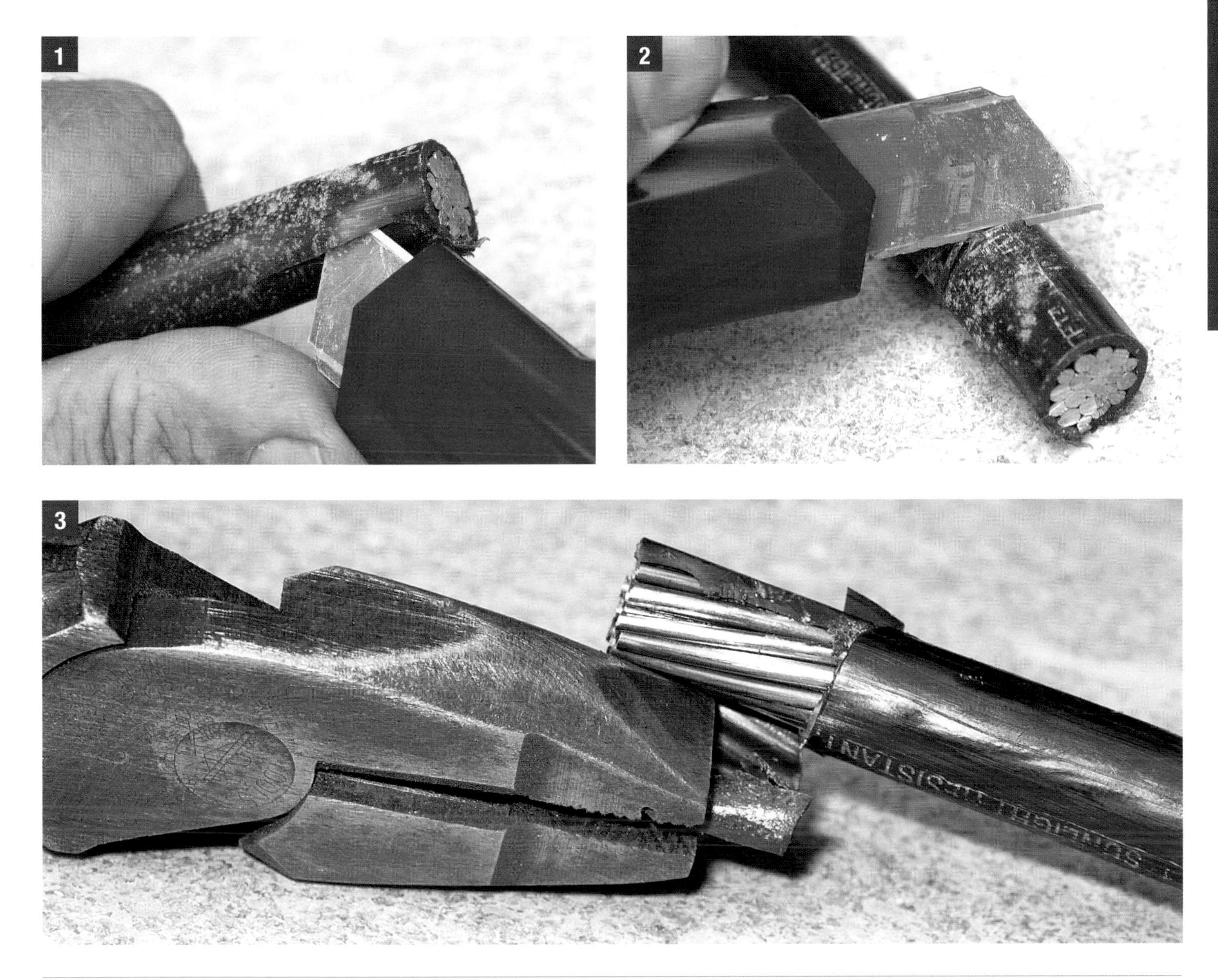

1 | Holding the utility knife parallel to the ground, slice into the insulation until you feel the blade on top of the conductors. Tilt the blade so that it slides over the conductors and cut the insulation to the end.

2 | Circle the cable with the utility knife cutting three-quarters of the way through the insulation.

3 | Using electrician's pliers, peel the insulation off the cable.

being eaten through by water that had accumulated inside the sheath of the service cable connecting the panel to the base. This is quite a common occurrence, and it is one of the many reasons why I won't separate the panel and the meter base.

Locating the meter base back-to-back with the service panel, or as close as is practical, simplifies the installation, reduces installation cost, and as indicated by the example here, makes for a better job. Remember, water will follow the cable any time the SE cable leaves the bottom of the meter base to enter the house. Meter bases are not watertight—they

Use 9 1/8-in.-long conduit to connect a meter base to a service panel through an 8-in. block wall (with a 1 1/2-in.-thick wood spacer behind the panel). Placing only the male adapters back to back shortens the conduit length to 2 1/2 in. Cutting 1/2 in. off each adapter hub further shortens the distance to 1 1/2 in. If the meter base and service panel are closer than 1 1/2 in., use a metal all-thread nipple.

Use a PVC saw for cutting plastic conduit. Its stiff, wide blade keeps the cut straight.

can leak through the "watertight" hubs on top or around the glass.

You can connect the meter base and the service panel in several ways:

1. Via flat, all-in-one style-U cable. If you use style-U cable, conduit is not always required between the meter base and the service panel. (Be sure to check local regulations.) However, if you don't use conduit, there's a risk that someone may drive a nail or screw into the cable. To make the interface, all you need is an NM connector in each box with a bushing over the threads. Run the cable between the two boxes.

2. Via cable inside a plastic or metal conduit. If you use individual cables, such as THHN/THWN, you must use some type of conduit to enclose the cable. I prefer plastic schedule-80 conduit rather than metal because it's easier to cut, easier to install, and less expensive. I also prefer running individual conductors rather than all-in-one (style-U) cable, because individual conductors are easier to bend and maneuver in the conduit and panels.

   The length of conduit nipple for a back-to-back run from the meter base to the service panel varies. For a block wall with a treated 2×6 spacer mounted inside the building, the conduit length needs to be around 9⅛ in. or just slightly less (slightly less and you'll never know; slightly more and it won't fit). Putting two hubs against each other yields 2½ in. to 2⅝ in. (cut the pipe that fits into the hubs 2⅜ in. long). For panels that are literally back to back, use a metal all-thread nipple and a grounding bushing.

3. Via 2-in. flexible liquid-tight conduit. If you've got a lot of turns to make and you're required to encase the cable in conduit, then liquid-tight conduit is the way to go. The flexible conduit and fittings cost more than PVC, but they're faster to install because there's no gluing or angled fittings to deal with. I use liquid-tight because the smooth, fittingless conduit makes it extremely easy for me to pull through individual cables.

A grounding bushing screws onto the metal conduit that exits the meter base and goes into the house. Note the ground wire (#4 copper for 200-amp service) connecting the bushing to the ground lug of the meter base.

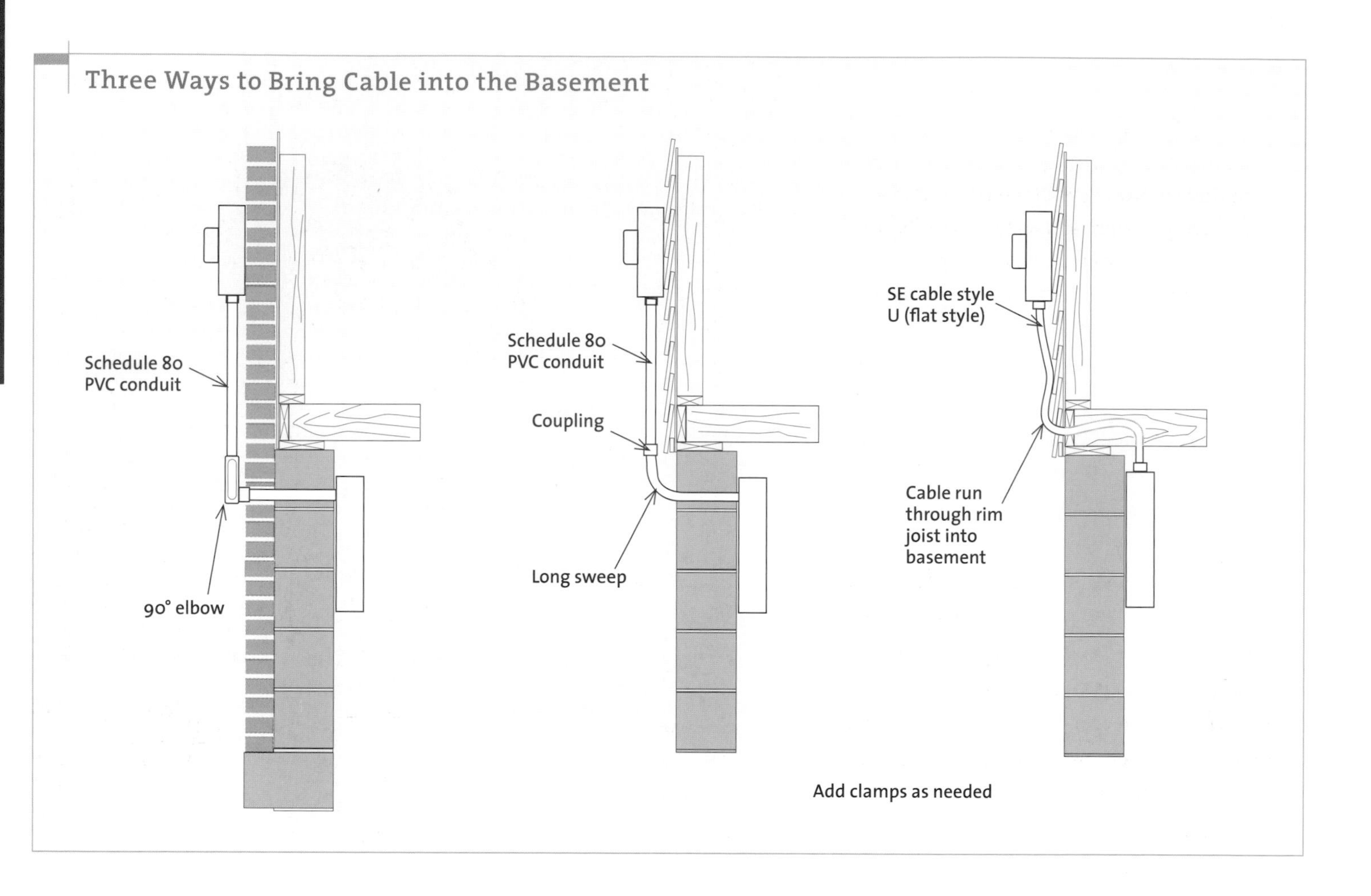

Connecting to the main panel in a basement

I don't recommend it, but if the main panel must be located in the basement, there are three ways to get the SE cable from the meter base to the panel. The first option is to run the SE cable (make sure it's rated to be outside and in direct sunlight) without conduit right into the basement. However, most codes don't allow you to run cable through concrete without some type of protective sleeve. To get around this, you can run the cable through the siding and rim joist immediately above the concrete and then turn down to the panel (some areas will still require conduit).

The most common method of running cable into the basement is to run schedule-80 PVC conduit immediately out of the bottom of the meter base, or cutoff panel, just far enough to get into the basement. The PVC normally travels straight down, then through a long sweep or a 90-degree—immediate-turn—elbow to go through the wall. (Remember to seal the hole around the conduit; otherwise, if the hole is below grade, the basement will flood with the first rainstorm.) Once through, you can either go directly into the main panel, or if you're still above, use an elbow to turn down.

Overall, installing the panel in the basement is not only expensive and time-

Work Safe • When large-diameter SE cables bend up inside the panel, they can get in the way of smaller wires and even prevent you from screwing on the lid. To keep such cables under control, pull them against the back of the panel with cable clamps and 5/16-in. hex-head screws.

consuming, it's also difficult to pull the individual conductors through the conduit, especially the elbows. In addition, water coming in around the meter itself (and sometimes through the alleged watertight connector on top of the meter base in aerial installations) will sometimes flow down the cables and into the main service panel, destroying the main breaker. I don't recommend putting the main panel in the basement unless it's absolutely necessary. If you are going to do it anyway, run the cable or conduit into the bottom of the panel. Once in, bend the conductors up and route them along the panel gutter to the top of the panel to connect with the main breaker. This will, at least, keep water out of the main breaker.

Cutoff panels

Many locales require a cutoff panel immediately adjacent to the meter base if the two are separated by 5 ft. or more. (The distance depends on your local restrictions. Some locales allow 10 ft.; in my county, there is no set distance.) The purpose of the cutoff panel is to protect the cable between the meter base and the service panel. Without it, there would be no overcurrent protection for this cable as it travels through your house. A single nail or screw could cause a meltdown. By the time the fuse on the transformer kicks in, your house will be on fire. Expect to pay around $500 for this switch—another good reason to keep the meter base and panel back to back.

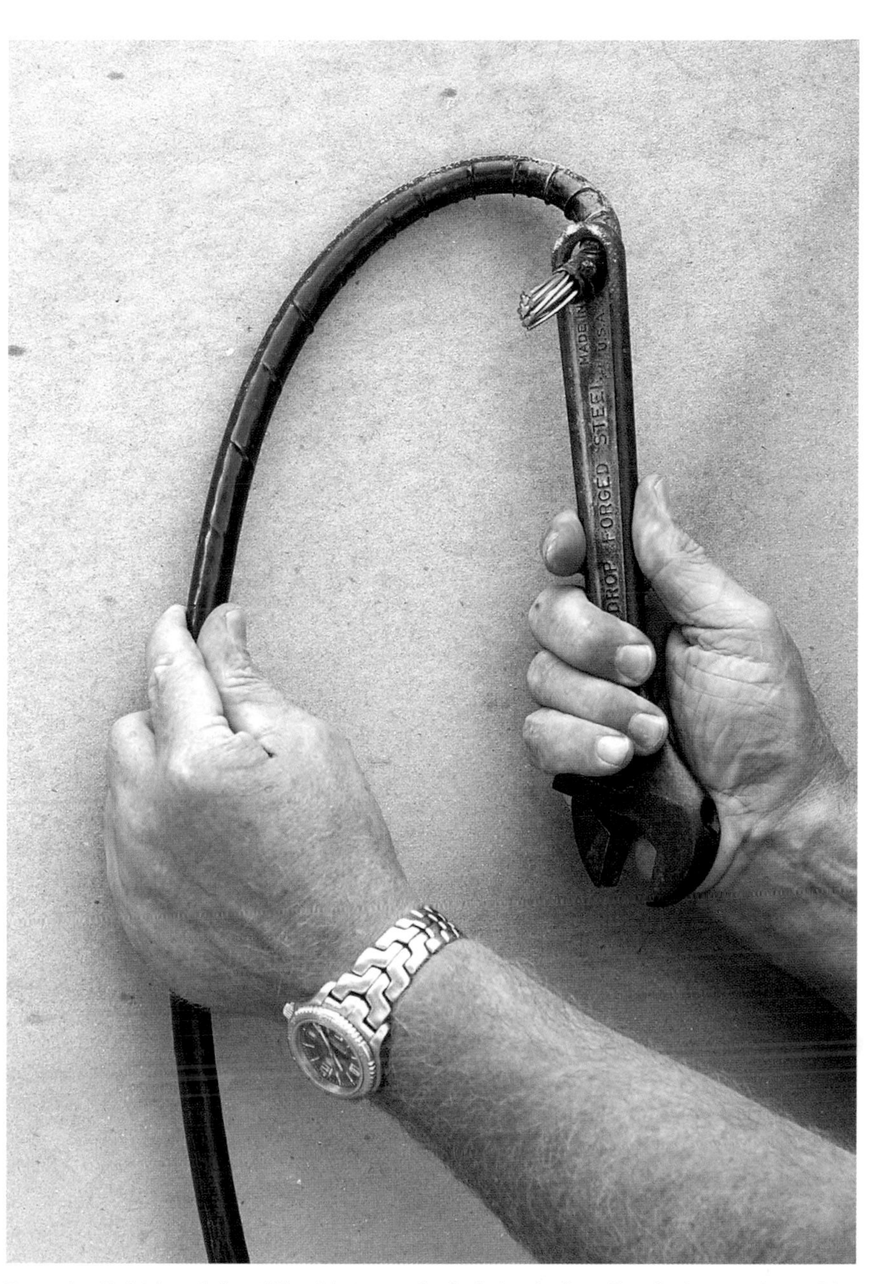

To make tight bends in stiff cables, use the hole in the handle of a large adjustable wrench for extra leverage.

Panels and Subpanels

A **The main service panel and subpanel** are the heart and soul of a home's electrical system. Through its outgoing wires flows the lifeblood of the electricity used throughout the residence. The main service panel has several important functions:

- It serves as a distribution center for all branch circuits.
- It houses all overcurrent devices that protect individual branch circuits.
- It houses a master switch that you can manually throw to cut off all outgoing electrical power from the service panel. (Some locales require the master cutoff switch to be adjacent to the meter base. If that is the case, there may or may not be an additional master switch in the panel itself.) The master switch also monitors the total amount of current the house uses and automatically cuts the power if it exceeds a specific amount.

Many installations fail, regardless of the class of the house, because some installers incorrectly assume that all main panels are

Main Service Panel

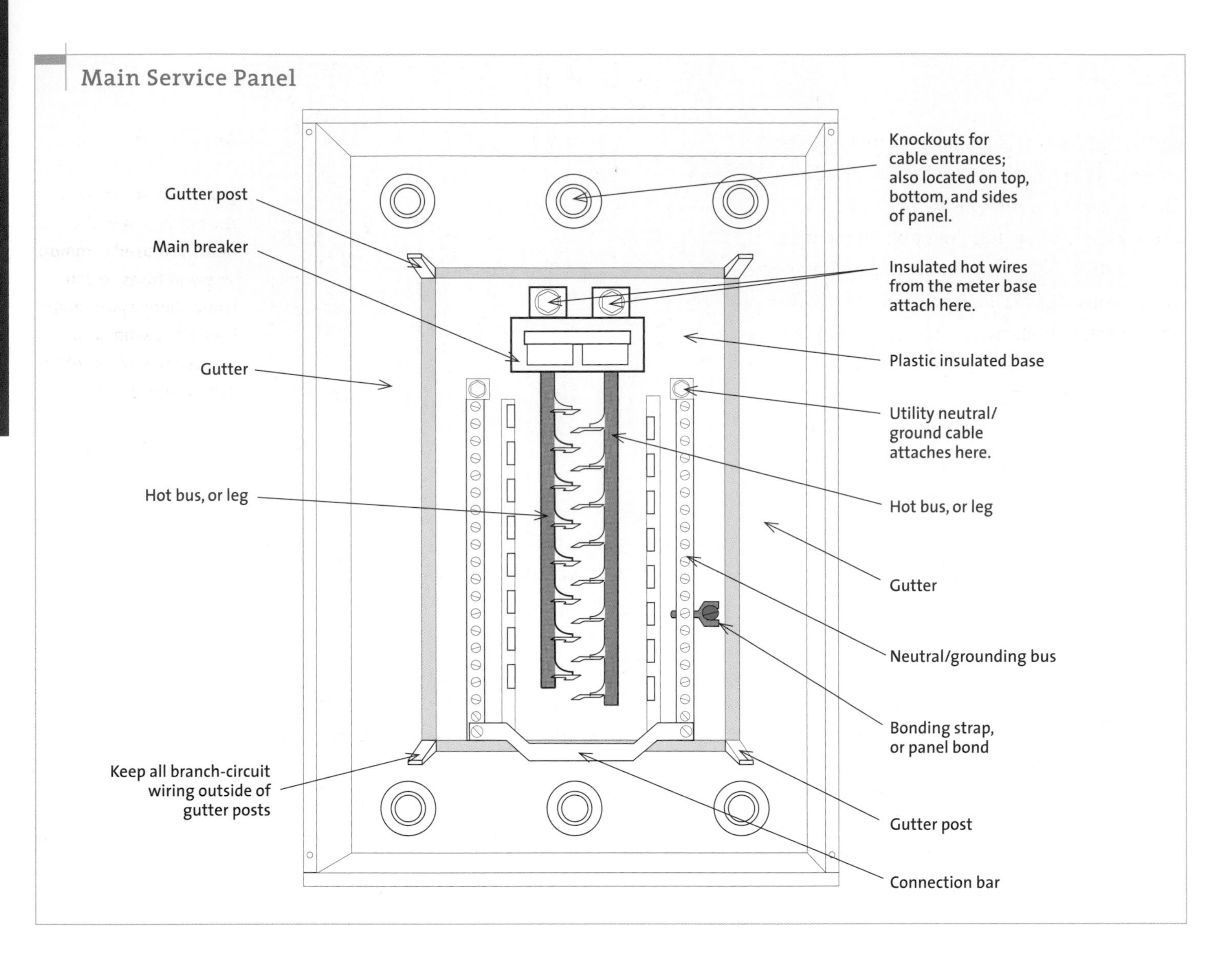

designed the same way and therefore install the cheapest one they can find. The fact is, all panels are not created equal: There are good designs and there are bad designs. It's also important to realize that even the finest panel is no better than the person who installs it. A good panel gives you only the potential for a good installation—it doesn't guarantee it. The installer must know the basics of a good installation and what pitfalls to avoid, as well as have the willingness to do the job correctly. In this chapter, I will help you pick and install a high-quality panel.

Elements of the Main Panel

The main service panel (also called the box, panel board, or load center) has many sections, and each one performs a specific function. The three most important sections of the main panel are the main breaker, the hot bus, and the neutral/grounding bus. To understand how the panel works, you must first understand what each section does.

Main breaker

The main breaker, also called the main disconnect switch or just "the main," is normally located at the top of the main panel. All the power that comes into the house goes through this switch. All adults in the home should know what this switch does and where it's located. If there is ever an electrical emergency in the house, or perhaps a fire, throwing this one switch will allow you to cut off all power. Although it can be used as a manual disconnect, the main breaker's primary purpose is overcurrent protection. It monitors the total amount of current coming into the house so that it cannot exceed the current-carrying capacities of the SE cable, meter base, and main panel. If this amount is exceeded, the main breaker disconnects the incoming power from the buses. This amount, in amps, is written on the breaker handle.

The two hex-screw terminals immediately above the main are for the two hot SE cables coming in from the meter base. The two plastic pieces immediately above the main and below the buses are called gutter posts. All branch-circuit wiring should remain outside the posts (wires cannot cross over the breakers from one side of the panel to the other).

Hot bus

The main breaker transfers power to the hot bus. The hot bus consists of two copper or aluminum strips, which are sometimes called legs; they are located immediately below the main and run down the center of the panel. Each hot leg has a row of tabs that allow for the insertion of circuit breakers. The bus takes the hot power and distributes it to the circuit breakers. Each leg is the same voltage, in reference to neutral, but acts as an independent power source. Simply put, the current in the hot bus comes to each leg from alternate sides

An early one-circuit main or branch circuit panel that fused both the hot wire and the neutral. It used common screw-in fuses so you could change your main breaker to what you wanted regardless of the consequences.

of the utility transformer. When measured from leg to leg, the voltages add up to 240 volts; the voltage for each leg, to ground or neutral, is 120 volts.

Neutral/grounding bus

The neutral/grounding bus provides a common return point for the current after the load has used it. It also provides a ground reference point. The neutral bus consists of two long aluminum strips with many screws; they are usually located on the outer edges of the hot buses. In a main service panel, the screws serve as attachment points for the neutral and ground wires (there will be a lot of them). In a subpanel, the neutrals and grounds are connected to independent buses—one grounding bus and one neutral bus. The large hex screw at the top, called the main neutral lug, connects to the neutral coming in from the utility meter base.

Even though the neutral is grounded at the utility pole, an additional ground is provided by the installer at the main panel. This is the ground-rod connection and it is normally a large diameter, solid-copper wire (4 or 6 gauge). For additional protection, a bonding screw is installed, connecting the neutral/grounding bus to the metal frame of the main panel. The bonding screw, called the panel bond, places the metal of the panel at ground potential so that it can never become a conductor if a hot wire touches it. In addition, all equipment grounding wires from every receptacle and every appliance, as well as all neutral wires (the white ones), connect to the neutral/grounding bus.

A well-designed panel has a bar that connects the two buses as one electrical unit. If the panel with this setup needs to be converted to a subpanel, which has its neutral isolated from the ground, the two can be separated easily by removing the bar. Not every panel has this handy option.

Picking a Panel

Before purchasing a main panel, you will need to know—or at least have a rough idea of—the number of circuits required for the house. Almost all houses can get by with 40 circuits for a 200-amp system, 60 for a 300-amp system, and 80 for a 400-amp system. If you need more than that per system, consider installing a subpanel.

The way to tell panels apart when buying them (other than by a part number or amperage) is by their circuit-quantity designation. That designation appears on the packaging as two numbers, such as 20/30 or 40/40. The first number indicates the total number of full-size breakers that the panel can hold if only full-size breakers are installed. The second number indicates the total number of circuits allowed in the panel. For example, a 40/40 panel can hold 40 circuits and all 40 are full-size—no dual breakers.

Panel size comparison. Both panels hold the same number of breakers, but the full-size panel on the left gives you much more wiring room than the split-tab panel on the right.

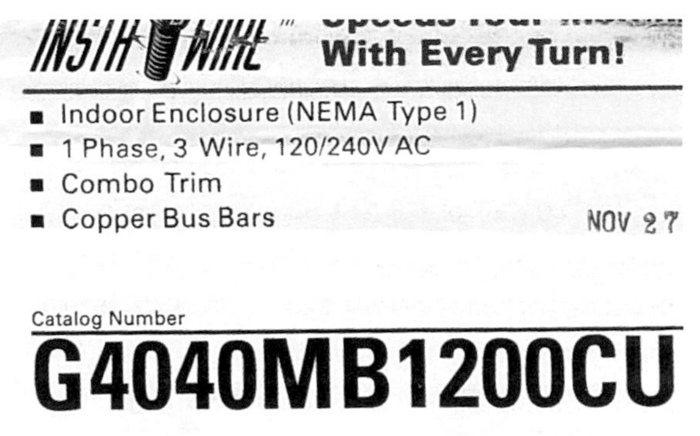

Check the specs on the box. The left side gives you its maximum rating (200 amps) and indicates that it can hold 40 full-size 1-in. breakers. The right side lists the part number and indicates that this panel is for interior use, is designed for 120/240 single phase, and has a copper bus.

If the second number is larger than the first, then dual breakers are allowed. For example, a 20/30 panel can hold 20 full-size breakers for 20 circuits. Utilizing the same 20 tabs, 30 circuits can be obtained if dual breakers are used at specific places within the panel. These types of panels normally have special tabs for dual breakers located on the bottom of the buses.

It is possible to cheat and overfill a panel by installing dual breakers where they don't belong, which you may be tempted to do if you bought the smaller panel to save some money. Experienced electricians have learned how to break the metal "stop" out of the back of a dual breaker to allow it to fit on a full tab. This makes it physically possible to put 40 breakers in a 20-circuit panel or even 80 breakers in a 40-circuit panel. Don't do this. You not only void the warranty of the breaker and panel but you exceed the maximum number of breakers for which the panel is designed, creating a potential hazard.

Bus design

When you go shopping for a service panel, you will have a choice of three bus designs:

- Panels with all solid bus tabs, called full-size panels.
- Panels with all slotted (or split) tabs.
- Panels with both solid bus and slotted tabs.

The advantage of full-size panels over split-tab panels is their size. A full-size panel has plenty of room for wiring and can only house full-size breakers, which I prefer. Split-tab panels can fit both full-size and two-in-one

Panels differ in their hot-bus tabs. The solid-tab panel (above) accepts only full-size breakers, whereas the slotted-tab panel (top) takes both full-size breakers and two-in-one breakers. Contrary to what you might think, the solid bus tab is the one you normally want.

This undersized panel was filled the day it was installed. (Those two blank spaces over to the right are not empty slots—they're fakes.) The owner now has to install a subpanel or upgrade his service to wire in his new spa.

Above Code • In most cases, I use full-size panels. A larger panel is easier to wire and looks neater when you're finished. And in electrical work, neatness counts. When you finish a wiring job, no matter how well you did the design or laid out the wires inside the walls, the only part that shows is the panel. When an inspector sees a messy panel, a red flag goes up in his or her mind. Based on the appearance of the box, the inspector may start wondering about the overall quality of your work. Odds are good that he or she will spend much more time inspecting the rest of the house.

The end terminals of this breaker had corroded so badly due to moisture damage that I had to cut the cable when the breaker died.

breakers. This may seem good at first, but having more breakers with less room can turn the panel into an overcrowded mess that is hard to work in.

In most cases, contractors use these smaller panels to save a few bucks. Ironically, I've found these panels even in million-dollar homes. There are times where a few extra dollars becomes money well spent, and a buying full-size panel is one of those times. The one time when I may not use a full-size 40/40 panel is when I am installing a 300-amp or 400-amp system. The reason is that if I plan to put several high-amperage circuits in the second panel—for example, two heat pumps—I may need only eight slots to max out the panel's amperage.

Mounting a Panel

Where you put the panel makes a big difference in its life expectancy. If it's at all possible, try to place the panel back-to-back with the meter base. A garage, for example, would be a perfect location: meter base outside, panel inside immediately behind it. Never install a panel in a bathroom or wet basement, because the moisture destroys it. Never install a panel in a closet, under the stairs, or in some other area that might contain flammable materials or be difficult to access. And never install a panel where you know the clearances around it cannot be maintained. For example, if a spa is placed two feet in front of a panel, the inspector will make you move either the spa or the panel. To prevent other moisture-related damage, avoid placing water and drain lines directly above the panel and make sure that clothes dryers are properly vented outside.

Panels are maintainable items. For that reason, you must make sure that you are able to comfortably stand in front of the panel in order to work. Thus, do not install a panel along stair steps. Maintain proper minimum distances, and the door must be able to open at least 90 degrees. Because you may need to get to the panel fast in an emergency, you don't want to be climbing over boxes or tables to kill the power. Most common problems are due to renovations made after installation, such as new overhead plumbing, counters mounted in front of the panel, shelving mounted across the front of the lid, and appliances (especially the clothes washer) set in front of the panel. One home I worked on had a spa mounted in front of the panel. Another had the panel mounted over a stove, and you had to crawl over the burners to get to the breakers.

Take Note • Some locales allow you to run a plumbing line over the service panel, provided you use a drip pan. Drip pans are available at heat-pump supply stores or they can be custom made by a sheet metal shop.

Clearances at the Panel

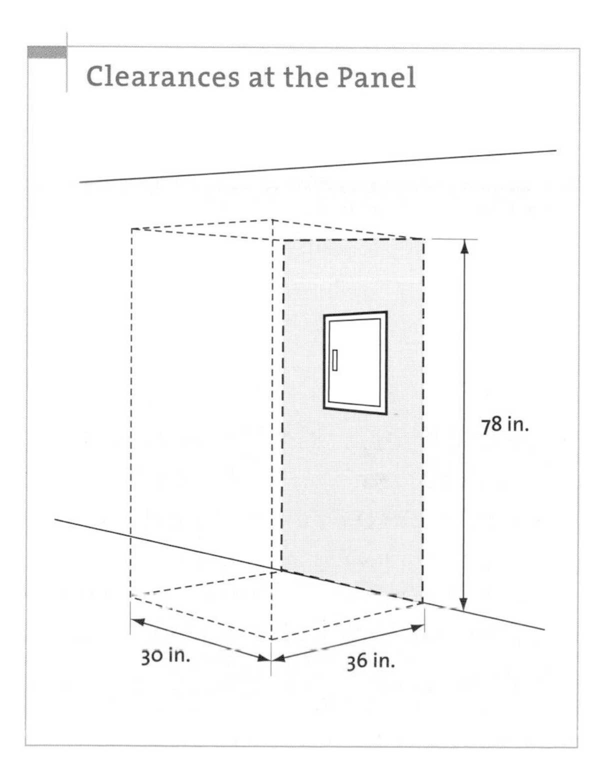

Mounting options

Service panels are designed to be mounted several ways. On a solid surface (as opposed to between studs), use the five small, prepunched holes in the back of the panel. First, hang the panel like a picture frame using the elongated hole in the top center of the panel. Next, level the panel, then install the four corner screws. On masonry, be sure there is at least a ¼-in. gap between the metal of the panel back and the masonry itself. The wood spacer must be pressure-treated. I normally use a treated 2x6

Panel Shopping List

All panels are not created equal. Some designs are great; others I would never use. What you want is an intelligently designed panel—one that is safe and wires fast, easily, and logically. Everything about the Siemens panel, shown here, suggests that an electrician designed it. Here is a shopping list of features that you'll want your panel to have:

- A fall-safe lid. With some panel lids, an electrician needs three hands—one to hold the driver and two to keep the lid from falling and smashing his toes. With a Siemens panel, it's okay to have only two hands. The lid has two metal tabs that ride on top of the box frame.
- A comfortable cutoff. I have made service calls where the only problem was that the homeowner couldn't get the breaker back on or off. This panel breaker has a wide handle that turns on and off vertically (side to side is harder to operate). You can use four fingers to pull it down and the palm of your hand to push it back up.
- A movable neutral lug. Being able to move the lug to where the cable enters the panel speeds installation, leaves much more room in the box for the branch-circuit wires, and eliminates having to run the large utility neutral all around the inside of the box. To move it, first remove the setscrew from the lug. Look inside and you will see a Torx-head screw holding the lug body onto the bus.
- Copper hot buses. Copper buses beat aluminum buses every time. Copper conducts electricity better than aluminum and has fewer corrosion problems.
- Lots of neutral screws in the right location—immediately next to each row of breakers. Note that they are neither against the panel side where they'd block the knockouts, nor at the panel bottom or top, but right where they belong—next to the breakers. Cutting all the wires to the same length saves time and makes the panel look neater.

- A system that allows several ground wires under one bus screw. Siemens buses are UL approved for up to three 12- or 14-gauge ground wires under one screw. Many manufacturers make you attach more than one wire per bus screw whether the panel is approved for it or not. You cannot, however, put a neutral and ground under one screw or two neutrals under one screw.

The back view of a panel lid. The metal tabs (in the photo at left) grab onto the panel frame to keep the lid from falling when the screws are withdrawn from the lid.

This bus screw is designed to secure up to three ground wires.

- Special neutral bus attachment system. These patented square/standard-drive screw heads speed up the wire attachment time dramatically because you can tighten the screws with a common 5/16-in. flat-blade screwdriver or a power driver and square-drive bit.
- Instant conversion from the main service panel to the subpanel. A Siemens main service panel has a tie bar at the panel bottom that connects the two neutral/grounding buses. Remove this tie bar and the panel is instantly converted to a subpanel. Use one bus for the grounds and the other for the neutrals.

A Torx screw is located under the utility neutral lug. Remove it to move the lug body.

(be sure to get one that isn't twisted). If you need to protect the wires, build a vertical box, or plenum, to house the cables once they are installed.

If mounted inside the wall, the panel must be attached by at least two of its four sides. If the panel is mounted between two 16-in.-on-center studs, you'll lose the ability to draw cable through the sides of the panel unless you cut the wood. When mounting a panel between 24-in. on-center studs, attach one side of the panel to the stud and run a crosspiece above or below the box for the second side. Remember to leave enough of the panel sticking out from the studs so it will be flush with the finished wall.

Above Code • Use screws, not nails, to mount the panel. Nails work loose over time and do not allow for minor adjustments. I always use 5/16-in. hex-head screws.

If the box is mounted inside a stud wall, you need a way to get all the cables through the top and/or bottom plate within the wall cavity (for cables going into the attic or basement). You can drill holes for each cable or group of cables, but that takes time and is both ugly and messy. Instead, create two large oval slots next to each other. Use the left slot for the cables entering the left side of the panel and the right slot for the cables on the right. Drill two 7/8-in. holes about 6 in. apart, then use a reciprocating saw to make a large oval cutout between the two. Coating each opening with some cable lube or liquid soap will make it easier to pull cable.

Opening the Plates

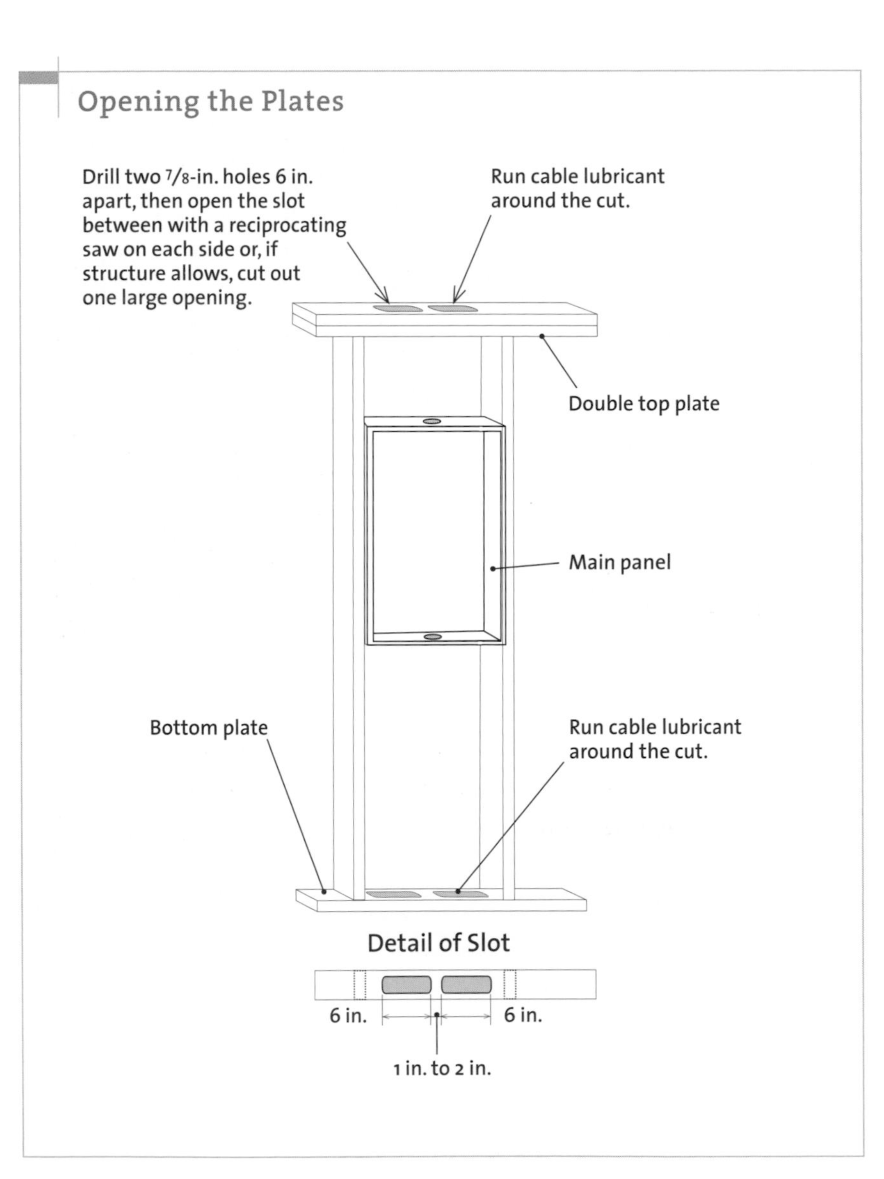

Roughing in wiring to the panel

After mounting the panel, rough in the cables to the panel location. If you are bringing the cables into a wall cavity with the two slots cut into the top or bottom plate, bring the 120-volt circuits into the left slot and the 240-volt circuits into the right slot.

Do not bring the cables inside the panel at this time. To save time, most electricians run

Removing Panel Knockouts

NM connectors are used whenever you're installing NM cable to a panel, a subpanel, or metal receptacles. Without the protection of the connector, the cable can be cut on a sharp metal edge or accidentally pulled out. Anyone can pop out the easy knockouts with a screwdriver, but what do you do about the stubborn ones?

1 Make sure you're opening the knockout in the direction it wants to open. Use a sharpened screwdriver to cut sideways across the attachment point.

2 Use a drill and a metal cutting bit to cut through the attachment point.

3 Use a fine-toothed sawblade in a jigsaw or a mini reciprocating saw (as I have here) to cut the attachment point.

Above Code • During the rough-in stage, write where the circuit goes on the sheath and wrap the end of the cable with colored tape: for example, yellow for the kitchen circuits, red for the lighting, blue for the bath, and green for the receptacles. Leave the 240-volt circuits as is. Color-coordinating the circuits makes it easy to identify them by group and bring them into the panel in an organized fashion.

all the cables through the house and leave the cut ends dangling around the panel with each cable's destination written on its sheath. Once all the cables are run, the electrician then ties them into the panel. Working this way is faster than wiring the cables one by one, but it forces you to sort through 30 to 40 cables. You can try searching for cables one by one or pick the first cable that's handy and bring it in with no organization at all. However, there is a better way.

Bringing Cables into the Panel

First consider cable management. You won't wire every panel exactly the same, but you can wire every panel in the same basic way. To keep things organized, I bring all the 120-volt circuits into the left side of the panel and the 240-volt circuits into the right. I further break down the left side into groups. I first wire the kitchen circuits into the upper left of the panel. Then, working downward, I wire the lighting circuits, then the bath, then the receptacle circuits. I like to wire the receptacle circuits so that the washer/utility room receptacle is the last breaker on the left side. Then I put the electric dryer receptacle opposite it, on the 240-volt side. Of course, you don't have to use my system, but make sure you have one.

Be just as consistent when bringing cables into the panel. That is, bring the cables that go to the left-hand circuits into the left side of the panel and the ones that go to the right-hand circuits into the right side of the panel. Do not crisscross the cables once they are in the panel—it looks messy and wastes valuable space.

Now we bring the cables into the panel. Regardless of the panel type, all cables must enter the box via circular knockouts that use NM connectors or conduit to keep the wires and cables from getting cut on the knockout's sharp edges. Any unused holes that got knocked out by accident must be plugged.

In case you're wondering, you are allowed to put more than one cable through an NM connector. Typically, the common ½-in. connector can take two 12- or 14-gauge NM cables. If the cable is #4 gauge or higher, the lip of the

Work Safe • Instead of investing in a set of electrician punches ($50) to open just a few holes in a panel, consider buying a metal-cutting hole drill ($10).

Filling a Panel

Shown here is a typical panel fill for an all-electric house. I have all 120-volt circuits on the left side of the service panel and have broken them down into four sections: kitchen, lights, bath, and receptacles. All 240-volt circuits are on the right. (The numbers listed here do not coincide with the numbers on the panel—they just show you how many breakers it takes.) As you can see in the list, even a large 40-circuit panel can fill up fast. Each house, of course, is different. There are also some appliances that I haven't listed, such as a freezer, sewage pump, sewage pump alarm, whirlpool, and central vacuum, each of which takes its own dedicated line. When your box starts filling up, as this one has, consider installing a 300-amp or 400-amp system to have more breakers.

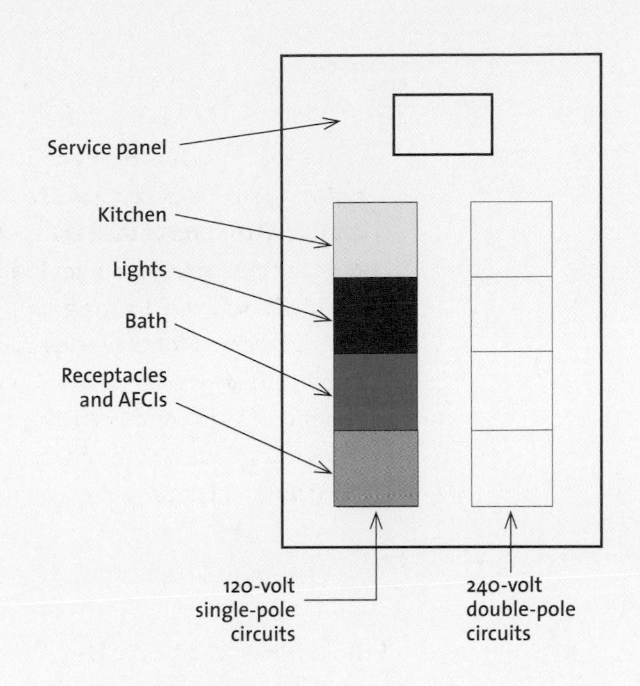

Left Side of Panel

(numbers do not correspond with panel numbering)

1. (20 amp) Dishwasher
2. (20 amp) Garbage disposal and instant-heat water dispenser
3. (20 amp) GFCI kitchen counter small appliance #1
4. (20 amp) GFCI kitchen counter small appliance #2
5. (20 amp) Refrigerator (optional)*
6. (20 amp) Kitchen, dining room, and pantry wall receptacles**

—

1. (15 amp) Smoke alarms, all 3-way switched lights***
2. (20 amp) Lights
3. (20 amp) Lights
4. (20 amp) Lights
5. Spare
6. Spare

—

1. Bath GFCI
2. Bath #2 GFCI or second receptacle in first bath

—

1. Garage receptacles—GFCI
2. Unfinished basement receptacles—GFCI
3. Outside receptacles—GFCI
4. Bedroom #1 and #2 branch circuit
5. Bedroom #3 and #4 branch circuit
6. Utility room (clothes washer receptacle)

Right Side of Panel

(numbers do not correspond with panel numbering)

1 & 2. Electric stove
3 & 4. Electric water heater
5 & 6. Heat pump outside unit
7 & 8. Heat pump inside unit
9 & 10. Bath #1 heater
11 & 12. Bath #2 heater
13 & 14. Water pump (for us country folk)
15 & 16. Spa GFCI (we country folk have these, too.)
17 & 18. Welder, tablesaw, or kiln (for do-it-yourself folk)
19 & 20. Electric dryer

*Code allows this to be a 15-amp circuit. I prefer to keep it 20 amp.
**The NEC allows these receptacles to be on the kitchen counter circuits. I do not. My Above Code system requires that their loads be taken off the counter circuits to allow more available load for the counter circuits.
***Here we are putting together all the 14-gauge loads that use 3 conductor cable. The 14 gauge can be used here because there is either no current on the load (smoke alarms) or minimal current (3-way switch loads).

***Note:* Codes may require all non-GFCI 125-volt, 15- and 20-amp loads to have AFCI protection.**

Work Safe • Don't overtighten the NM connector screws against the cable—just barely snug them. Otherwise, the connector will cut right across the wires and give a direct short.

Work Safe • If a large number of cables are coming into the panel, use the rear knockouts first. If you don't, the front NM connectors will get in the way of the ones being installed in the rear.

connector or conduit needs a bushing to protect the cable's insulation from abrasion. Never bend any wire or cable into a sharp 90-degree angle once it gets through the connector or you'll damage it.

The next problem is identifying circuits once you strip off the sheath. If you want, you can leave a couple of inches of sheath intact on the cable after it enters the panel so that you can write on it where it goes. There are fancy numbered sticker systems, but they don't work any better than freezer tape. Of course, you don't have to identify the circuit if you can remember where it went once you strip off the sheath. But if you have to leave in the middle of the job, you may forget. And most people forget just due to the sheer number of cables entering the panel.

To route the wires to the breakers, run them along the outside edge of the panel frame until the wire gets close to its breaker, then make a gradual bend over to the breaker. Do not make sharp 90-degree bends—this can damage the wire by creating a stretch area on the outside of the bend where the metal pulls apart, as well as a hot spot on the inside of the bend where the metal compresses and overheats. Install the neutral wire in the neutral bus screw closest to the breaker at which its accompanying hot wire terminates. Do not combine two neutrals under one screw and do not put a neutral with a ground wire.

In the old days, the white wire of a 120-volt circuit always went to the neutral bus, and its accompanying hot wire (normally black) always went to the breaker or fuse. Not anymore. With the advent of GFCIs and AFCIs, some of the neutrals go to the breakers as well. It is imperative that you read the side of these breakers to determine which wire goes where before the breaker is installed in the panel; otherwise, you may reverse the neutral and load connections on the breaker. Once the breaker is installed, you can't read the printing on its side that tells you which wire goes where.

On 240-volt circuits, both the hot and the white wire go to the breaker. In this case, you need to put a piece of black tape on the white wire to show that the wire is not a neutral but a hot conductor of a 240-volt circuit.

Panel bonding

Bonding refers to connecting two conductors to form a complete current path—as in bonding the grounding bus to the metal frame of the panel. A bonding jumper or screw normally makes the connection. Once the bond is in place, if a hot wire touches the panel frame, the fault current flows through the bond jumper to the bus, allowing the breaker to kick. If the panel bond is not present, a

The bonding screw grounds the panel so that it can never become a conductor if a hot wire touches it.

complete circuit for current flow does not exist and the breaker will not kick. In that case, the frame stays hot and can electrocute someone.

In addition to the panel bond, any metal conduit extending from the service panel into the meter base must be bonded to the panel (or meter base) in case a hot wire faults against the conduit. This is normally done via a grounding bushing that screws over the sharp threaded end of the conduit. The bushing grounds the conduit and protects the cable against the sharp edge of the conduit's threads.

Subpanels

A subpanel's most common use is to provide a secondary panel some distance from the main service panel. For example, if you have a long ranch house with the kitchen, baths, and utility rooms on the opposite side of the house from the service panel, you will need to run a couple dozen or more cables from one side of the house to the other. A better option is to run one large cable and terminate it in a subpanel adjacent to those rooms. In this case, the subpanel saves labor and material costs.

Neutral Current Can Be Lethal

Even though a neutral is not considered a "hot" conductor, the same amount of current that flows through the hot wire flows through the neutral wire. If the 120-volt load draws 10 amps of current, you have 10 amps leaving the breaker on the black hot wire going to the load. After it goes through the load, there will be 10 amps returning to the panel on the white neutral. Either wire can kill you if you put yourself in series with the circuit.

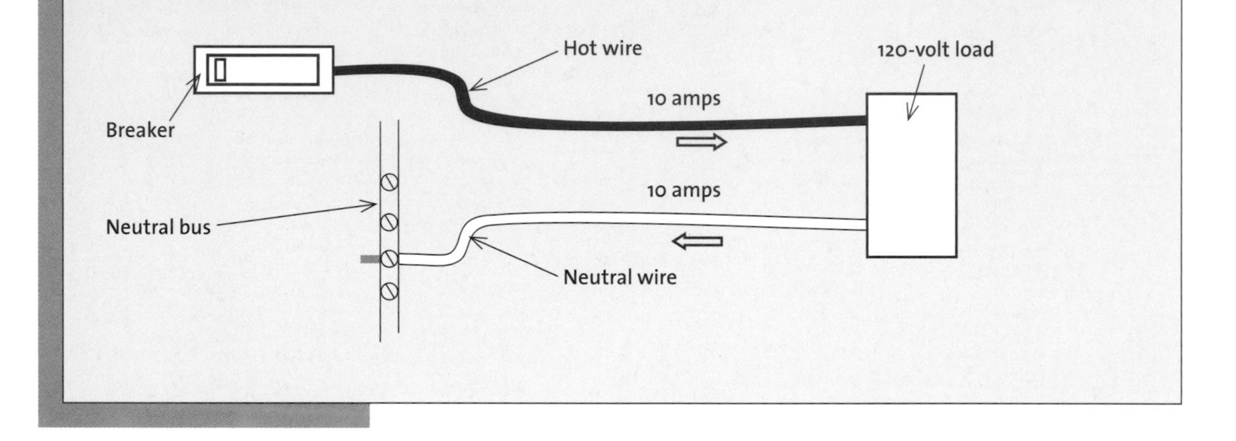

Above Code • You are not required to put a main breaker within a subpanel, because the feeder cable from the main service panel or cutoff switch is always protected by the breaker at the main panel. However, I recommend installing a main breaker in a subpanel. Without a one-switch-controls-all breaker within the subpanel, you will have to go to the main panel to kill the power. If the panel that houses the cutoff breaker is on a different floor, at a distant location, or perhaps outside in the rain, you might get lazy and not turn off the power to the subpanel, thereby taking a few unnecessary risks. In the event of an emergency, cutting off the power at a nearby subpanel may save a life.

Occasionally, subpanels are wired immediately next to a main service panel to solve circuit-shortage problems. This happens when the original panel is too small (probably a low-bid item) and the house needs more circuits (not necessarily more amperage) than the panel can supply. Some areas require a cutoff panel adjacent to the meter base on the exterior wall. If that is done, then the main service panel becomes a subpanel because it is downstream from the cutoff panel.

Wiring a subpanel

Subpanels require a four-conductor cable: two insulated hot conductors, one insulated neutral, and one ground. The two insulated hot wires provide the 120/240-volt power, and the insulated neutral return provides the

Picking and Installing NM Connectors

NM connectors such as these are inserted into the knockout holes of the panels.

NM connectors come in various sizes. All are meant to allow smooth entry of the cable into the panel.

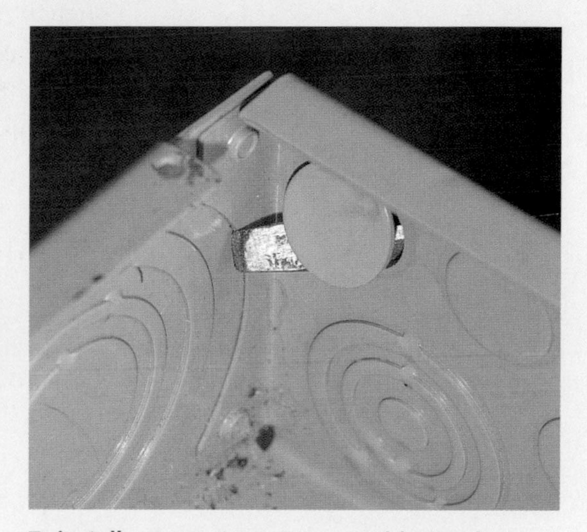

To install a connector, use a screwdriver and knock the center out of the prepunched knockout. Some knockouts can be bent in only one direction.

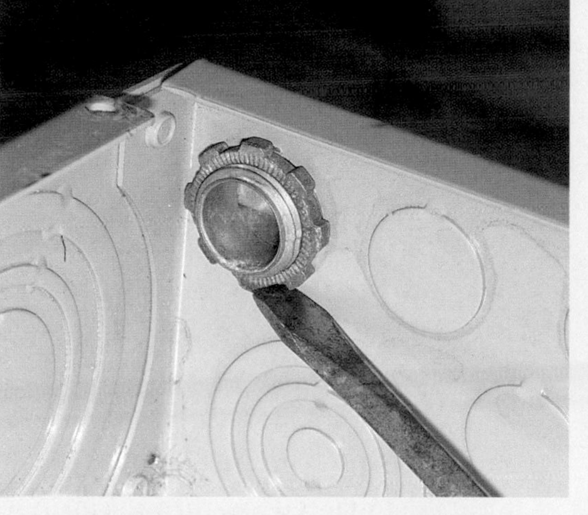

Insert the NM connector through the panel, then tap-tighten it with a screwdriver and hammer. Orient the screw heads of the connector so they are facing forward.

Metal or plastic bushings are required on any sharp-edged thread extending into the panel and on any conduit carrying #4 or larger cable.

A grounding bushing grounds the steel conduit that enters the panel. This is required on any metal conduit between the meter base and the service panel (both sides). To get around this expense, use nonmetallic conduit.

return path for all the 120-volt circuits. You don't need a neutral return for the 240-volt circuits. Because 240-volt circuits already have two wires, that's all that's needed for current to flow. Current flows from one wire, goes through the load, and comes back via the other.

In any subpanel, grounds and neutrals must be kept separate from each other. The ground wires must go to a bus that is bonded to the panel's metal frame. The neutrals must be on a bus that is isolated from the frame and all ground wires. If the grounds and neutrals are intermixed, you will have neutral current flowing on all the ground wires and everything that is grounded, such as water lines and ductwork.

People sometimes ask why it is required to separate grounds and neutrals at the subpanel and not at the main panel. After all, if current can flow on the grounding system at the subpanel, then why can't it flow at the main panel, too? The answer is that current is always trying to get back to where it came from—the utility meter. In the main service panel the current returning from the neutral conductor wants to go to one place—the transformer. The only way it can do that is via the SE utility

Work Safe • Long sheet-metal screws are frequently used to replace lost blunt-end panel screws. Don't do it. If the sharp-tipped screw punctures the SEC cable, you may not live to regret it.

Wiring a Subpanel

In a subpanel, the ground and neutral buses are separated to keep the neutral current off the grounding system. A main breaker is not required.

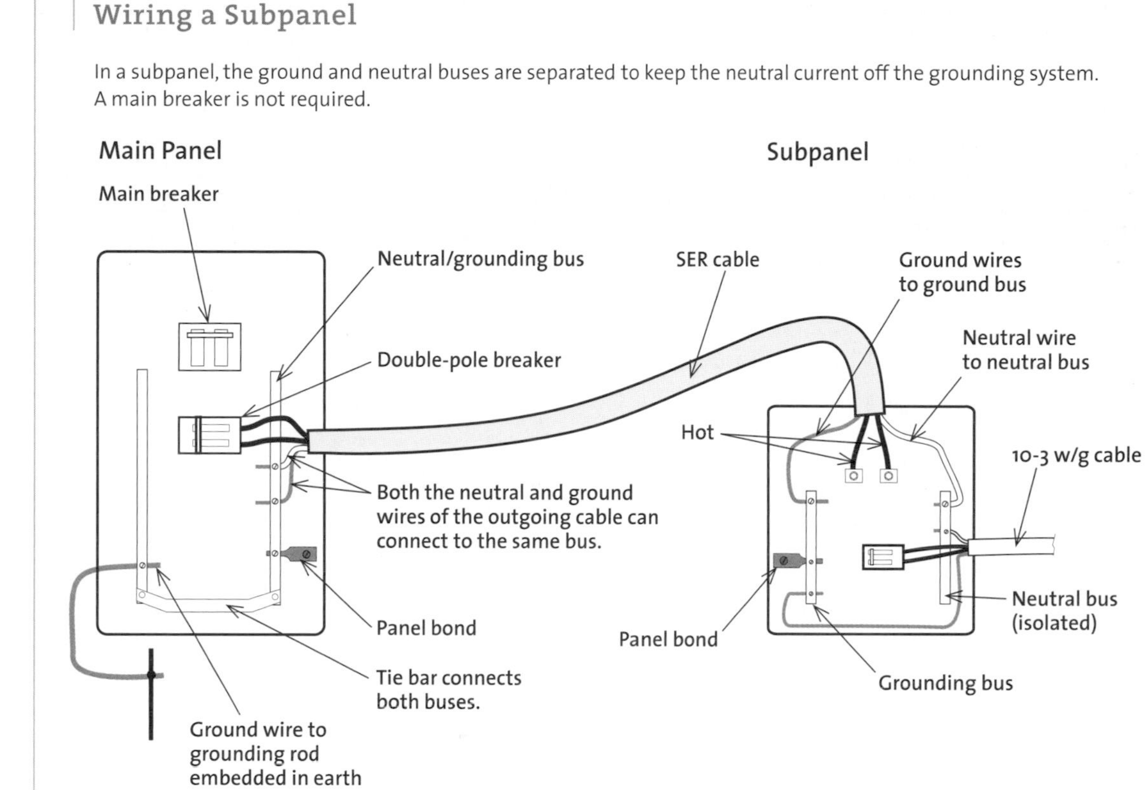

neutral cable. It can't go back there via a house ground wire; thus, no current flows on the grounding circuit.

The current at the subpanel also wants to get back to the utility transformer. But in this case, the subpanel is a distance away from the utility transformer. If we were to tie the grounds onto the neutral bus at the subpanel, the neutral current could also flow on the ground wires to get back to the main panel (as well as the utility neutral), because they also terminate at the same point in the main service panel.

If the subpanel is in an exterior building, such as a shop, add a couple ground rods to the system as well. That is, ground the subpanel just like you would a main service panel—just make sure you tie it into the ground bus, not the neutral bus. This will bleed off any stray electromagnetic voltage that may build up on the metal frames of the shop equipment (especially during thunderstorms). It's true that the stray voltage would bleed off via the ground wire going back to the main system ground at the service panel, but that route takes you through a lot of splices that have resistance. This is especially important if you are using any surge arrestors. It's better to keep surges out of the house, rather than send them to it.

Sizing the subpanel feeder cable

If the subpanel is in the service-entrance system (a meter base, a cutoff panel next to the meter base, and a breaker panel, which is now

Converting a Main Panel to a Subpanel

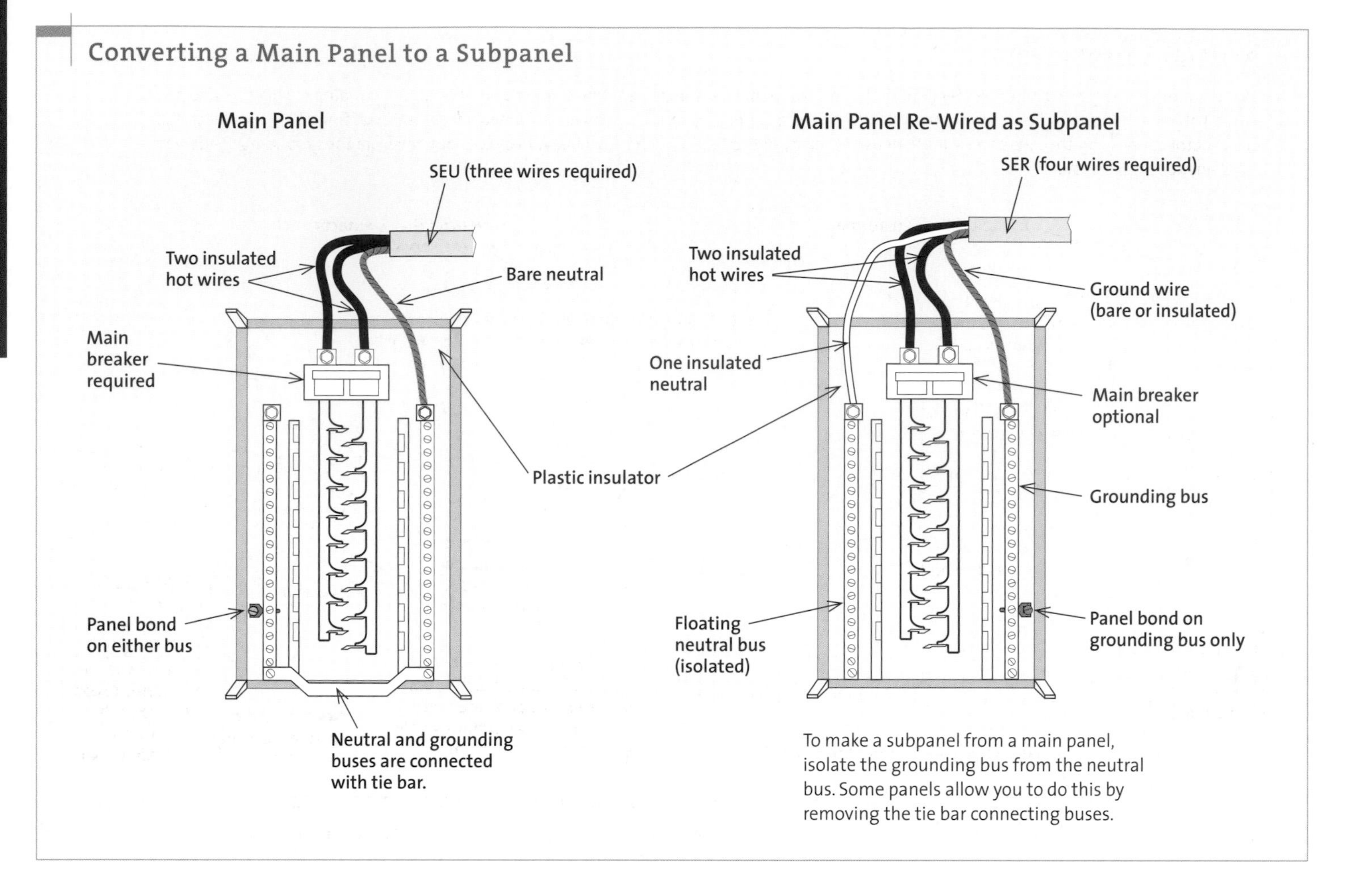

To make a subpanel from a main panel, isolate the grounding bus from the neutral bus. Some panels allow you to do this by removing the tie bar connecting buses.

Work Safe • If you are using a screw as a panel bond and it is among a number of other screws and hard to see, paint the head green so the inspector will have no problem finding it.

a subpanel), the cables feeding both panels are sized as SE cables. If there is a cutoff panel downstream from the main service panel, size the cable that feeds it the same as you would for a utility service cable for amperages of 100 and above. If the subpanel feeds off the main service, as in the long ranch house example, still use the SE-cable sizing chart for currents of 100 amps and above. For amperages less than that, follow Table 310-16 of the NEC. For example, if the subpanel is fed via a 30-amp breaker from the main panel, use 10-3 with ground NMB. For a 100-amp feed, you can use #4 copper cable. Always remember, the feeder cable that feeds a subpanel must be a four-conductor cable (hot, hot, insulated neutral, and ground).

Neutral Current Flow

When neutral and ground wires connect on a subpanel, subpanel neutral current flows along all equipment-grounding conductors and any metal objects that touch them, and then back to the service panel. This route is parallel to the neutral current that flows back via the neutral conductor connecting the two panels. If the neutral breaks, all the neutral current will flow back to main panel via the grounding conductors and any touching metal.

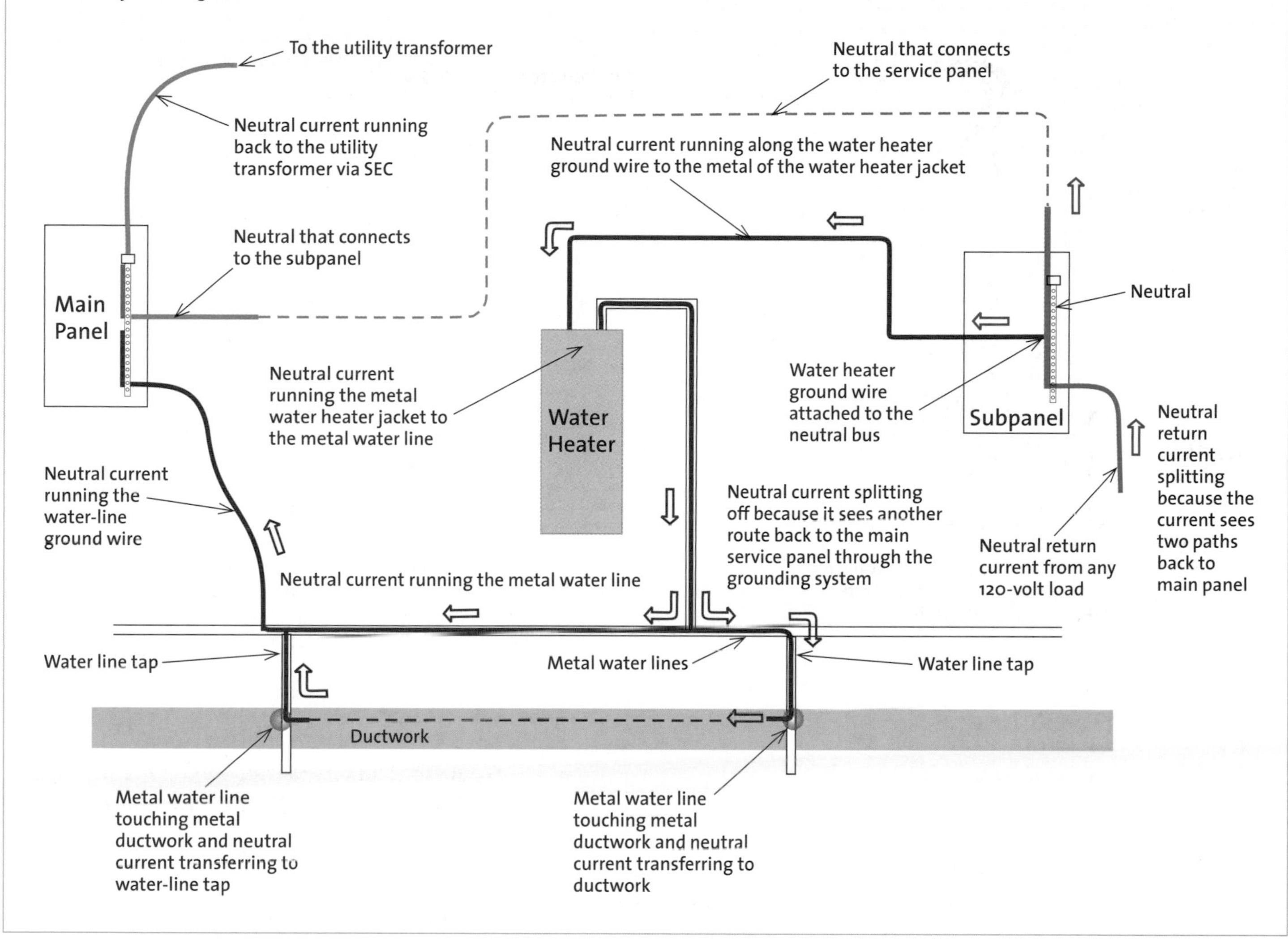

Above Code • If possible, always use a copper conductor to feed the subpanel. Copper is a better conductor, which means that you can run a smaller cable, and that makes for an easier pull. Also, copper doesn't corrode as readily as aluminum.

Balancing the Load

A balanced load is wired so that current on one leg is equal (or close) to current on the other, so current cancels out in neutral. If two circuits—one with a freezer pulling 5 amps and one with a refrigerator pulling 5 amps—are put on the same phase, or leg, current in both the neutral and the service-entrance hot leg will add up to 10 amps. If the same circuits are put on opposite phases, they will still add up to 10 amps in the hot leg but will cancel out to 0 amps in neutral.

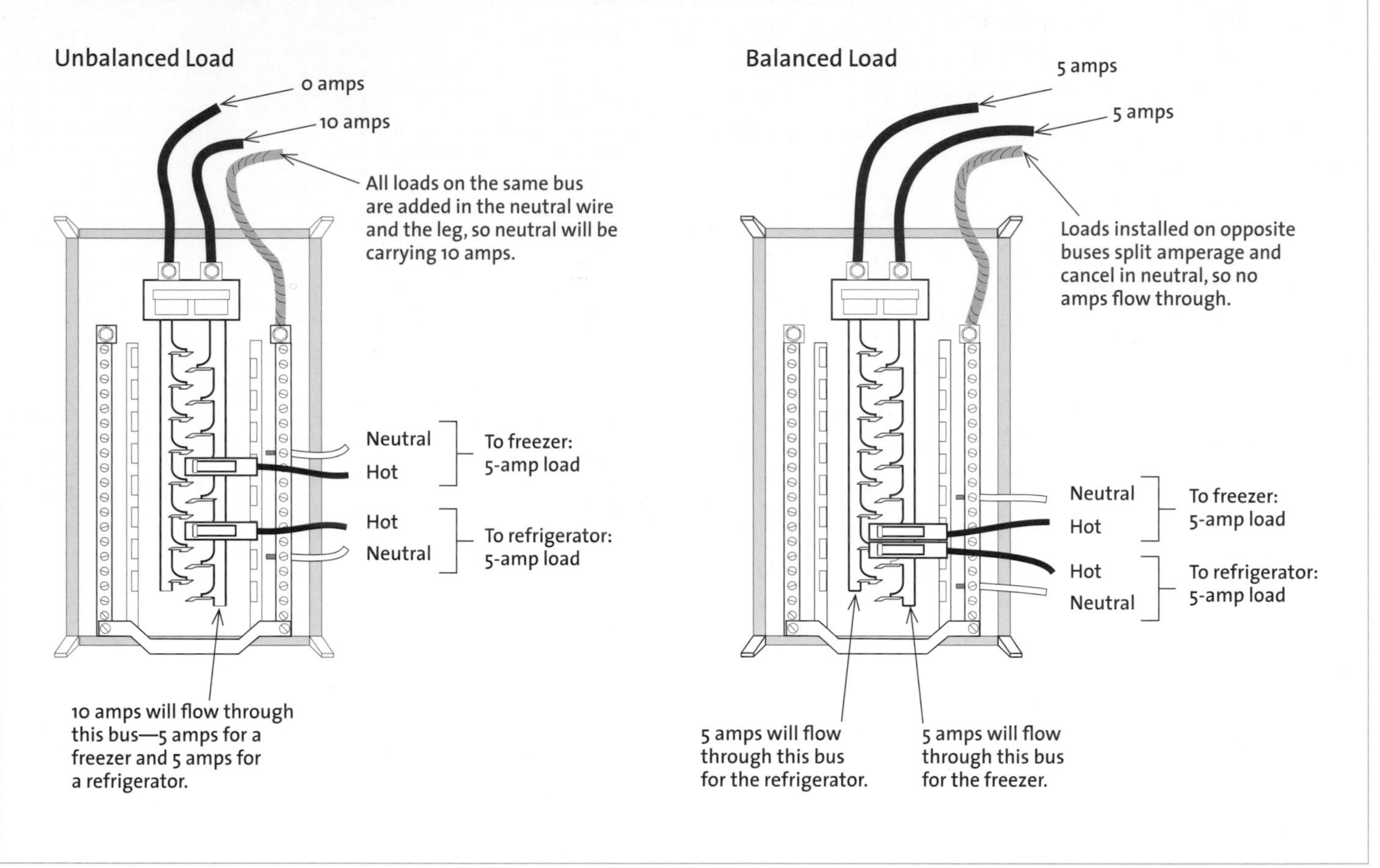

Work Safe • Never paint or put wallpaper over the cover of the panel. The panel should be conspicuous in case someone needs to turn off the power quickly.

Balancing the Load

In theory, and as much as is practical, the load needs to be balanced as it leaves the panel and enters the service-entrance wires. A balanced load is wired in the panel so that the current on one 240-volt leg is equal to the current on the other; this way, the current cancels out in the neutral. Balancing requires you to analyze the circuits to determine the load they will serve. After that, you must employ some common sense.

Balancing lowers the heat buildup in the wire terminals and allows you to put more load on the panel. Let's assume you have a 200-amp panel and main breaker. If you wired the load so that 300 amps of current are pulled through only one leg, the main would trip. However, if you split, or balanced, the load, the current would be reduced to 150 amps on each leg—well within the 200-amp rating of the main.

In most main panels, the tabs on each leg of the hot bus are arranged in opposite phases, so balancing occurs automatically. As single-pole breakers (those that use only one hot leg of the panel) are installed down one side of the panel, they automatically get put on alternate phases because of the tab/bus design. If a double-pole breaker (one that uses both hot legs of the panel) is installed, the current will automatically balance because the breaker, by design, fits on opposite phases.

Inside the house, if two loads of any kind will be on at the same time, put them on opposite phases. For example, if two circuits—one with a freezer pulling 5 amps and one with a refrigerator pulling 5 amps—are put on the same phase, their current on both the neutral and the service-entrance hot leg will add up to 10 amps. If the same circuits are put on opposite phases, 5 amps will flow through each hot leg and will cancel out to 0 amps in the neutral. This is why the neutral is allowed to be smaller than the two hot wires. Even if the neutral load doesn't cancel, it will be reduced by the amount of the smallest load. That is, if one load pulls 10 amps and the other 6, the resultant current flow through the service-entrance neutral will be 4 amps if the loads are on opposite phases, 16 if they're not.

If you're building a shop with a 240-volt feed and two 120-volt machines will be running at once, put them on opposite phases so that the current will cancel in the service-entrance neutral cable and in the neutral wire leading to the shop. If the shop only has a 120-volt feed, there is nothing you can do. It takes two opposing currents to cancel. With only one current, nothing will cancel.

Work Safe • When removing an old paint-encrusted service panel, run a utility knife around the edge of the lid to break the paint-to-wall seal.

CHAPTER 5

The Art of Grounding

Grounding is the foundation of the electrical system. An improperly grounded electrical system is a hazard to both life and property; it can damage or destroy appliances and even kill those who use or come in contact with them. One of my customers is alive today only because of a grounded flue pipe. At my insistence, he grounded it upon installation. During an extremely violent thunderstorm, a ball of lightning came out of an outlet and flew across the room as if to hit him straight between the eyes. At the last minute it changed course and diverted into the grounded flue pipe. It is imperative that residences, especially those with computers, answering machines, and other high-tech sensitive equipment, have an extremely good grounding system. Otherwise, these expensive machines will wind up in the high-tech junk pile.

In this chapter, I explain how to install a proper grounding system—not just a code grounding system, but also my Above Code grounding system. The fact is that the NEC doesn't care how high the ground resistance is as long as at least two rods are driven. If there's a high ground resistance, a surge won't be diverted into ground, but instead into that delicate electronic gear

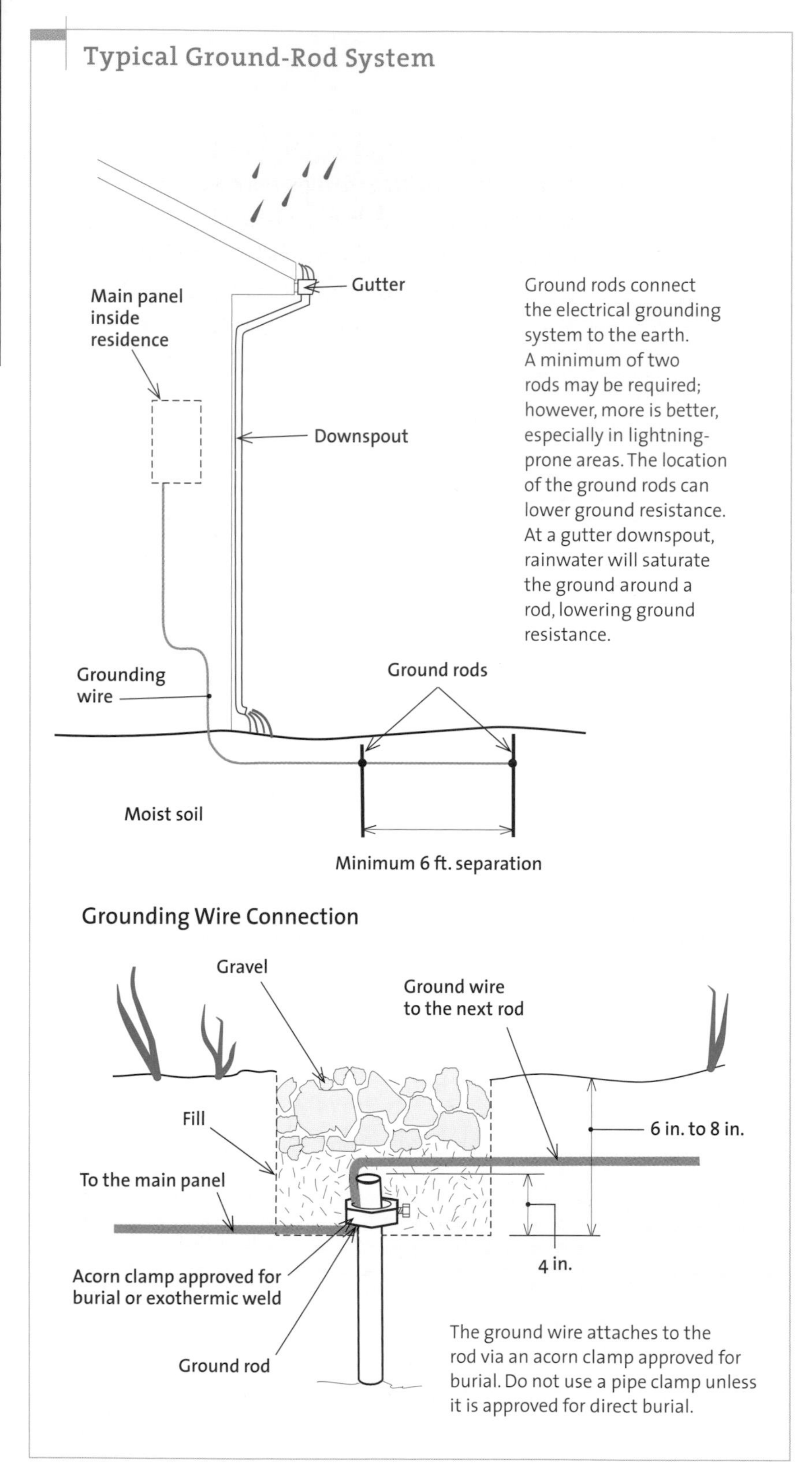

mentioned earlier. In comparison, my system gives you the lowest possible ground resistance, keeping both life and property as safe as possible.

Be selective when purchasing grounding material and take your time installing it—one wrong piece of equipment or one installation mistake can cause the entire system to fail. A grounding system is like a string of old-style Christmas tree lights. All the bulbs—in this case the connections—are in series. If any one is loose, burned out, corroded, or installed incorrectly, the system won't work.

A Grounding/Bonding System

Grounding and bonding is a process designed to keep all non-current-carrying conductors at a 0-volt potential. (A non-current-carrying conductor is a material that can conduct electricity but normally does not, such as the metal frame of a washer or dryer or the metal housing of an electric drill.) Bonding all the non-current-carrying metal together and taking it back to the panel grounding bus protects the user against electrocution and also protects the user and the appliance against the buildup of induced voltages, while providing a safe discharge path for surges.

Grounding/bonding is basically a two-part system. The first part consists of a wire present in all the branch-circuit electrical cables that connects the frames of all the appliances, tools, and metal outlet boxes back to the grounding bus of a fuse or circuit breaker panel. This wire is called an equipment grounding conductor. It is the bare wire in the NM cable and the green wire within a conduit system. It could also be the armor of an AC cable if the metal conduit

Methods of Grounding

There are many methods of grounding, such as using structural steel, rebar, or well casings. However, ground rods are used for most residential applications. But sometimes rods cannot give you a low-resistance ground due to low moisture in the earth or excessive rock and little soil. Below are methods of grounding from the worst to the best:

- Code minimum (unless you can prove 25 ohms or less ground resistance with one rod). Two rods are considered the bare minimum for today's electronic equipment.
- My Above Code requirement. Eight rods, 8 ft. to 20 ft. apart, depending on the available area. I also tie in the building's metal well casing, if present.
- A ground ring (#2 or larger copper bare cable circling the house). Ground rings work well but aren't that practical. In most residential situations, you cannot open a trench around the building, because it would be in the way of the rest of the building construction. You also risk cutting other utilities if they are already in place.
- A circle of ground wire under deep-driven pillars or posts. If deep support posts are going in, take advantage of the opportunity to obtain a low-resistance ground. Form a long continuous circle about the size of a dinner plate and stick it under the post.

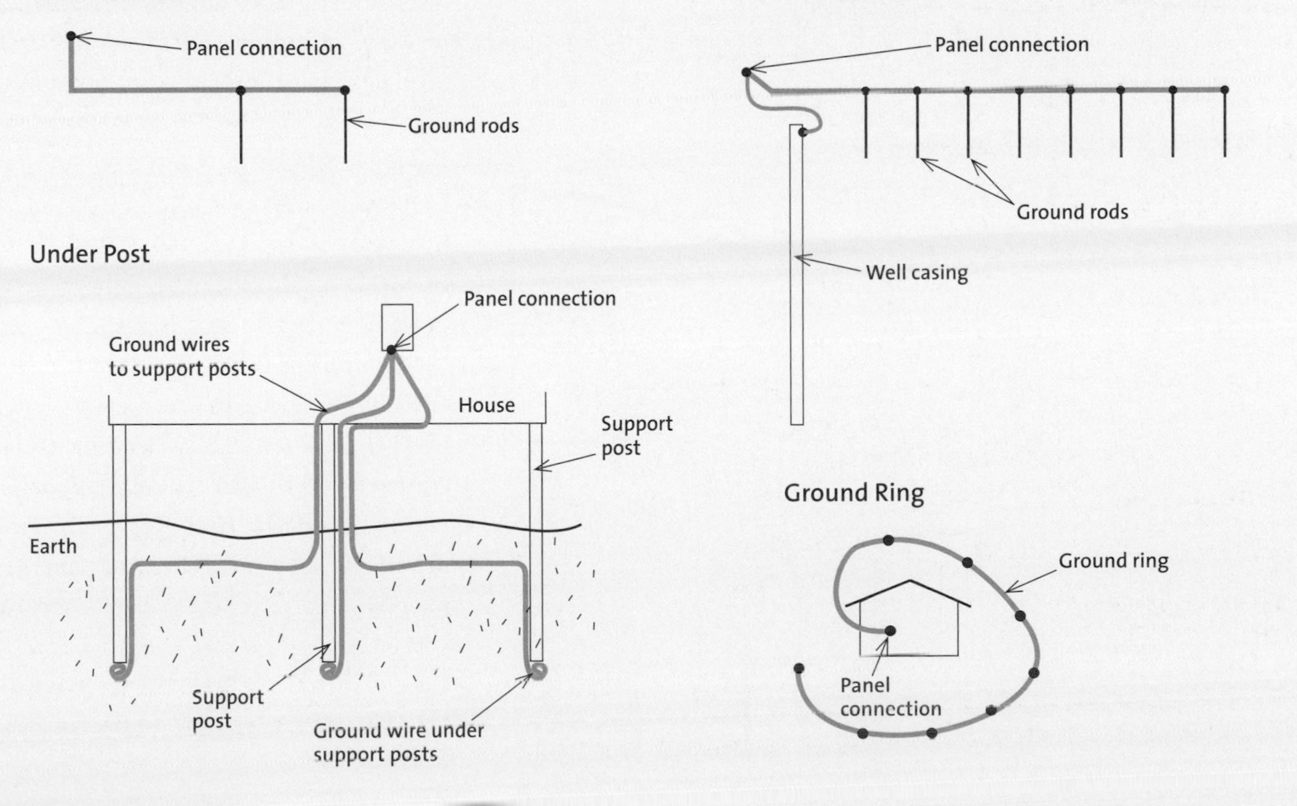

runs all the way back to the service panel. The second part is the earth grounding system. It is composed of a large copper wire, called the grounding electrode conductor, and a series of ground rods (or equivalent) that are buried in the earth.

Protection through Grounding

Grounding offers us life safety, equipment safety, and longer equipment life. The equipment-grounding conductor works like this: If a hot wire touches a grounded appliance frame, a grounded tool frame, or the side of a grounded metal outlet box, the fault current will ride back to the service panel via the equipment grounding conductor, complete the circuit, and trip the breaker. In this case, the ground rod and earth ground play no part in this system.

The earth grounding system comes into play whenever there is any static voltage buildup on an appliance's frame. A typical example would be a lightning strike. The strike puts out a massive electromagnetic force pulse that induces a voltage into metal appliance frames. The equipment-grounding conductor takes this surge back to the main panel, where it transfers to the grounding electrode conductor (ground wire), which sends it to the ground rods to bleed harmlessly into the earth. This same large ground wire is also used to transfer surge pulses (caused by the utility company opening and closing a circuit, a tree falling on a line, a car hitting a pole, etc.) to the earth that have been stripped off the line by arrestors mounted at the panel or at the point of use—like your computer's surge protector.

Ground Fault on a Grounded Appliance

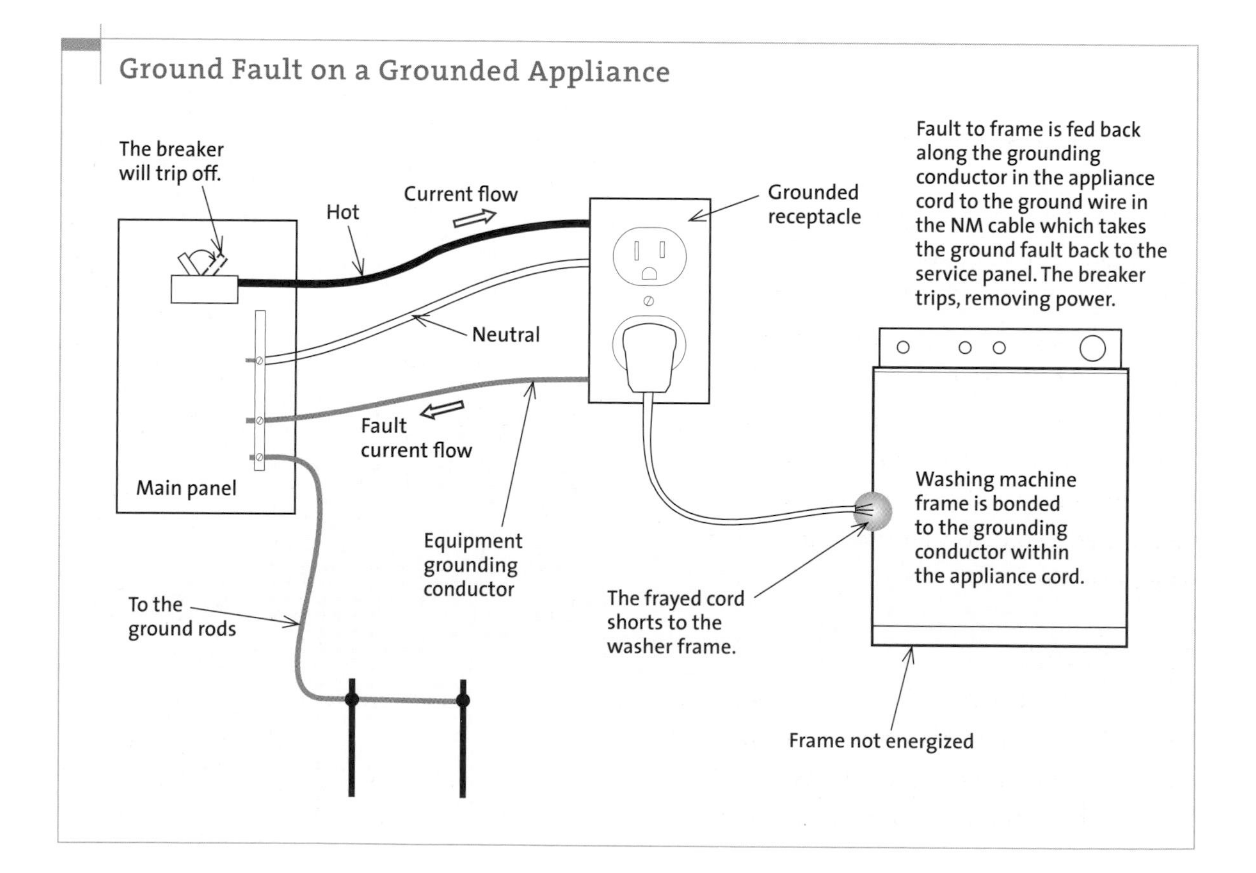

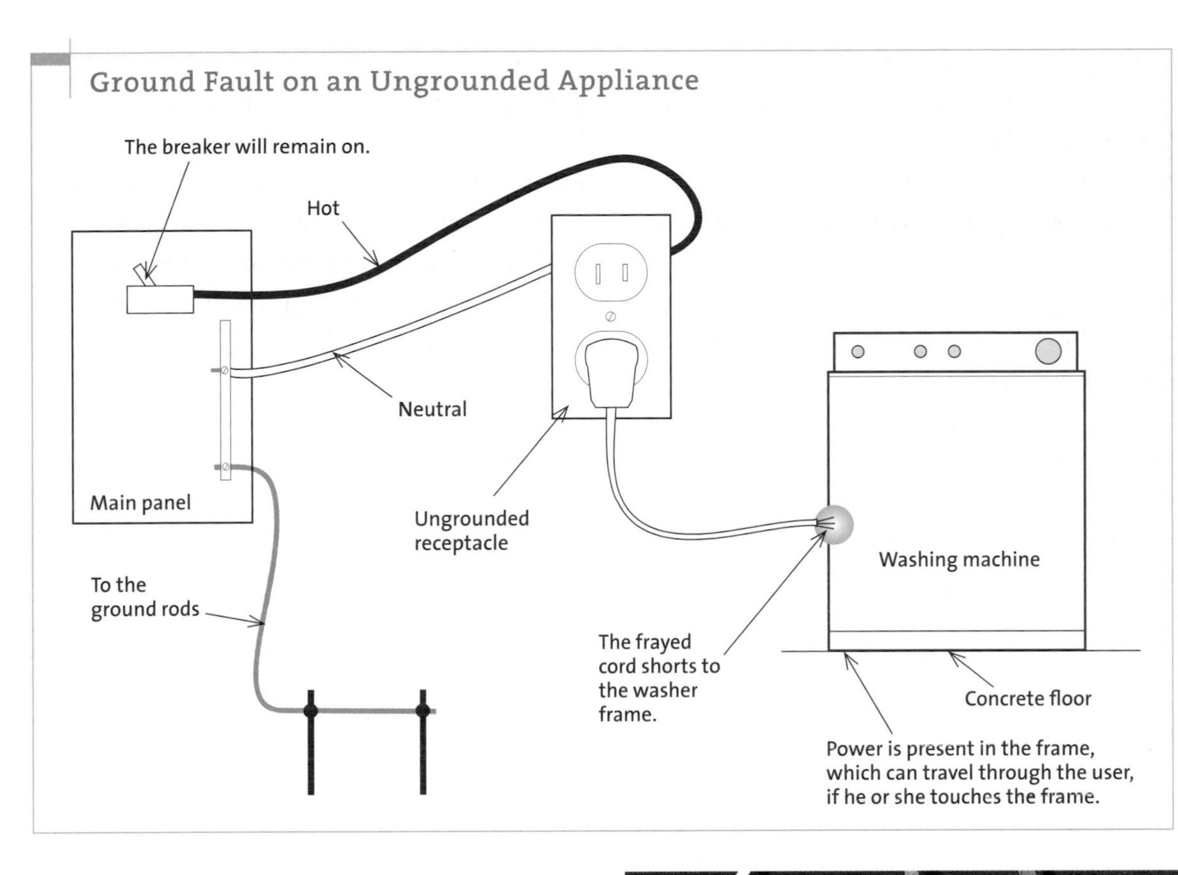

A deadly ground fault. A too-tight NM connector has cut into the cable at the pressure switch (top left of photo). The multimeter reads 120 volts to ground on the water line.

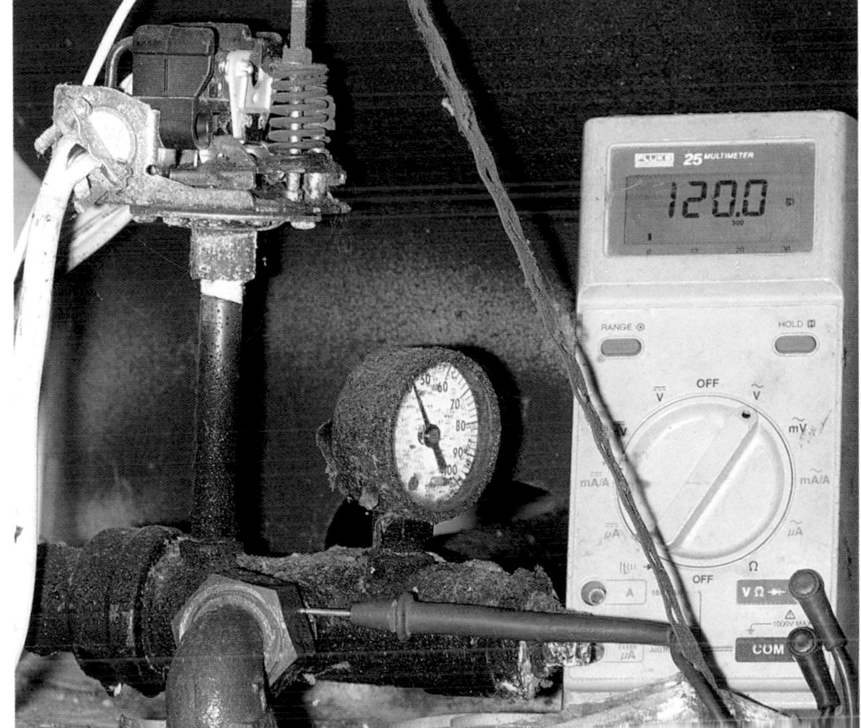

None of what grounding has to offer will make an appliance work any better if these problems aren't occurring. Indeed, electrical gadgets can work just fine without proper grounding. Perhaps that's the real reason grounding took so long to enter residences: It was hard to convince people that an appliance needed something when it was working just fine without it. Grounding is akin to a personal insurance policy. You never know when you'll need it, but you sure want to have it when you do.

Ground fault

If a hot wire accidentally touches the grounded frame of a tool or appliance and energizes it, a ground fault has occurred. This is one of the more common types of tool/appliance

An Ungrounded Can Opener

Some time ago, I was troubleshooting a problem where a customer was getting shocked every time she leaned over the sink to wash the dishes (the sink was nonmetallic). I traced the problem to a metal trim ring around the sink. When I lifted the electric can opener off the ring, the problem went away. Turning the can opener over, I discovered that one of the screws attaching one of the rubber feet to the opener extended beyond the rubber pad. A hot wire inside the opener had shorted to the metal frame on the opener's bottom, and because this was a two-wired (ungrounded) appliance the short was transferred through the screw into the sink trim ring. If the ground prong were present, then the short would have returned back to the main panel's neutral/grounding bus and kicked the breaker. Note that the earth is not in the current flow path during a ground fault.

malfunctions. For example, let's assume that the cord of a grounded washing machine has frayed (the washing machine is sitting on a wet basement floor) and the hot wire is touching the metal frame, resulting in a ground fault. Once the frame is energized, the fault current will flow back to the main panel via the equipment grounding conductor and the circuit breaker will trip off. In this situation, the fault current does not flow to earth ground to dissipate. It remains on the low-resistance wiring to the breaker, allowing the breaker to trip off.

If an equipment grounding conductor is not present, if a cheater plug is used (I see this all the time), or if there is a bad connection somewhere along the equipment grounding conductor, the breaker cannot trip because there is no way for the fault current to get back to the main panel to complete the circuit. The frame of the washing machine will remain energized, and a person who touches it can receive a serious shock. What's happening here is that the person's body becomes an electrical conductor to complete the grounding conductor path.

A ground fault doesn't have to be direct, either. Though the initial fault or short to the appliance may be direct, the routing back to the panel may be of high enough resistance that it lowers the current below the level needed to kick the breaker off. Although the current may not be high enough to trip a breaker, it could still be lethal to both you and the appliance.

Unwanted voltage and current

If an appliance frame isn't grounded, the 0-volt potential (on its frame) that we need for safety cannot be maintained. Without grounding conductors to direct the unwanted current away and an earth grounding system to dissipate it, the stray voltage can build up to a damaging potential—arcing over to us if we touch it or through the appliance itself.

Stray current resulting from induced voltage can come from any current-carrying source (for example, current through one wire can induce current into another). Induced voltage also comes from any continuous voltage source, such as an overhead high-voltage line. When a good grounding system does not exist, a power line can induce current into a home's metal roof or metal framing. I've seen this problem in mobile homes that are close to high-voltage transmission lines. The siding's metal skin can build up some very high voltages and give you quite a shock. To avoid this type of voltage buildup, the metal must be well grounded into a system of ground rods.

Surge current comes primarily from lightning. The magnetic field created by a lightning strike, even if it's over a mile away, can induce voltage into any current-carrying conductor (such as house wiring) or non-current-carrying conductor (for instance, metal water pipes and appliance frames), producing a short-lived high-voltage spike. If the induced voltage and current find no designed path from the appliance to ground, it will find its own, such as through sensitive electronic equipment, or even arc through the air. The resulting shock to humans can vary from a minor jolt to electrocution.

Surge arresters

Surge arresters suppress voltage surges (that's why they're also called suppressors) by directing excess voltages and currents into the grounding system and thus into the earth. But neither a high-quality surge arrester in the main panel nor a point-of-use arrester at the appliance will work properly without a high-quality grounding system to dissipate the surges. Without that, the surges simply go right into the equipment you are trying to protect. And the odds of it picking that new large-screen TV rather than an old black-and-white set are almost 100 percent.

Choosing Grounding Materials

Now that we know what a grounding system is and what it can do for us, the question is how to build one. Choosing the right materials can sometimes make the difference between a properly working grounding system and one that doesn't work at all.

Ground rods

Ground rods connect the grounding system to the earth. Manufacturers make many different types of grounding rods, but the two most common choices are galvanized steel and copper clad. Both types will do the job.

Diameter Code requires a minimum of a ⅝-in.-diameter rod. (Listed rods can be ½ in.) The next steps up are ¾ in. and 1 in. The latter two are only available at large electrical wholesalers. In general, I don't think the larger-diameter rods are worth the extra expense.

Avoid the little 4-ft.-long, ½-in.-diameter rods that are sold at electronics stores and discount houses. These mini-rods aren't legal for anything. I've noticed that people use these

Work Safe • It's imperative that you bury the entire ground rod. Leaving one end of the rod sticking out of the ground is a trip and lawnmower hazard.

rods to ground individual items, such as a TV antenna. Don't do it. The antenna, the telephone system, the cable TV system, and everything else must be tied into the main house grounding system at the service panel.

Length The rods must be at least 8 ft. long; if you can find them, they do come in 10-ft. lengths. Special clamps are also available; they allow you to stack ground rods on top of each other. Deeper is better, but it is debatable whether or not the extra length is worth the extra expense and effort. In most cases, the extra depth does not significantly lower the ground resistance, unless it gets you into a moisture bed. It's better to drive a second rod (NEC requires a minimum spacing of 6 ft.).

In cold areas of the country with deep frost lines, it may pay to go deeper. Low temperature yields increased ground resistance within the frost-line area. Ground rods must be long enough to go deeper than the frost line. In that case, extralong rods may have an application. If possible, consider coupling ground rods together or taking advantage of a metal well casing.

Above Code • Ground rods (also called grounding electrodes) are normally driven into the earth close to the main panel or utility meter base. Code demands two rods unless you can prove a resistance of 25 ohms or less (many homes only have one rod, which is not enough in most cases). I install eight rods because I'm always after the lowest ground resistance I can get. In theory, doubling the number of rods should reduce the ground resistance by half. In reality, it varies and can be significantly more. Regardless, the bottom line is the more rods the better. Because ground rods are cheap, normally less than $10 each, there is no reason to obtain anything other than the best ground possible.

Other grounding options

At one time it was common practice to use a metal plumbing system in lieu of ground rods. Many homes are still grounded that way. If this is the case with your home, change it immediately. Code does allow metal plumbing lines to be used as a ground rod if three conditions are met: a ground-rod system is also present, the pipe is in contact with at least 10 ft. of earth (which is impossible to know), and the connection to the pipe is within 5 ft. of where the pipe enters the building. However, should a plumber replace the metal pipe with a plastic one, he'll nullify the ground. To be safe, I always think of the metal plumbing system as something that needs to be grounded, not something to ground to.

Rebar is also sometimes used instead of a ground rod. The footer, connected into the grounding system via rebar, is an effective low-resistant ground. To use it, the rebar must not have a nonconductive coating (in my section of Virginia, nonconductive coating looks like light-green paint) and must be in contact with at least 20 ft. of concrete. In addition, the concrete footer must be in direct contact with the earth, as opposed to having a plastic barrier underneath it.

Another problem with using rebar is connecting it to the ground wire. An acorn clamp connection, which simply presses the copper wire against the metal of the rebar, rusts quite quickly and doesn't provide a good long-term ground. If you want to use the rebar as a grounding electrode, you should use an exothermic weld system; unfortunately, these welding devices can be difficult to find and tricky to use. Because it's not possible to verify that all of the conditions are met, and because of the problems associated with the rebar/clamp attachment, I rarely use rebar in lieu of a ground-rod system.

Ground wire (grounding electrode conductor)

Because aluminum corrodes so easily, I use only copper ground wire. Although the ground wire from the service panel to a ground rod never has to exceed #6 copper, I use #4 copper most of the time, because it has a significantly larger diameter and thus has less resistance, especially on long runs.

For 200-amp service, you need #4 copper to connect to the metal plumbing, rebar (if used as a grounding electrode), or anything else you may want to bond into the grounding system. For a typical eight-ground-rod system, you need around 60 ft. of wire, assuming you don't have to make a long run to get into that moist (low ground resistance) valley.

Clamps

Make sure that the clamp you use to connect the ground wire to the ground rod is approved by UL for direct burial. The common clamp used is called an acorn clamp. Do not use hose clamps or the toothed clamps commonly used to attach wire to pipes inside the house—they are not approved for direct burial and quickly rust and corrode.

Acorn clamps are the most common clamp used and are approved for direct burial. Note that the ground wire is in the back of the clamp—in the V—not under the screw.

I once measured the clamp resistance of a large number of grounds for a utility cooperative that was trying out some new equipment to measure clamp and ground resistance. The average resistance through their corroded toothed clamps was 140 ohms (that's a lot of resistance). A several-thousand-amp surge (a typical lightning pulse) through such a high-resistance connection can develop several thousand volts. This high voltage is felt throughout the grounding system, and it will arc through sensitive electronic equipment if it finds a lower-resistance path to ground. For some unknown reason, this normally occurs through the most expensive electronic gear, such as a color TV. With the approved clamps, you can expect a ground clamp resistance of close to 0 if they are installed properly.

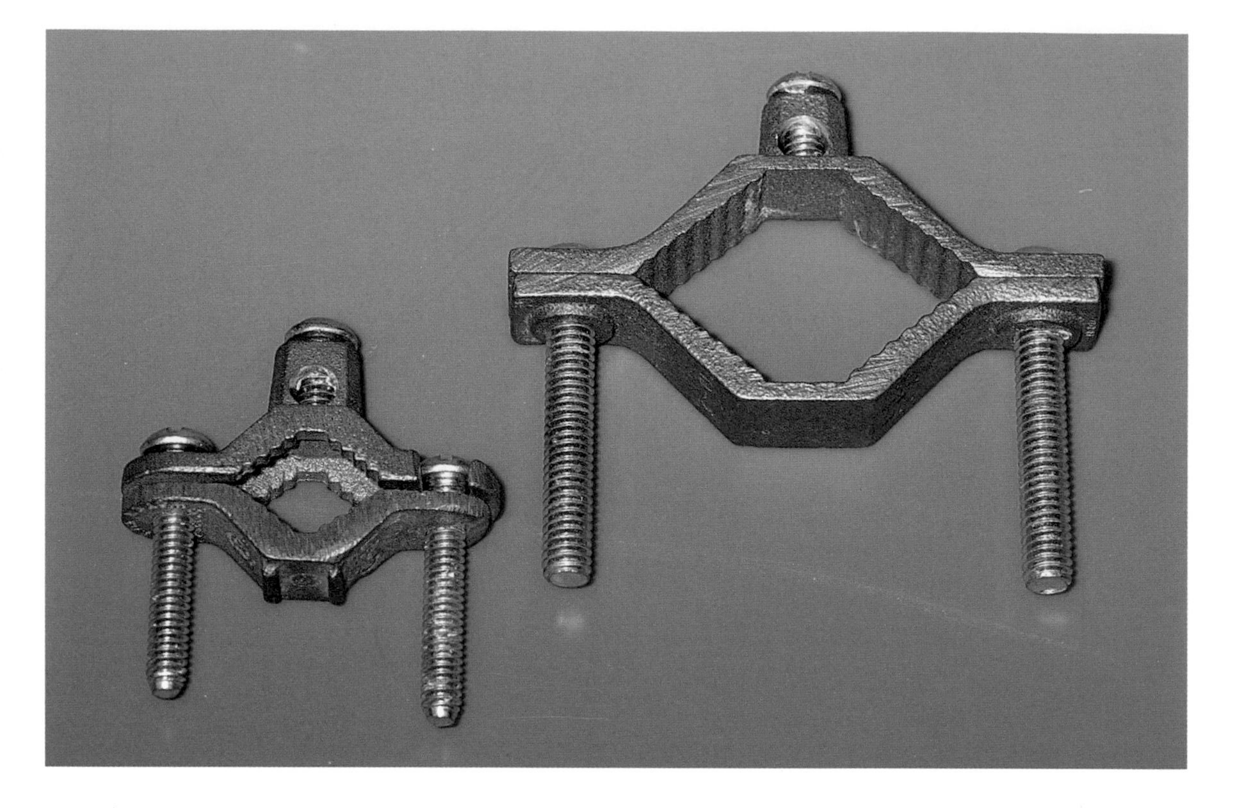

Brass-toothed clamps. Use these to grab onto the water lines inside the house to ground them, not onto the ground rods outside.

You are allowed to skip the clamp system entirely and instead use an exothermic weld system to connect the ground wire to the rod. Such a system actually welds the wire to the rod, so you never have to worry about a bolt coming loose on the clamp. However, I no longer use this system due to problems with the wire snapping off the rod.

A Low-Resistance Panel-to-Earth Ground

Installing a low-resistance panel-to-earth ground to instantly drain away unwanted current flow and keep the ground system at 0 volts is not an easy task. Absolutely everything is working against you in obtaining a low-resistance ground: rocky soil, low moisture in the earth, poor conductivity in the soil, human error, and so on. Thus, your objective in designing and installing a grounding system is to overcome these problems to the point where life and equipment are no longer in jeopardy. In common residential construction, a good grounding system starts with the actual earth connection itself—the ground rods.

Installing ground rods

Once you have all the materials, the question is where to drive the first rod. Though the first one or two rods should be driven reasonably close to the main panel to appease the inspector, the rest should be driven where the ground is moist and can provide low ground resistance. A good location close to the house is normally around the gutter downspouts and the drip edge. To obtain a lower-resistance ground, consider running the ground wire to a creek bed, dry pond, or wooded valley. When driving rods into a creek or pond bed, do not use an AC-powered hammer drill (for obvious reasons). Use a sledgehammer or a battery-powered hammer drill. Don't cheat when you install the ground rods. What I mean by cheating is driving the first rod into the loose fill adjacent to the footer or support post—so loose that you can easily shove the rod into the ground bare-handed. If the rod goes in easily, odds are you do not have good, tight contact with the earth and will wind up with a higher ground-to-earth resistance.

In rocky areas, you may be lucky to get down 2 ft. If you can pull the rod back out when you hit a rock, do so. If not, cut it off just under the final grade and use it as a ground, but keep driving more rods. The more rocks you hit, the more rods you will have to install. If you have no significant depth at all, consider laying the rods in a trench and bringing in dirt to cover them. Covering the rods with rock will do you no good—you need soil.

Rods must be in contact with moist earth for their full length to get a good low-resistance ground. This ground rod was installed in the rock fill around the footer because it was easier to drive down. The result will be a higher ground resistance as compared to a rod that is completely surrounded by soil. Also, only one wire is allowed per clamp.

Driving Ground Rods in Rocky Areas

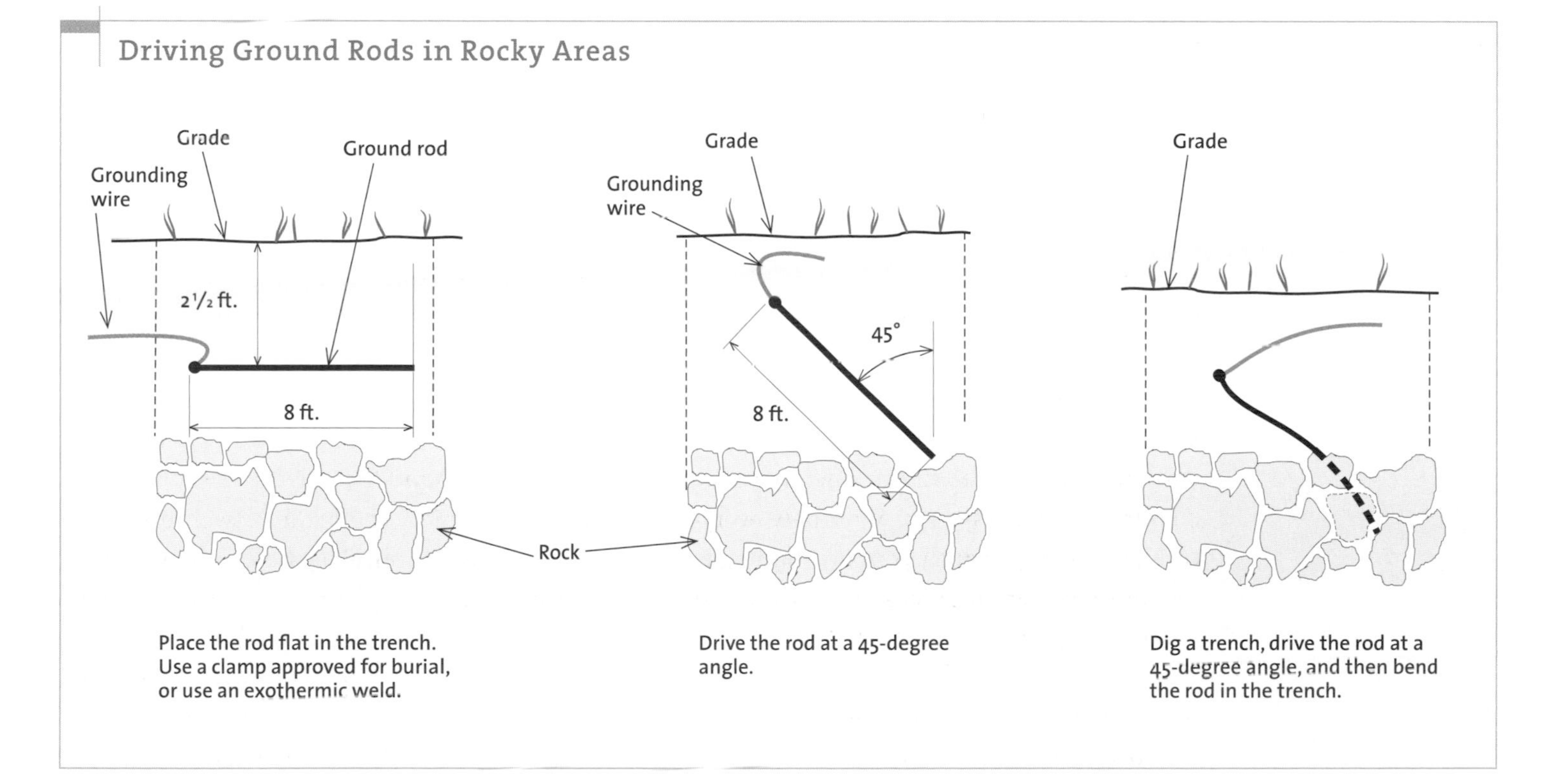

Place the rod flat in the trench. Use a clamp approved for burial, or use an exothermic weld.

Drive the rod at a 45-degree angle.

Dig a trench, drive the rod at a 45-degree angle, and then bend the rod in the trench.

If you use a sledgehammer to drive the rods, you'll need to steady the rod as it is hit. Never use your hands. Instead, have someone steady it with a board that has a hole drilled in it or a hook made out of plastic pipe.

For rocky areas, I highly recommend using a rotary hammer to drive the rods. In normal soil conditions, it can drive an 8-ft. rod in less than a minute. In rocky areas, the many hammer blows it produces can split a small- or medium-size rock. If you hit a large boulder, you're still out of luck.

If you're unable to drive the rod all the way in because you encounter solid rock, you've got three options other than just cutting the rod. You can drive the rod at an angle of up to 45 degrees from vertical or lay the rod in a trench (minimum 2½ ft. below the surface, or grade). Another option is to dig a trench and drive the rods at 45 degrees until you hit rock. At that point, bend the rod flat into the trench.

The distance between the rods is very important. Code requires a minimum of 6 ft., but I feel this is not enough. You want the distance between the rods to be at least equal to the length of the rods (normally 8 ft.) to as much as twice that distance. What you are trying to avoid is the "sphere of influence," where the electrical field around one rod affects the electrical field around another. This increases the resistance of the electrical current flowing through the soil.

Once you've found a suitable location, use a post-hole digger to dig a hole 6 in. to 8 in. deep for each rod. Drive the rod, leaving about 4 in. above the bottom of the hole. As mentioned earlier, the easiest way to drive a ground rod is with a hammer drill (you can rent one). The most common way to drive a rod is with a sledgehammer. The problem with using this method, however, is that the rod swings as you

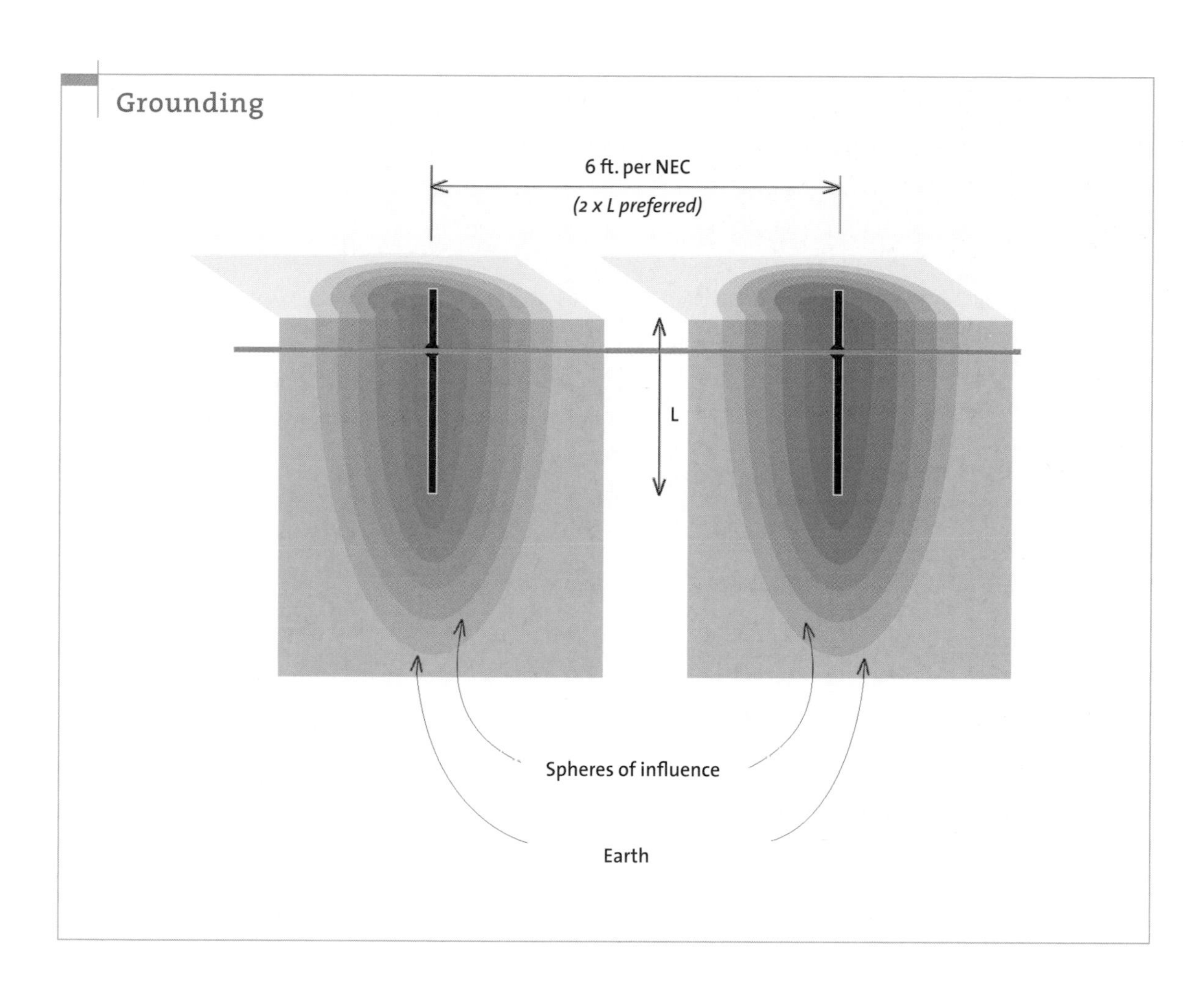

try to hit it. To help hold the rod steady, drill a 1-in. hole through scrap board and put it over the rod or make a hook using PVC pipe. Try not to mushroom the end of the rod, or you'll have to file it to fit the clamp over it.

Running the ground wire

Once all the ground rods are driven, open a ditch between them with either a pick or a mattock. Lay the ground wire in the ditch and cut it to length. Cut the wire about 1 ft. longer than what it lays out to be, just to be on the safe side. Do not cut the wire anywhere along the run—it must be continuous from beginning to end, regardless of the number of ground rods you plan to install. (This is why I buy the wire in rolls and then unwind the length I want.) Slip the acorn clamps onto the wire. Slide them along the wire to the approximate location of the ground rod, then slip the clamp onto the rod as you push the wire into the ditch.

Work Safe • Smooth all the sharp bends out of the ground wire as you unroll it along the ground. This allows the acorn clamps to slide easily along the conductor.

For protection, the wire must also be buried (at least 6 in. deep) along its run and enter the wall cavity as soon as possible. If the wire along the outside of the residence is not subject to damage, it can be run without conduit as long as it is stapled flat against the wall approximately every 2 ft. (the wire can be painted to match the building's color scheme); if the wire can be easily damaged, run it through conduit.

Attaching the clamps

The clamps need to be properly fastened to the rods. To do this, run the wire parallel to the rod for about 1 in. before and after the clamp. Do not run the wire into and out of the clamp from the side; doing so will result in a loose, high-resistance connection. If you are using an exothermic weld system, be careful not to

Work Safe • For a better ground connection, use two acorn clamps and apply thread lock compound on the bolts to keep them from working loose.

Work Safe • Do not use electrician's pliers to tighten the ground clamp—they can't get it tight enough. Use an end wrench, an adjustable wrench, or a socket wrench.

This ground wire is entering and leaving the clamp from the side (as opposed to dead center down the V back). In a few weeks or months, the ground wire may slide to the center of the V, creating a loose, high-resistance connection.

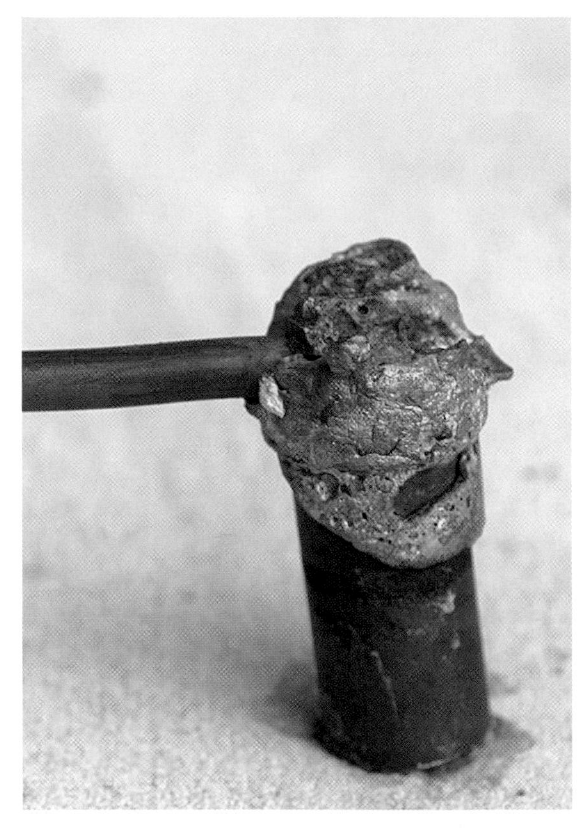

A missing ground wire. The wire that used to run to another rod has snapped off the right side of this one. The wire was made brittle by the exothermic weld, and when bent, easily broke free of the connection. When this happens, you must cut off the top of the rod, as I did here, and start from scratch.

bend the wire close to the weld. The heat produced by the welding process makes the wire brittle and easy to snap, which is why I no longer recommend this system.

Cover the ditch once the clamps are attached. As you fill it in, push the wire down to ensure that it's at the bottom. If aesthetics allow, fill the ground-rod hole with gravel to allow the rain to flow through the gravel and saturate the soil around the rod for an even better ground. If you located the hole close to a gutter, then the gutter water will saturate the earth around the rod.

Bonding the panel frame

The main panel frame, being metal, will always have a panel bond, which is simply a wire or metal strap that connects the panel into the grounding system. An inspector normally makes a point to verify its presence. Some panels use a screw that is turned down through the grounding bus and into the panel's steel; others have a bonding wire that runs from the bus through a prepunched hole in the steel frame. Although the inspector may only check the bond in the main panel, all metal box frames in the electrical system need to be bonded into the grounding system, including the cutoff boxes for heat pumps, water heaters, and submersible pumps.

What Needs Grounding

When it comes to grounding, remember this rule: If it can become accidentally energized—through an induction or a fault—ground it. I've been shocked by sheet-metal ductwork, water lines, drain lines, and metal siding—if it's metal, it's bitten me. Therefore, I require all ductwork to be grounded. Even if a heat pump or furnace frame is grounded by its own equipment-grounding wire, the ductwork may not be. Many times there is a nonconductive flex joint between the unit and the ductwork that acts as an insulator. I also recommend grounding the rails of garage-door openers to prevent lightning surges from building up and destroying the electronic opener. The following is a list of things that I recommend be grounded.

Take Note • Soldering the wire to the ground rod does improve the connection, but you must also use an acorn clamp in case a lightning surge melts the solder.

Metal water pipes

To ground metal water pipes, make the primary connection from the pipe to the grounding bus within the main panel and connect it to the pipe on the house side of the main valve. You'll also need to run bonding jumpers where needed. The bonding jumper is a separate section of wire with clamps on each end; it allows the grounding conductor to be run around anything that could interrupt the grounding circuit. For example, you need to run a jumper around a water heater because some water heaters are now made of nonconductive material. Also, water heaters get replaced, and you don't want to disconnect the ground at any time. If a jumper is not there and if fault current is flowing in the water pipes (this is quite common), you can be electrocuted when you change the water heater. If all pipes are metal, including the feeder from the city connection, install a jumper around the meter.

Bonding Jumpers

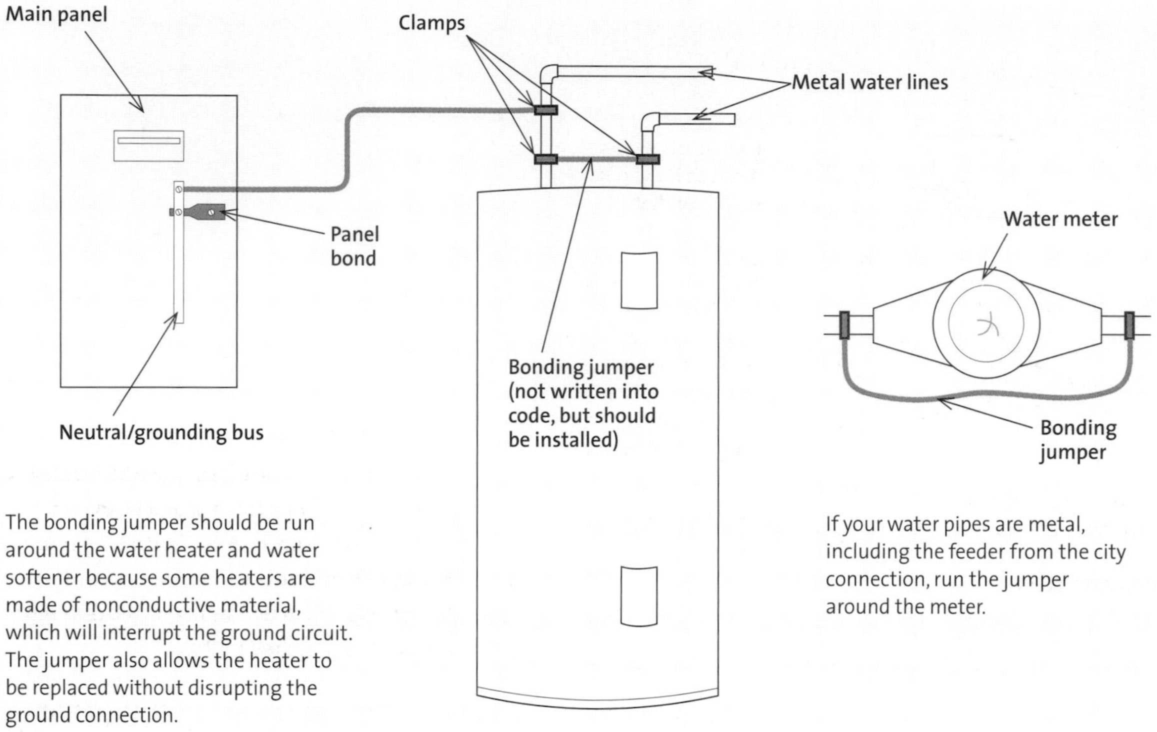

Main panel

Clamps

Metal water lines

Panel bond

Water meter

Neutral/grounding bus

Bonding jumper (not written into code, but should be installed)

Bonding jumper

The bonding jumper should be run around the water heater and water softener because some heaters are made of nonconductive material, which will interrupt the ground circuit. The jumper also allows the heater to be replaced without disrupting the ground connection.

If your water pipes are metal, including the feeder from the city connection, run the jumper around the meter.

Metal conduit

All metal conduit within the house must be grounded. I normally don't depend on conduit to carry the ground, because there is too much of a chance that a slip-on fitting will loosen and open the grounding system (screw-in fittings make better grounds). I normally run a grounding wire within the conduit as much as is practical. Consider using plastic conduit if you don't want to worry about grounding it.

In a residence, metal conduit is most commonly used in the garage and basement. Short sections of thin-wall conduit protect wire as it runs from receptacles and switch boxes vertically up to the floor-joist system. To ensure a proper ground, connect the metal conduit tightly to a metal receptacle box and, in turn, ground the box via the grounding wire in the NM cable entering the box. The particular box you want to use must have a single knockout hole as opposed to multiple ones. The punched metal surrounding a multiple knockout is not a reliable conductor of fault current (thus the use of a grounding bushing and a jumper to bridge the gap).

Receptacles and switches

Receptacles and switches in nonmetallic boxes can be grounded in several ways. The standard way is to simply connect the incoming bare grounding wire onto the outlet grounding screw. If there are two or more cables, cut

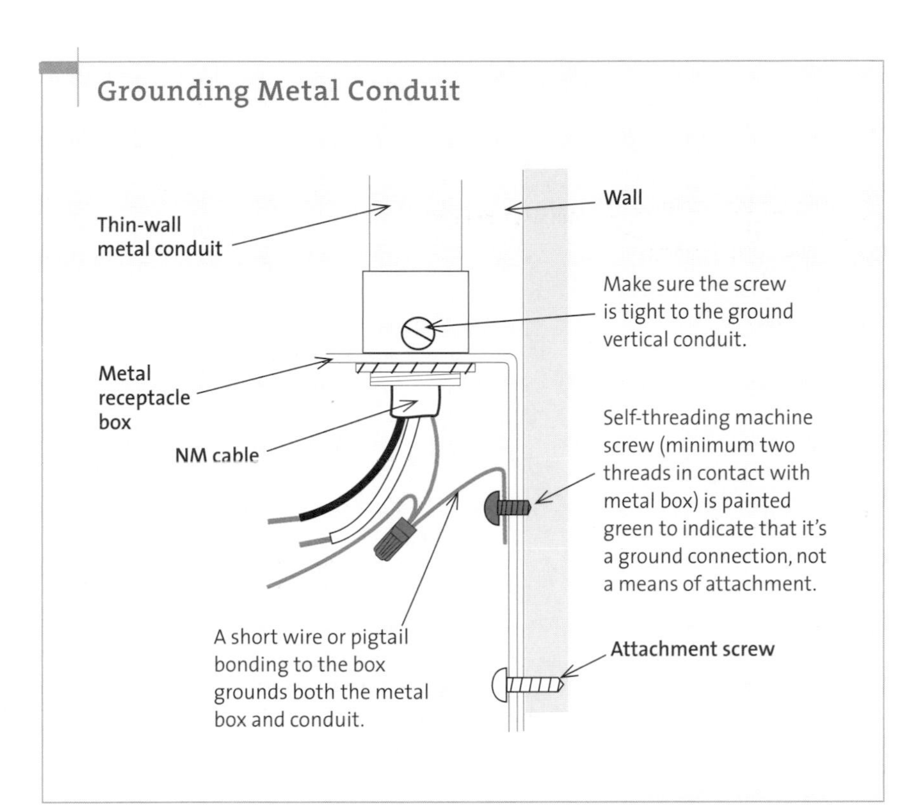

This box has prepunched holes for both single and multiple knockouts. A single knockout hole can be used to ground conduit via a grounded box. Multiple knockouts need a grounding bushing, because fault current may not get through the area of the punched metal.

the ground wires from the cables to the same length and add a 6-in. bare wire pigtail to the bunch; this will serve as a jumper from the splice to the receptacle. Using broad-nosed electrician's pliers, twist the ends of the wires together in a clockwise direction for about 1½ in. to 2 in., then twist on a wire connector. The wire connector does not have to be green.

Some inspectors require crimped ground wires in the outlet boxes (the NEC does not). Although their intent is good (to provide the best ground connection), the end result is not. Your chances of having a loose ground splice due to a bad crimp are greater than having a loose ground due to a bad wire-connector splice. Personally, I don't crimp my grounds because I want to be able to add or remove one later if I change the wiring design. You can't do that with a crimp system—you have

Grounding Receptacles in Nonmetallic Boxes

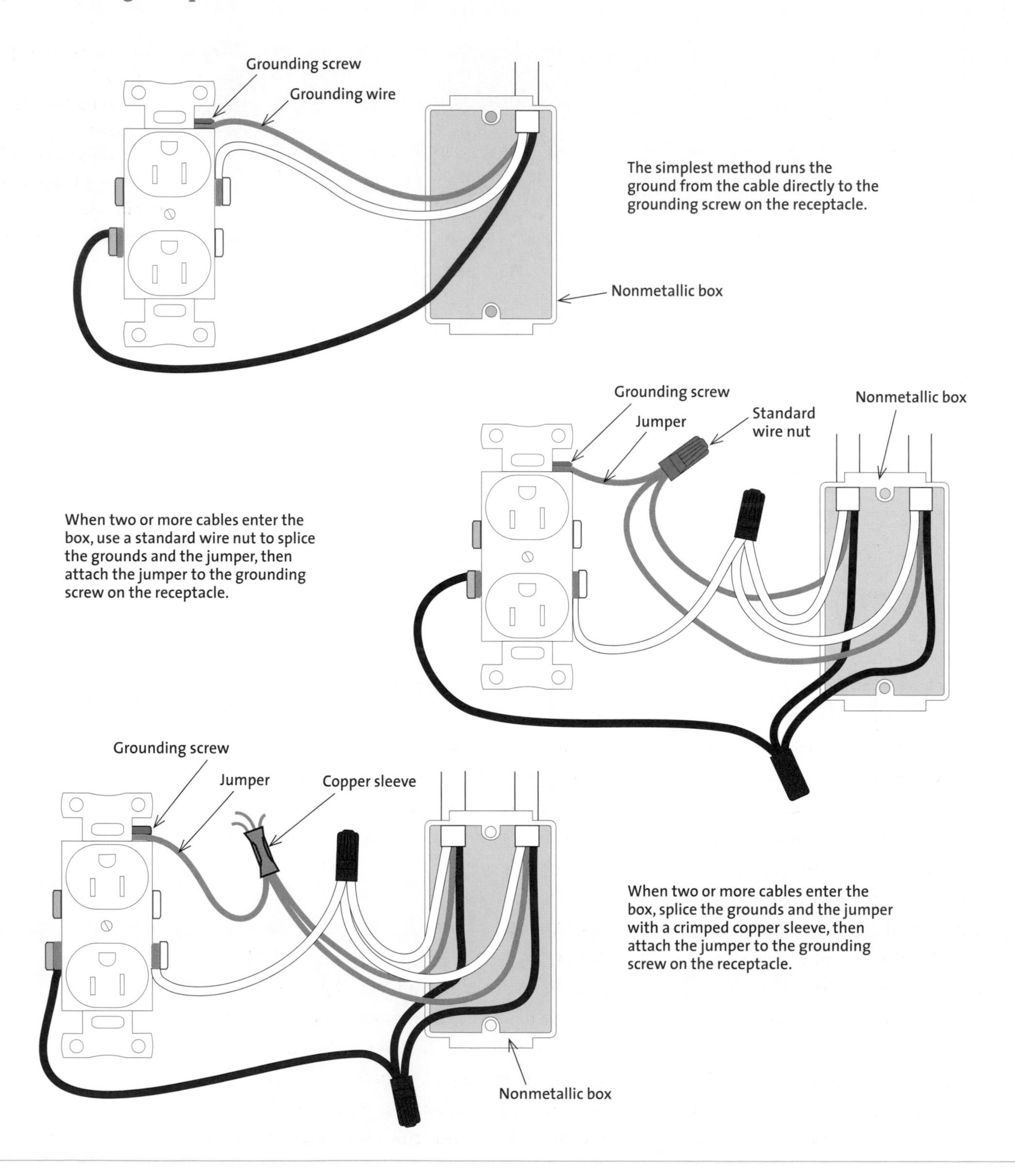

A four-way crimp system. **I don't recommend crimping ground wires, but some inspectors require it. If you must crimp, be sure to use the proper tool; don't use pliers or a common crimper, which just mashes the wires together in one place.**

to cut every ground wire at the crimp. In my opinion, if wire connectors are good enough to splice hot conductors, they are certainly good enough to splice ground conductors. If a crimp system is used, you must use the proper crimping tool: a four-way or four-point crimper. Another method utilizes a green grounding wire connector that has a hole in its end. Forget using the hole—it's more trouble than it's worth. If you want to use a green wire connector like a typical wire connector on the splice, go ahead. However, the color green is not required for the splice.

You can ground receptacles in metal boxes via the grounded metal of the box in lieu of installing a wire jumper to the ground screw. To do this, you must have direct metal-to-metal contact between the box and the metal ears of the receptacle. Some receptacles come with special clips on one ear just for this purpose. This grounding rule does not apply to metal boxes where the receptacles attach to a removable cover and not to the box.

Metal boxes

A metal box must always be grounded. The ground can come via a piece of grounded conduit if the box knockout is a singular hole (you cannot obtain a ground through a multiple knockout) or via an equipment-grounding conductor. In a residence, you normally use the latter. The conductor terminates in a splice

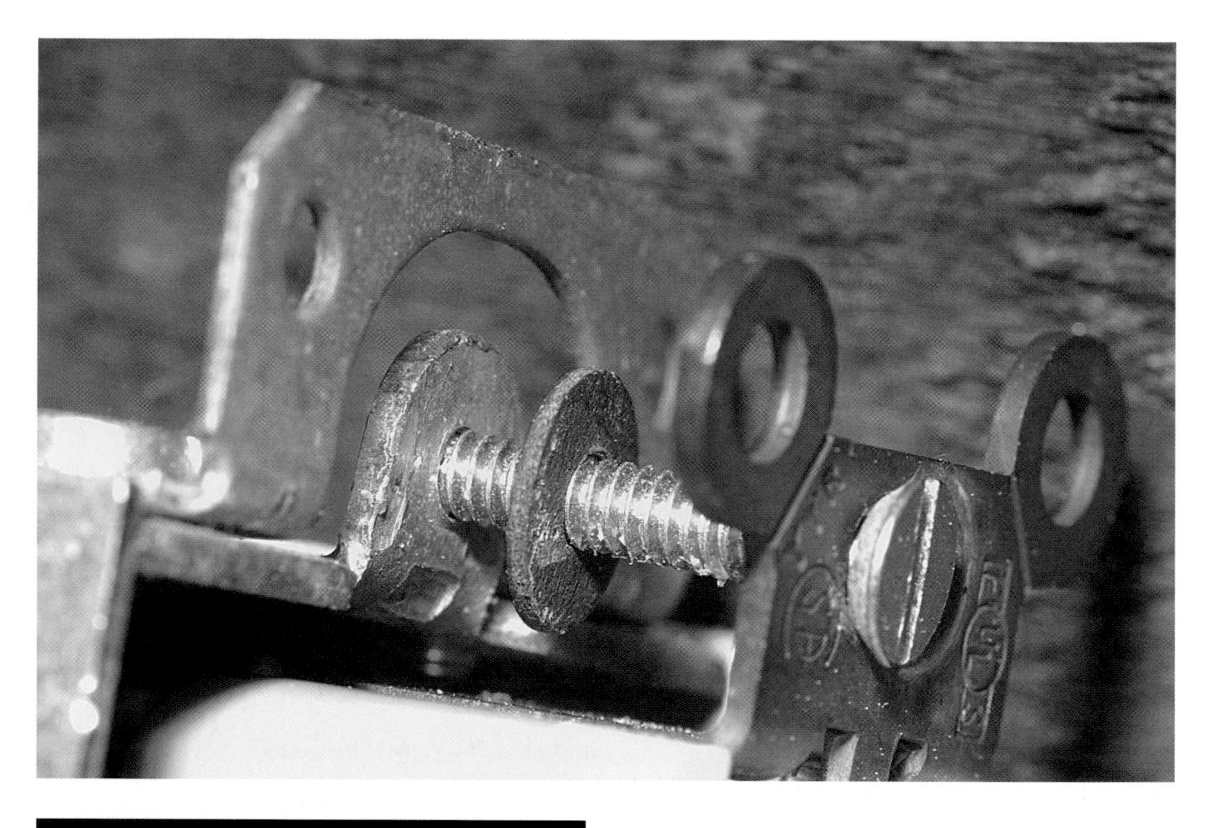

To have yoke-to-box ground contact, you must remove the insulator bushing. This one is orange. However, code now limits such use to surface boxes only. Thus, this is a violation for this box.

High-quality receptacles have a built-in spring-type grounding strap that automatically grounds the receptacle when the attachment screw enters the grounded metal box.

where a jumper leaves to ground the box via a screw. These jumpers are called pigtails; they can be purchased with screws already attached if you don't want to make your own.

There are a few things you cannot do. First, you cannot attach the ground pigtail to the box via the screw that secures the box to a stud or a wall (it's against code); in this case, use a separate screw. Second, you cannot run the jumper to the receptacle rather than to the box and then think that the box is grounded through the receptacle. And you cannot use a sheet-metal screw to ground the box (it doesn't have enough threads).

Appliances

Until 1996, the NEC allowed some appliances, in particular electric dryers and stoves, to use the neutral as the ground as well, and you could get away with using a three-prong plug. Now, all stoves and dryers must use an equipment-grounding wire that is separate from the insulated current-carrying neutral.

Many smaller appliances are grounded via a three-prong plug and cord (the roundish

Work Safe • If you're thinking of plugging a grounded appliance into an ungrounded receptacle using a cheater plug, don't do it. The cheater plug is simply an adapter that allows you to insert a three-prong plug into a two-prong outlet. Cheater plugs cannot ground an appliance unless the receptacle box is metal and there is some type of conduit providing a ground all the way back to the main panel. (This is rare in modern residential situations, but it does occur in older houses wired entirely with BX or Greenfield cable.) If the appliance does not have a proper ground connection and a ground fault occurs, someone could be electrocuted.

prong is the ground connection). However, for an appliance to be truly grounded via the plug and cord, a grounded three-prong receptacle must be installed with an equipment-grounding conductor connecting it back to the main service panel grounding bus. To make sure the receptacle is grounded, check it with a plug-in tester.

The biggest problems come when you have new appliances and old wiring. If you buy a new appliance that has a grounded plug (or a hard-wired appliance that has a green wire in its splice box), it requires grounding via the manufacturer's instructions; otherwise, you have the potential of electrocution. The most common violations occur with common appliances, such as dishwashers and water heaters. There is no grandfather exemption. If you have an older house with ungrounded wiring, you must either run a ground wire back to the panel or run new grounded cable.

Double-Checking the System

Grounding connections are like Christmas-tree lights—one problem, and nothing works. If you had a ground-resistance-testing machine, called a megger, you could check the overall system, but most of us don't have one. Here is a checklist to help you to verify that everything is in working order:

- Don't just wrap the ground wire around the rod or use hose clamps. Use an acorn clamp and install it properly.
- Leave the first hole open so the inspector can observe the clamp connection.
- If you can't get the ground wire inside the wall immediately, be sure to clamp it tight against the siding.
- Use copper ground wire.
- Don't use wire less than #6 to the rod, and use #4 copper for everything else. Make sure all connections are tight.
- Do not place the ground rod(s) under a paved driveway.
- Bond the metal gas and water lines to the panel with #4 copper.
- Bond the panel's grounding bus to the metal frame of the service panel.
- Attempt to place as many rods as possible in a moist area.
- Use a grounded bushing on metal conduit connecting the service panel to the meter base.
- Add a metal well casing, if there is one, into the grounding system.

Submersible pumps

If you've got a well, its submersible pump and metal casing must be grounded. The metal well casing, around 6 in. in diameter and driven through soft soil and into bedrock, is arguably the best ground medium available. That is, it has the lowest resistance to ground of anything around. The NEC requires that you simply attach the equipment-grounding wire from the pump to the metal casing. In high-lightning areas, it's best to supplement the code-mandated ground connection by running a 4- or 6-gauge copper wire from the casing to the ground-rod system. I also install a Tytewadd surge arrester at the ground attachment to provide surge protection for the pump.

Structural steel

Exposed structural steel, such as a building frame, should be bonded to the grounding system. The metal siding, if any, receives a ground through its contact to the beams. The bonding jumper needs to be accessible where the jumper is connected to the frame. Although you can connect the bonding jumper to the ground rods or a grounding electrode conductor, I prefer to connect it directly to the main-panel grounding system so I don't have to worry about any bolted connections coming loose. The conductor is sized according to tables within the NEC (for 200-amp service, use 4-gauge copper). Although it's not required, metal siding and metal roofs on buildings with no structural steel can be grounded to the main panel.

Grounding Multiple Panels

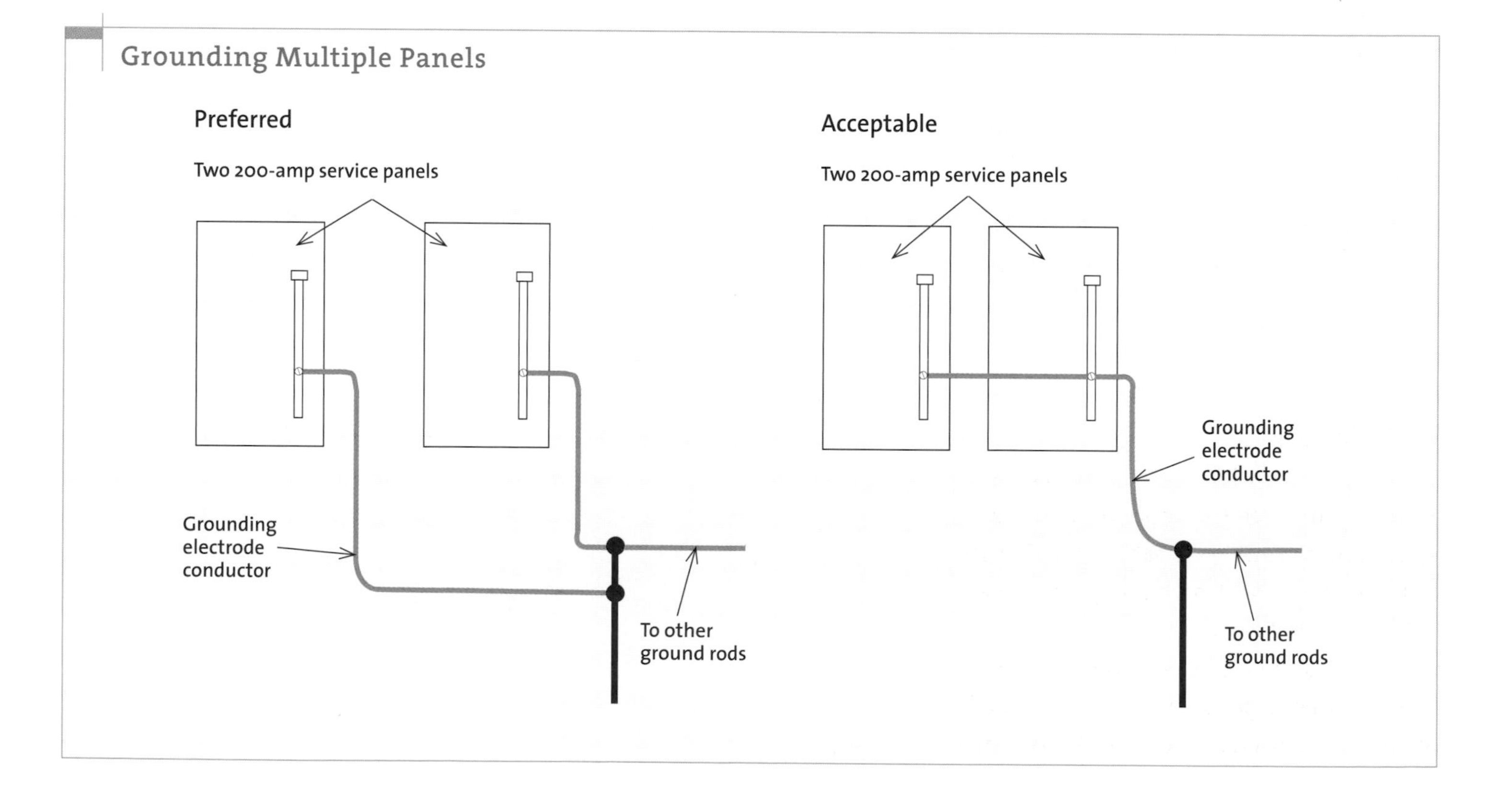

Work Safe • Work Smart • Think Ahead
Don't drive a separate ground rod and run a separate ground wire solely to ground a gas line. You do not want more than one ground-rod system.

Gas pipes

The NEC forbids the use of a gas pipe as a grounding electrode. Whether or not gas pipes need to be grounded depends on the local code; however, the NEC requires metal gas lines to be bonded to the house grounding system. I always ground the gas pipes because it makes good sense. Simply run a #4 copper wire to the metal pipe on the house side of the turn-off valve. You can use the equipment-grounding conductor of the cable feeding the appliance. If you ground the furnace, for example, and the pipe is screwed into the furnace, it is automatically grounded and you don't need to run a separate ground wire. Be aware that this equipment-ground wire is much smaller in gauge than a #4 copper ground wire, so you'll be safer with the #4 in addition to the equipment-grounding conductor.

Multiple panels

It's common knowledge that a service panel must be connected to a ground rod or an equivalent system. But what do you do with multiple panels, which you may have in a 300-amp or 400-amp system? The common method is to run a grounding electrode conductor from one panel to the next and have one panel connected to a ground-rod system.

The problem with this logic is that the noise and surges that are stripped off the line in one panel are sent to the next on that same wire. Normally, there may not be a problem, but if you have some sensitive electronic gear in the second panel, there may be trouble. The noise and surges may not destroy equipment in the second panel, but they could overstress them and shorten their life expectancy. And if a big surge—a several-thousand-volt blast stripped off the line by a surge protector in a nearby panel—does come through, the last thing you want to do is send it to another panel. You want it to go straight to earth. The only way to do that is with a direct connection. Therefore, the best way to ground multiple panels is with independent ground wire systems. That is, each panel has its own grounding electrode conductor connected to the same ground-rod system.

Wiring Room by Room

Once you've installed the service-entrance system, passed your inspection, and gotten power from the utility, it's time to install the house wiring system. In other words, now that you've run the really big cable, it's time to run the little ones. If you know what you're doing, you can do the job yourself in about a week. The work will go easier and faster if you can enlist a helper to drill holes in the studs and pull cable through the studs and joists.

Planning is key. First, know what materials will be needed and have them on site. This includes the light fixtures. Store all materials in the center of an unused room. Storing them in the center allows access to the stud walls. Second, locate all outlet locations and determine how you're going to get to them. Third, pull the cable. Through trial and error, all electricians develop tricks and methods to get them through the problems of getting the wire there; I've included as many of these techniques in this chapter as I could.

Stocking Up

A typical 2000-sq.-ft. house uses approximately twelve to fifteen 250-ft. rolls of 12-2 gauge wire, four to six rolls of 14-2, and three

What Comes Around, Goes Around

In previous editions of *Wiring a House* I suggested that you color code your cable with spray paint so you never mix up the gauges—and so you can tell from across a room what gauge is in what wall. As of this edition, I no longer suggest painting cable as long as you use cable that is already color coded.

Here's why: Before I wrote the first edition of *Wiring a House* I called every cable manufacturer to suggest that they color code their cable sheaths by gauge. I got no response. Later, after discovering that one of those manufacturers, SouthWire®, had begun color coding by gauge, I called the company to inquire. I was told that the head honcho who made the decision got the idea from a wiring book that illustrated painting sheaths to designate gauge. The book was *Wiring a House*.

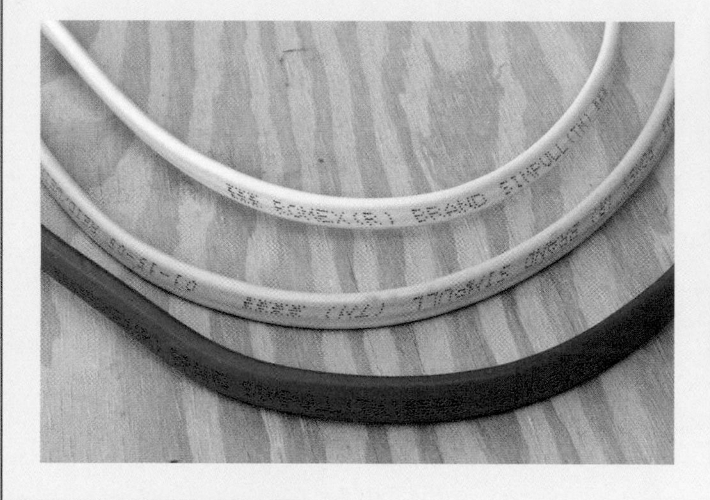

Most cables today use color-coded sheaths to advertise their gauge: white, 14 gauge; yellow, 12 gauge; orange, 10 gauge.

Take Note • All cables have a ground wire even though it may not be noted. For example, "14-2" means a 14-gauge, two-wire conductor with ground; "12-3" means a 12-gauge, three-wire conductor with ground; and so on. All conductors are copper unless otherwise noted.

to four rolls of 14-3. Try to avoid using 12-3 because it is very expensive. However, I do use 12-gauge for all the receptacle circuits even if code allows 14. The 14-3 will be used for interconnecting the smoke alarms and for wiring the three- and four-way switching circuits. Now that bedroom smoke alarms have to have AFCI protection, these rooms may now be fed via 14-gauge cable, assuming there are no heavy loads that require 12 gauge.

You'll need UF cable if you are running any outside circuits. If the house has an electric dryer and a water heater, you'll also need 10-3 for the dryer and 10-2 for the water heater. You can use leftover 10-3 for the water heater; just cut the red wire dead (cut it back to where the sheathing has been trimmed). Last, you'll need 6-3 copper for an electric stove.

Outlet Box Location

Overall, code requires at least one receptacle within 6 ft. of a door and then every 12 ft.

Code-Minimum Placement of Receptacles along Wall Sections

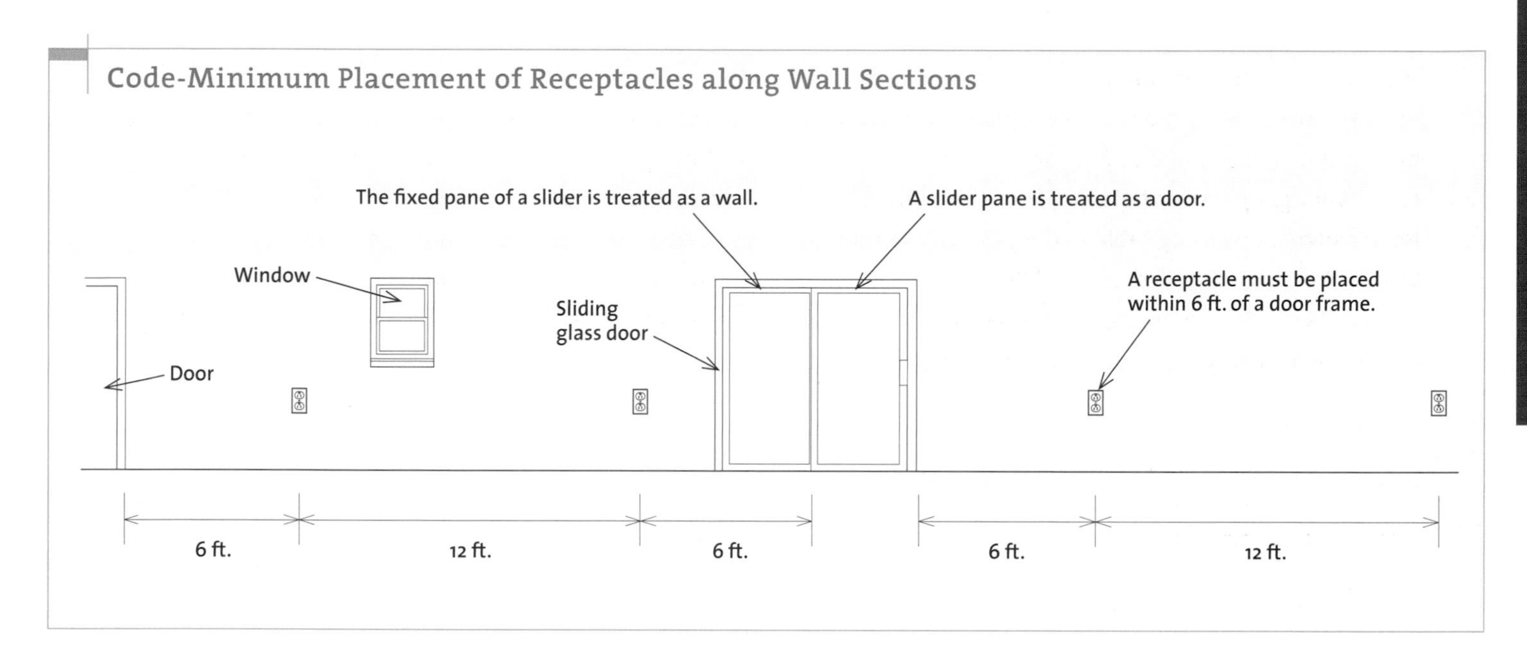

thereafter. The fixed-glass section of a slider door and fireplaces should be treated as a solid wall. Any wall 2 ft. or wider requires an outlet.

Code allows certain details that I try to avoid whenever possible. For example, code allows standard receptacles to be placed in the floor face up, but you shouldn't do that unless you have to. Floor-mounted receptacles are a catch-all for liquids (from spills or cleaners), trash, and dust. This buildup will destroy the receptacle or may start a fire. If you must have a floor receptacle, invest in a commercial unit that has a cover or a tombstone design (it looks like a tombstone rising out of the floor), which keeps the receptacle vertical.

Above Code • Code allows switched receptacles in lieu of switched lighting in some rooms. Ignore that. For room lighting, I always prefer switched overhead fixtures. True, switched receptacles save the material and labor costs of installing an overhead light, but doing so commits the homeowner to using one light on that one end table forever. Overhead lights not only provide better overall illumination, they also allow you to move the furniture around.

Planning outlet and switch locations

Building a house is a stressful time for most couples. One way to prevent blowups is to have both partners lay out the locations of receptacles and switch boxes together. Otherwise, if one partner picks the receptacle and switch locations alone, he or she should be prepared to hear, "Why didn't you put one here?" or "What about my lighted curio cabinet?" To that, the only answer is, "Whoops."

Once the framing is done, I advise couples to walk through their home with a big black marker and put an X on the stud where they want the outlets. If you want an outlet on a particular side of a stud, put an X on that side. Don't worry about the height of the outlet

Mark the proposed door swing on the floor and have the head carpenter approve it. This eliminates any responsibility on your part if the door is mounted in the opposite direction.

Take Note • As you start marking the receptacle box locations, you may begin to wonder how many outlets you can have on one circuit. In commercial work, we are limited to 10 outlets on a 15-amp circuit and 13 on a 20-amp circuit. However, there are no such limits in residential work—the number is infinite. Common sense, though, tells us to stay close to these numbers or to simply put the receptacles in different rooms together.

box just yet. Take your time—bring a couple of folding chairs so you can sit and talk about where the furniture is going and where the outlets will best be placed.

Don't forget to mark the switches. To do this, you must know which way the door is going to swing. Sometimes it will be obvious, as when the door opens against a wall, but when the door opening is dead center in the wall, check the blueprints and get a verbal confirmation from the head carpenter, then mark the direction on the floor to avoid changes later.

The light switch should be on the wall adjacent to the latch side of the door. Never mount the switch outlet box too close to the edge of the frame, as it may get in the way of the trim. To avoid this, put the box on the next stud back from the door-frame stud, or scab one or two pieces of scrap wood onto the doorframe and mount the box to that.

Now consider switch and outlet heights. Switch outlet boxes should be 48½ in. to 48⅝ in. from the top of the upper plate. Assuming that the drywall is going to be installed from the ceiling down, placing the outlet box at this height allows for easier drywall installation. The drywall installers will only need to cut the bottom sheet. Unless you have a specific reason not to, place common receptacle boxes one hammer length up from the floor. There is no code spec on height here, and using the hammer as a height reference makes everything uniform and speeds things up—just make sure you always use the same hammer.

Using a hammer as a height gauge keeps all receptacle boxes at the same level and allows for quick installation.

Kitchen

There are perhaps more receptacles in the kitchen than in any other place in the house—thus, there are more rules that apply here than in any other place in the house. First, I'll discuss the NEC method, or the minimum code approach. Next, I'll go into my Above Code method, which I feel is a more practical approach, considering today's abundant kitchen appliances.

The NEC kitchen Code requires two 20-amp circuits to feed kitchen receptacles. How these two circuits are distributed around the room is immaterial, as long as both circuits are on the countertop. Any receptacle that feeds the countertop must have GFCI protection.

Along the counters, the receptacles must be spaced so that no one spot is more than 2 ft. from an outlet. Practically speaking, this means that when starting from a counter edge, stove edge, or sink lip, you must have an outlet within 2 ft. and then an outlet every

Code-Minimum Placement of Kitchen Receptacles

Having receptacles this far apart normally doesn't provide enough outlets for appliances.

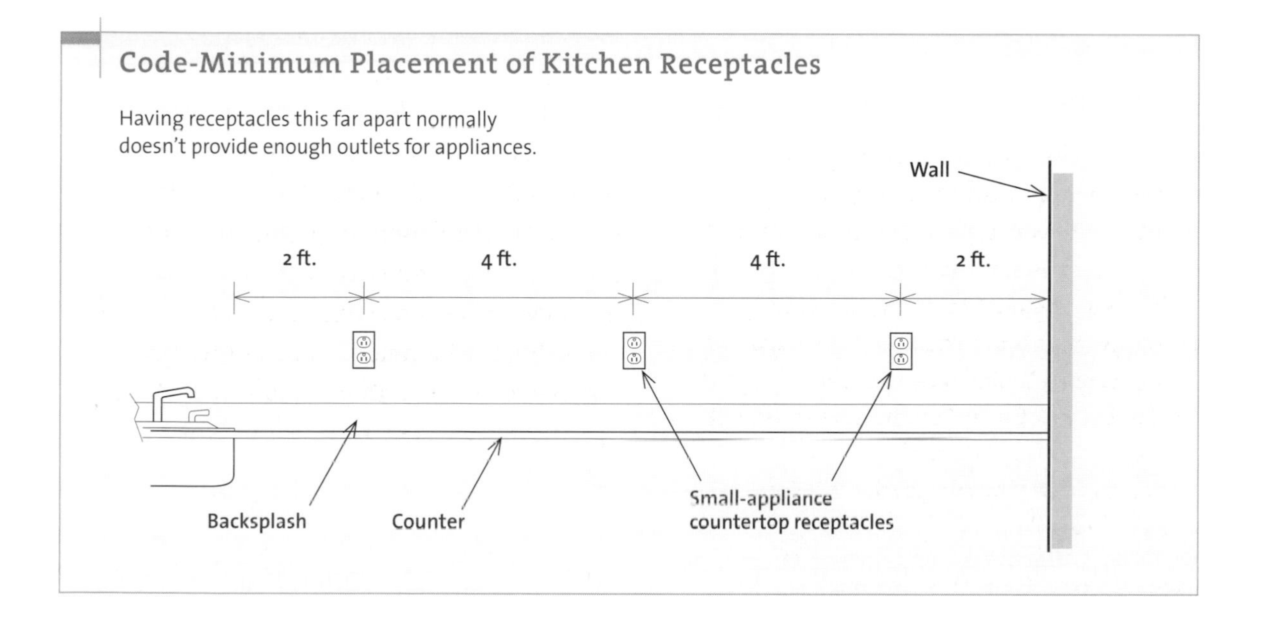

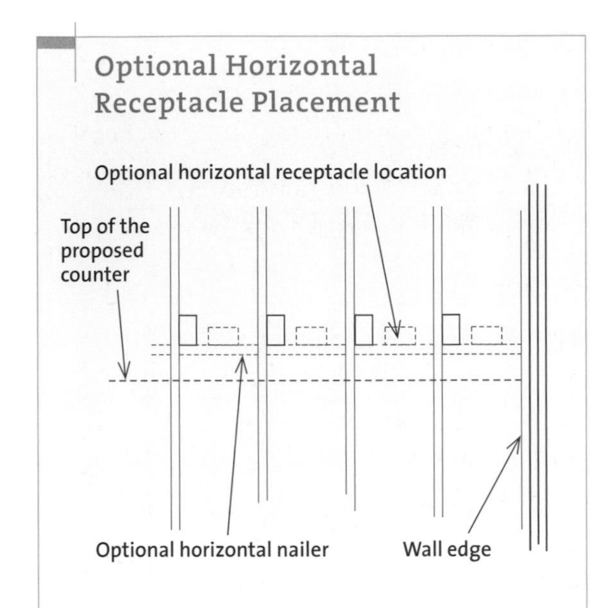

Take Note • In codes for an NEC kitchen, the dining room, breakfast room, and pantry are wired together. In this case, all references that say "kitchen" also mean the dining room, breakfast room, and pantry.

4 ft. thereafter. Also, outlets cannot be mounted face up in the countertop because water will spill into them. If receptacles are placed above the counter, keep them within 20 in. of the counter. If receptacles are in an appliance parking garage, they do not count as the required number of receptacles.

Kitchen receptacles must be, in general, reserved for plug-in appliances. For example, loads such as an outside receptacle outlet, in-room lights, or fixed-in-place appliances (for instance, a dishwasher) cannot tap into the receptacle circuits. However, there are a few exceptions. A receptacle installed solely for an electric clock and receptacles installed to provide power for supplemental equipment and lighting on gas-fired ranges, ovens, or counter-mounted cooking units are allowed.

Islands and peninsulas must have at least one receptacle outlet if they measure 1 ft. by 2 ft. or larger (if you don't want to install a receptacle, make the measurement 1 in. shorter in either direction). You are allowed to put the receptacle outlet in the counter base but not more than 12 in. below the countertop, as long as the countertop lip does not extend more than 6 in.

Fixed-in-place appliance loads include items such as electric stoves, dishwashers, and garbage disposals. Use #6 copper (50-amp breaker) for the stove. The dishwasher is normally a 12-gauge dedicated circuit. Garbage disposals can be run on either 14 or 12 gauge, depending on what the manufacturer requests. The refrigerator is also considered fixed in place and will need its own receptacle. In this case, the refrigerator can be tied into the kitchen circuit or can have a separate 14-gauge cable. The lights (switched lighting is required) can be tied into any other general-purpose receptacle circuit in the house, such as the bedroom or living room, or put on a separate lighting circuit.

The common along-the-wall outlets must follow the same 6-ft. rule that applies to the rest of the house: Nowhere along the wall can you be more than 6 ft. away from an outlet. This means there must be an outlet within 6 ft. of a door and every 12 ft. thereafter. The fixed glass of a slider is considered a wall.

The Above Code kitchen The NEC method is good, but I don't think it adequately reflects the trend of today's homeowners, who are using more heavy-wattage appliances. To see this in action, all one has to do is attend a bridal shower and note all the kitchen appliances. And as most of us tend to forget, the NEC requires a minimum code and we are allowed to exceed it. Here is my version for the kitchen.

First, unload the countertop appliance circuits. Take the two 20-amp, code-mandated circuits and isolate them to the countertop. Next, bring in an additional circuit for the common wall receptacles. Thus, the kitchen will have a total of three 20-amp receptacle circuits: two GFCI circuits for the countertop (I use panel-mounted GFCIs because of their longer life expectancy) and one non-GFCI circuit for the common wall receptacles. The refrigerator can be tied into the 20-amp wall receptacles or can have its own 20-amp circuit. Along the backsplash of the countertop, place a receptacle on each stud or every other stud. With 16-in.-on-center stud spacing, this places a receptacle every 16 in. to 32 in. If the wall has 24-in.-on-center studs, place a receptacle on each stud.

If you don't want the stud wall layout to dictate where the receptacle will be placed along the countertop, you can make your own design. To do this, put horizontal nailers in between the studs along the countertop. The height of the common kitchen counter is approximately 40 in. at the backsplash. Installing the nailer bottom at 42 in. will place the receptacle just above the backsplash. In this situation, I prefer to install the receptacles horizontally, which makes a more pleasing line with the horizontal counter. You can now plan the distance between the receptacles to match your needs.

For lighting, place a wall-mounted switch at each entrance to the dining/kitchen area. If three-way switching is needed (for example, if the kitchen has more than one entrance), these switches can be fed from a 14-gauge, 15-amp light-duty lighting circuit (the circuit that feeds the smoke detectors). In addition to overhead lighting, consider adding a light at the sink and installing undercounter lighting.

Bathroom

You may think that the bathroom, being so small, would be quite easy to design and wire. No such luck. Next to the kitchen, the bath is the most complicated room to get right. As I did in the kitchen section, I'll first go over what code requires, then offer my Above Code version.

The NEC bathroom Code requires that a bathroom have a 20-amp GFCI-protected receptacle circuit—either a GFCI breaker or receptacle. In the past, this receptacle had to be isolated or, at most, tied in to other bath receptacles (an infinite number of receptacles in an infinite number of baths). As if the infinite number of

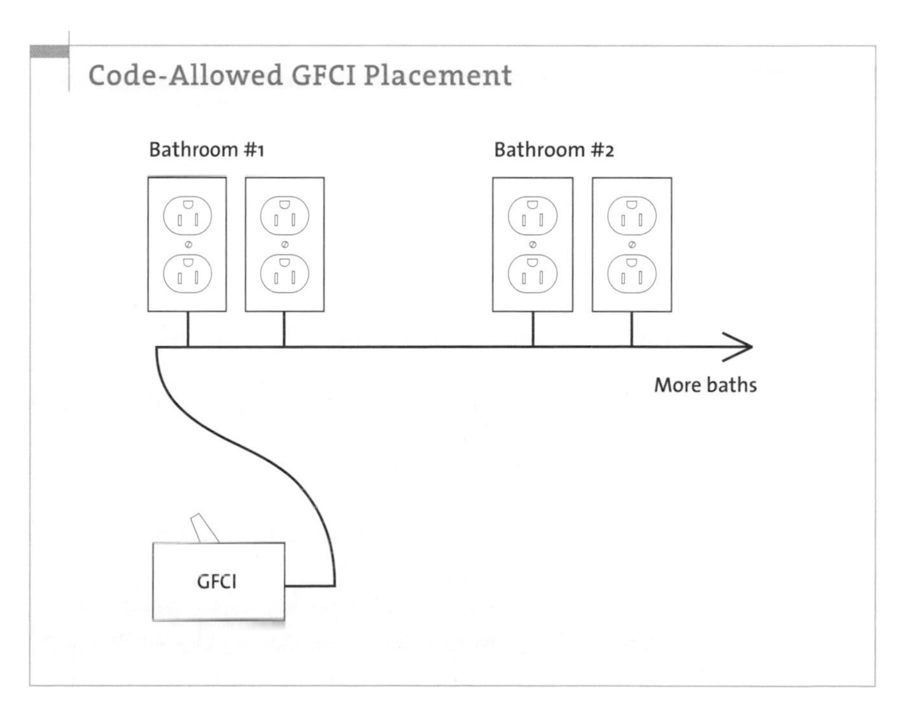

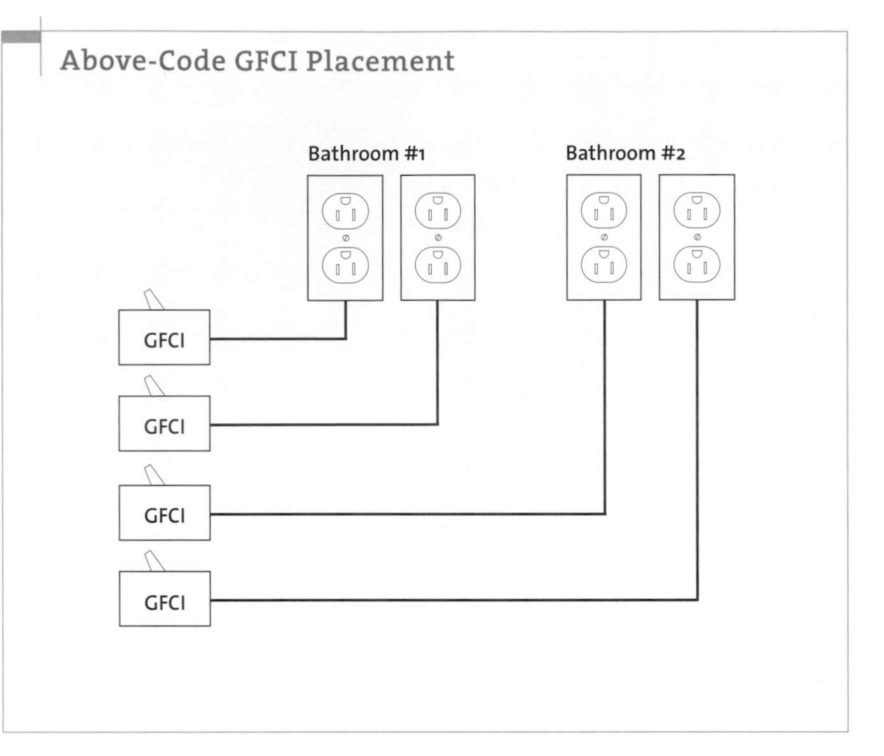

receptacles wasn't bad enough, the 1999 code made it even worse. As long as the receptacle doesn't tie in to any other bath, code now allows you to load it up with anything and everything in that bathroom (whirlpool tub, lights, heaters, and so on).

Code also requires switched lighting at the point of entrance and a receptacle within 36 in. of the sink lip (the receptacle must not be in an adjoining room). Double sinks do not need two receptacles as long as the sink lip of both is within 36 in. of a receptacle. If the bath doesn't have an operable window, it needs a vent fan. If the fan is in the shower/bath area, it must be listed for wet locations and must be GFCI protected.

Last, no hanging fixture or lighting track can be located over or within 3 ft. of the tub, unless it is more than 8 ft. above the tub.

Take Note • Beware of a receptacle in a medicine cabinet, especially an old medicine cabinet. If one is present, do not count it as the required receptacle. Many times they aren't even grounded. I always cut these receptacles dead and don't worry about them.

The Above Code bathroom I much prefer a common-sense version of the NEC, keeping the receptacle unto itself. Allowing a circuit to take on more and more high-wattage loads as the years go by, such as hair curlers and hair dryers, is not a wise decision. I recommend isolating the bath receptacle from all other bath loads. In addition, I recommend that when there's more than one receptacle, each one should be wired on its own circuit. This may seem like expensive overkill at first, but it isn't. It is quite common to have two people (both spouses or two teenagers) working on their hair at the same time. Two heavy-amperage hair dryers cannot work simultaneously without their own circuits.

To wire an Above Code bathroom, start by bringing in two power cables: one for the receptacle and one for the lights, fan, and so on. If you want a heavy-duty heater, bring in a third. Bring the power cable for the lights into the light switch outlet box by the door. Typically this is a double box–one switch for the fan and another for the overhead lights. The lights by the mirror, if any, are powered by a switch on the wall adjacent to the sink (run a jumper over to it from the door switch box).

The power cable for the bath receptacle is a home run from the service panel directly to the receptacle outlet box adjacent to the

Heating a Bathroom

No one likes a chilly bathroom, especially when dripping wet. You can always crank up the main thermostat—if you like a tropical climate and don't mind paying sky-high fuel bills—but a space heater is a better solution.

There are a lot of space-heating options, ranging from portable heaters (most of which are illegal to use in bathrooms) to in-floor radiant systems (wonderful but pricey). I advise my customers not to use overhead fan/heater/light combination units. This type of heater throws heat overhead; the heat stays near the ceiling or is sucked out by the vent fan. Electric baseboard units are not a good choice, either. They have a tendency to rust, especially when mounted close to a shower or toilet.

I generally recommend a fan-assisted, in-wall heater with an integral thermostat (240-volt, with its own circuit). This type of heater circulates hot air, which helps absorb moisture and prevents the temperature from settling in different layers (which can lead to frozen pipes).

sink. Bring in an additional home run for each receptacle. A whirlpool tub (see also p. 293), if present, requires its own GFCI-protected circuit starting from a GFCI breaker and terminating at a hard-wired splice at the tub's motor. Don't forget that the tub must be installed so that the motor can be removed, if needed.

Hallways and stairs

Code requires one receptacle if a hallway is longer than 10 ft. Stairs with six steps or more require switched lighting at each end and at a landing that opens to a door. Code doesn't require receptacles on stairs, but I think it's a good idea to install an extra receptacle either at the top, middle, or bottom of the staircase and at one of the landings. This will give you a place to plug in a vacuum cleaner, night-light, or Christmas lights for the railing.

Bedrooms

Bedrooms require switched lighting via either an overhead light or a switched receptacle. Always go with overhead lighting; otherwise, you will be stuck with a table lamp on that end table in that one spot forever. Once you mark the light outlet box, determine the location of the bed, dressers, end tables, and any other specific piece of furniture. If you think you're an average person and don't need anything other than the few outlets required by code, the wiring is fairly straightforward. But I've never met an average person. I advise my customers to consider everything that they may want in the room and to wire it accordingly. For example, will the room house a computer or stereo system? Also consider track lighting above the bed, overhead lighting, and paddle fans.

Rex's Dream Bedroom

The design of a home can bring a relationship together or tear it apart. Although I don't have the space to go into detail about complete house design, I'll give you some of the more popular electrical designs for bedrooms that can help reduce tension and increase comfort. Remember that no one has to live in a cookie-cutter room—ask for what you want.

Outlets

The rule here is that more is better. It has been my experience that you need a receptacle outlet within 2 ft. of each room corner (one on each corner wall, except where there is a door), and if the room is large enough, another one in the middle. If you have windows and there is no receptacle close to either side, consider installing an extra outlet adjacent to one or both sides of the window. Having an outlet in each corner gives you the flexibility of putting furniture almost anywhere you want. The windowsills can now accommodate electric candles, Christmas stars, and other animated toys you may want to display during the holidays.

Install a receptacle quad on each side of the bed adjacent to the end tables for all your creature comforts: clock, radio, TV, transformer for the laptop computer, complete stereo system, portable CD player, ion generator, sleep machine, half fridge, fan, VCR, DVD player, and so on. You'll never run out of things to plug in—only receptacles to plug into.

Lights

If you're fairly certain that the bed won't be moved, I recommend installing a switch on each side of the bed to control the overhead or wall (sconces) lighting—that's a four-way switching arrangement. This way, you can dim the lights to whatever mood seems appropriate without leaving the bed. If you have lights below an overhead fan, they can be controlled by remote.

Reading in bed is sometimes a point of contention between spouses. To be able to read at night and not send your significant other out of the room (or worse, have to endure constant suggestions to turn out the light), consider installing track lighting on the wall above the bed and wired or recessed lighting with dimmers in the ceiling. With both installed, the light can be concentrated on just one side of the bed.

Above-Code Placement of Bedroom Receptacles and Switches

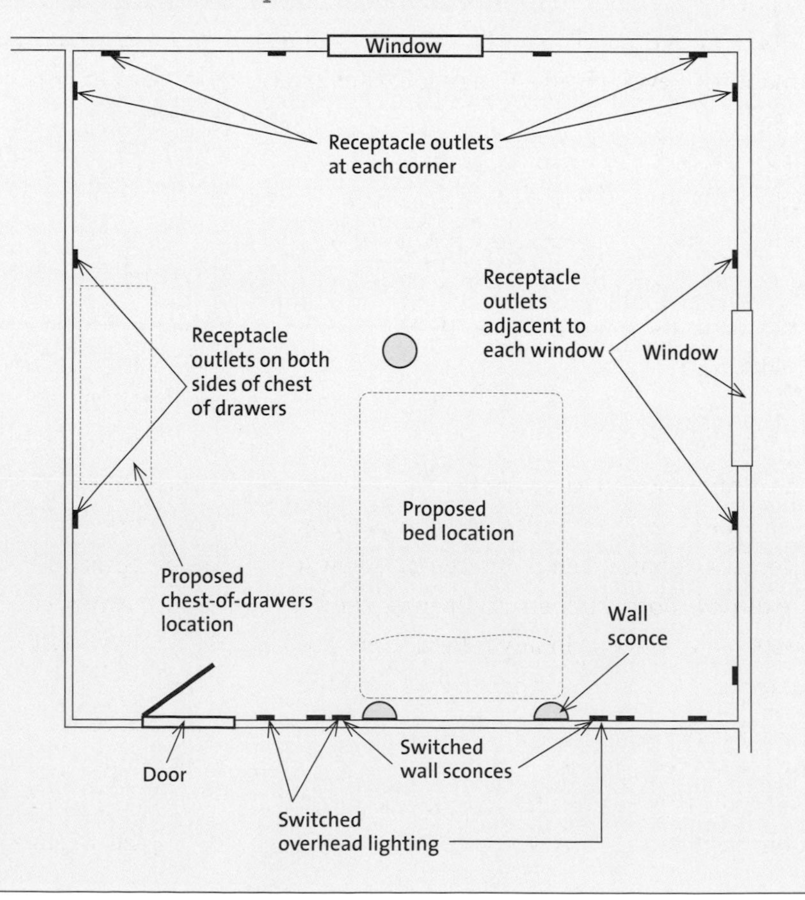

With the latest code changes, all bedroom 120-volt outlets must have AFCI protection. In this case, an outlet is defined as any 120-volt cable that comes out of the wall or ceiling and leads into a smoke alarm, ceiling fan, light switch, light, 120-volt receptacle, etc. All bedrooms can be on one AFCI breaker, but doing so will make troubleshooting very hard if the AFCI kicks.

Closet lights

You don't have to install closet lights, but if you do, the rules are quite simple. First, always use an enclosed bulb. Second, install the light immediately above the top of the closet door trim. This is about as far away from anything flammable as you can get. Keep all light fixtures a minimum of 12 in. away from all flammables, even though code allows 6 in. in some situations. Never install a hanging fixture in a closet.

Whether closet lights are included on the bedroom AFCI is up to the local jurisdiction. To be on the safe side, assume the closet is part of the bedroom and give the closet light AFCI protection by putting it on the bedroom circuit.

Dining room

The dining room needs switched lighting at each entrance. The standard rule for placing receptacles applies here. All receptacles within the dining room, kitchen, and pantry cannot be shared by circuits in any other rooms. For example, if a bedroom shares a wall with the dining room, it cannot tap off one of the dining room's receptacles.

Typically, you want the overhead dining room light dead center in the ceiling. Rather than guess at dead center, take advantage of laser technology by shooting a beam from floor center to ceiling. By measuring from corner to corner, you can find dead center of the floor. Place a vertical beam laser on the floor where the strings cross and you will find ceiling dead center.

Living room

The living room needs switched lighting, preferably at each entrance. The lighting can be overhead or a switched receptacle, but overhead is always preferred. In addition to the standard receptacle placement, it's a good idea to place receptacles on both sides of each window and corner, as well as on both sides of the fireplace. Be sure to place extra receptacles for the stereo, VCR, television, and any other electronic gear that may be used in the room. It's a good idea to put the electronic gear on its own 20-amp circuit. For finding ceiling dead center for the overhead light, see the section on the dining room.

Garages and unfinished basements

Attached garages and unfinished basements are required to have at least one GFCI-protected receptacle and light (an unattached garage does not require these, unless power is brought in). Garages and basements may have workbenches, so be sure to install a few receptacles at workbench height (around 48 in.). Any accessible general-purpose receptacle must be GFCI-protected, unless the receptacle is for a dedicated purpose, such as a freezer, or is inaccessible, such as a garage-door opener. A dedicated receptacle for an appliance must be a single receptacle unless both parts of the duplex are used.

Typical Minimum-Code Receptacle Layout

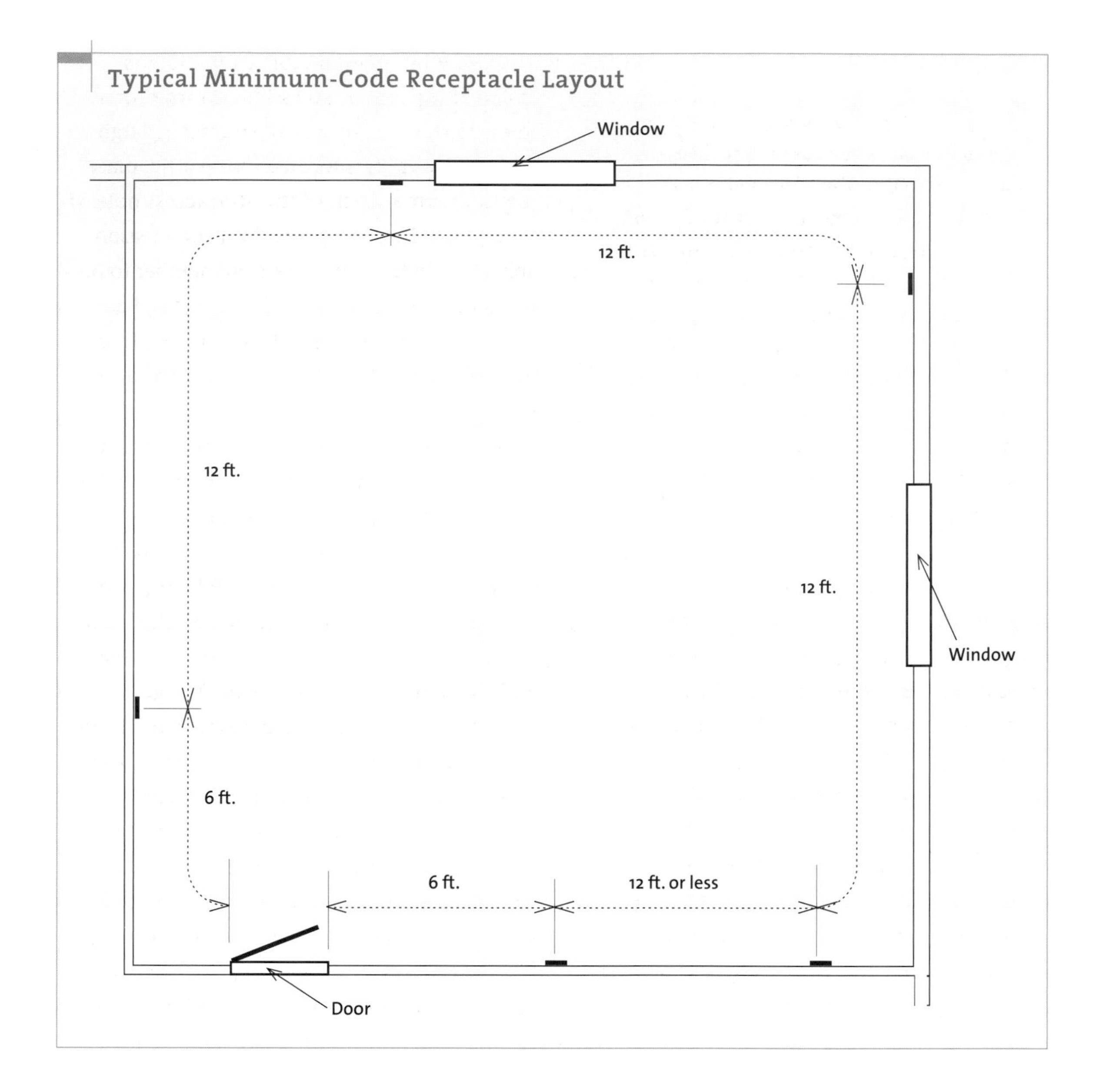

If the garage shares a wall with a room in the house, never install back-to-back receptacles within the same wall cavity. In a garage, the drywall is a fire barrier. If you cut a hole in the fire barrier in the garage and then cut another hole inside the house in the same wall cavity, you open up a fire entry into the house. It's especially important to avoid installing a panel in this wall, too. This is a common and expensive mistake.

Laundry room

In addition to an electric dryer (which will need its own 240-volt circuit), laundry rooms require at least one 20-amp circuit that cannot be a shared with any other room. Although

Above Code • The question always comes up as to what is considered finished and what isn't. Officially, "finished" means habitable. So if someone is currently living in or can live in the basement, then it is habitable and the GFCI requirement does not apply. In my opinion, electricity doesn't care who's living where; it's only looking for an easy path to ground, which means that the floor material is the most important consideration. If the basement floor is bare concrete, then the receptacles need to have GFCI protection.

only one receptacle is required, you'll probably need a few more. Scatter at least two or three around the room. For example, determine where you plan to iron in the room and make sure you have a receptacle nearby. A gas dryer will need 120 volts and can plug into any of the duplex receptacles. You'll also need switched lighting as you enter the room. Put this circuit on any other lighting circuit in the house, but not on the utility receptacle circuit. If the washer and dryer are located in an unfinished basement, the utility receptacle cannot double as the one receptacle required for the basement.

Attics, crawl spaces, and outside

If there is maintainable equipment in the attic or crawl space, it must be accessible and there must be a switched light at the point of entrance that illuminates the equipment and the path to it. The light fixture should not be a bare bulb. It is too easy to hit your head on the bare bulb and injure yourself and too easy for a hot filament to fall on flammable storage items.

Even if no maintainable equipment is inside, if the attic is accessible via stairs or a foldable ladder it should have switched lighting. If the switch is located in the attic, place it within arm's reach of the attic access hole. If it's outside the attic, place the switch reasonably close to the attic access. Remember to protect all cables within 6 ft. of the access, because this is where people tend to set boxes. Never drape run cables on the top edge of the attic floor joists.

There are several ways to run wires through an attic. The most common is to run the wires over the joists. This saves wire but can be against local code. It's also dangerous, because if you plan to use the attic you'll be tripping over and stepping on wires. One method is to nail a running board right next to the cable and staple the cable to the board's edge.

The best method for attics is to run all wires on the faces of the joists until you get to the low edge of the attic where the rafters rest on the top plate. There, nail a 2×12 across the joists, run the wires on top of the board, and drill through the board to get into a particular joist cavity. Once all wire is run, nail on a sideboard toward the open area of the attic. The advantage of this method is that you open up all the flooring as usable storage area—and you can lay down plywood over the joists for walking.

You are required to have two outside receptacles: one in the front of the house and one in the back. If you have a deck, consider installing outlets on both sides of it. These come in handy for extension cords, Christmas lights, cooking equipment, radios, TVs, and so on. If you have a long wraparound deck, consider installing a receptacle outlet every 12 ft. These receptacles must have GFCI protection.

Maintaining Structural Integrity

No matter where the cable is run, the structural integrity of the wood must be a prime consideration. That is, do not drill a large number of large-diameter holes throughout the framing member to pass cable through. Local codes dictate whether or not the structural integrity of any framing member—stud, plate, or joist—is being violated. If you have any doubts, it's a good idea to check with the local inspector or lead carpenter before drilling or cutting. Some common codes for retaining structural integrity are illustrated here.

Notching and Boring Studs

Drilling Multiple Holes

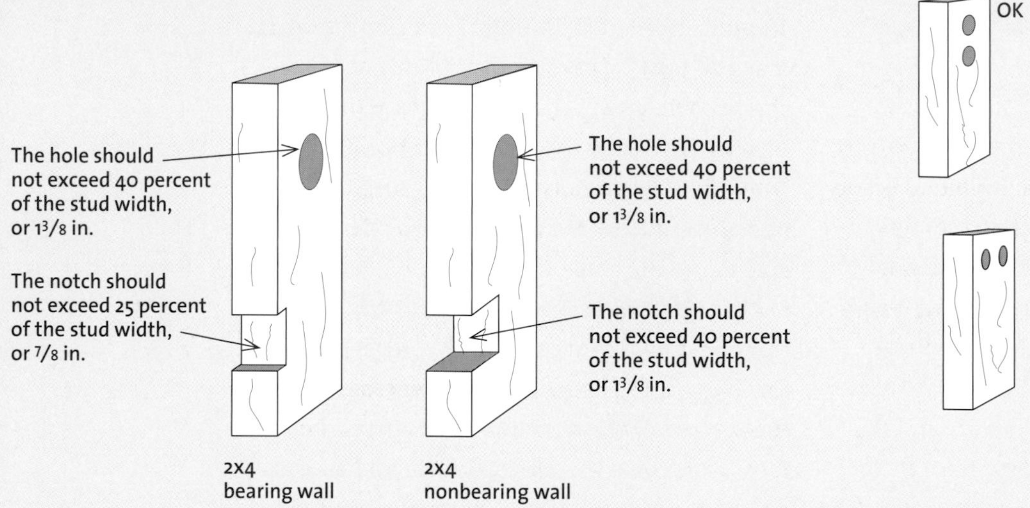

Protecting Cables

When a stud is notched or bored on its edge, code requires a $\frac{1}{16}$-in. steel plate to protect the cables from screws and nails.

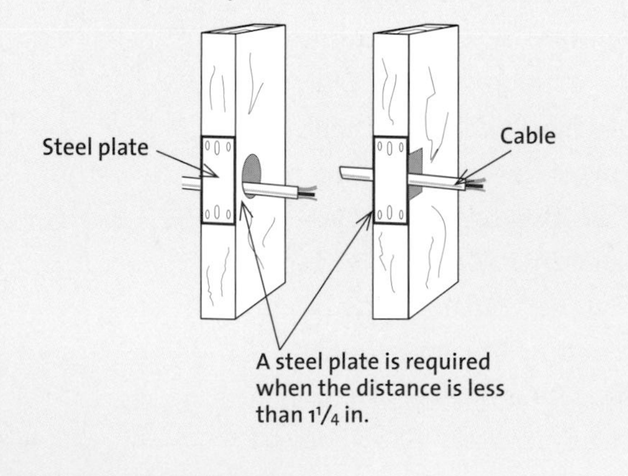

Notching and Boring Joists

Holes must be a minimum of 2 in. from the joist edge and cannot exceed one-third of the joist width.

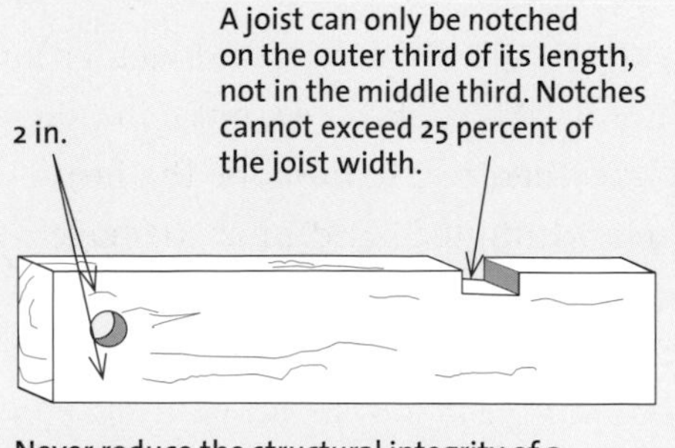

Never reduce the structural integrity of a load-bearing wall, floor, or ceiling.

Outside outlet boxes can be flush-mounted on the finished siding or recessed within the wall. Flush-mount boxes must be listed for outdoor use; recessed boxes can be the common nail-on types. In most cases, surface-mounted boxes are easier to install, because you don't have to cut the siding around the box opening. All you need to do is drill a hole through the siding for the wire and attach the box with a few screws.

The 2008 code wants any outside receptacle to be rated Weather Resistant. At the time of this writing these receptacles are not widely available and your area may or may not be enforcing that part of the code. I would check before spending the extra money.

Required outside lighting is limited to grade-level access points for all doors—excluding garage doors. The lighting can be indirect, such as an overhead floodlight, or a light right at the door. Neither light has to be switched.

Pulling Cable in New Construction

After you have all the boxes mounted, it's time to pull the cable. You can route cable through the crawl space or the attic and through, or over, studs and joists. For neatness and safety, try to keep the home runs along the outer walls.

Don't bundle groups of cable together for long runs. It may look neater, but if bundled tightly together for 2 ft. or more, the heat generated within the bundle may damage the wire's insulation. This means that you may have to lower the amount of current flowing through the wires (called derating) to lower the temperature and thus protect them. For instance, for a bundle of 10 to 20 conductors (five to ten 12-2 w/g cables), a 12-gauge cable must be fused at 15 amps and a 10-gauge cable at 20 amps (the NEC gives guidelines for derating conductors). To avoid having to make these adjustments, don't bundle the cables together. It will make your life easier.

I once went to a job that had failed inspection, and the owner was paying me to fix another electrician's problems. It was the neatest job I had ever seen—even the stranded neutral had been braided like a girl's pigtail. The electrician ran all the wires—10 cables (20 or more current-carrying conductors)—through the same large holes and bundled everything tightly with tie wraps for the entire run. It looked great, but it failed inspection because the cables, especially the ones going to the kitchen and laundry, could not be derated.

Through basements and crawl spaces

If a basement will not be finished, 8-gauge and larger NM cable can be run across the bottom edges of the floor joists, or you can nail a running board under the joists and staple the cables to the board. Cables smaller than 8 gauge that are run across the bottom of the joists must be stapled to a running board.

If the basement will be finished (either now or in the future), run the cable through the floor joists or along one of them. When running cable along a joist, staple it every 3 ft. to the center of the face to avoid the nails and screws that will be driven from below when the ceiling is installed. When boring holes to run cable through joists, follow the guidelines shown in the sidebar on the facing page. Crawl spaces have no rules for running cable along the joist edges because you can't trip over them.

Unrolling Cable Correctly

Before you pull the cable into the walls, unroll the cable as you would a bait-casting reel. Wrap one end around a stud and walk backward, spinning the cable roll in your hands and unrolling enough cable so there is slack to pull through the studs. Doing it this way allows a much easier pull because the cable is flat. Pulling the cable out of the center of the box, as it is designed to do, makes the cable spiraled and harder to pull through the drilled holes.

Unrolling cable like a bait-casting reel will keep it from spiraling.

Letting the cable spin out will leave it in spirals—making it impossible to pull it through the studs.

Here's how to pull cable from the main panel, through a basement or crawl space, and to an outlet:

1. Lay the box of cable in front of the main panel and unroll the approximate amount you'll need. Try to unreel the cable from the outside of the roll so that it will remain flat, as opposed to taking it from the center and having it loop in spirals. Next, feed one end of the cable through the lubricated slot in the bottom plate.
2. Pull the cable through the holes in the joists or across the joists (the method you choose depends on whether the basement will have a finished ceiling) to the wall cavity in which the cable will terminate.
3. Drill through the bottom plate of the interior wall. Pull enough slack to the location of the receptacle or switch.
4. Staple or attach the cable wherever needed in the basement (along the face or edge of the joists or on the running board). Attach it every 3 ft.

5. At the main panel, cut the cable with enough slack to go from the large hole in the bottom plate to the top of the main panel.

6. Mark the cable to indicate where it goes.

Through attics

Once you've planned the wiring route through the attic, you can begin to pull the cables. Here's how to do it (this method can also be used for joists or trusses):

1. Lay the box of cable in front of the main panel and unroll the approximate amount you will need from the outside of the roll. Work out the coils so that the cable lies flat. Next, feed one end of the cable up and over the first ceiling joist. Don't insert the cable through the large hole in the double top plate above the panel yet. Doing so will make it much harder to pull it over the top of the joists and could damage the cable because of the sharp 90-degree turn it has to make immediately above the top plate.

2. Pull the cable over the joists to the stud-wall cavity you need to access. Wiring home runs (the cables from the main panel) along an exterior wall allows you to keep the wiring out of harm's way. To get cable to an interior partition wall, pull through enough slack to make the run along the edge and then along the face of the ceiling joist to the cavity.

Running Cables in an Attic

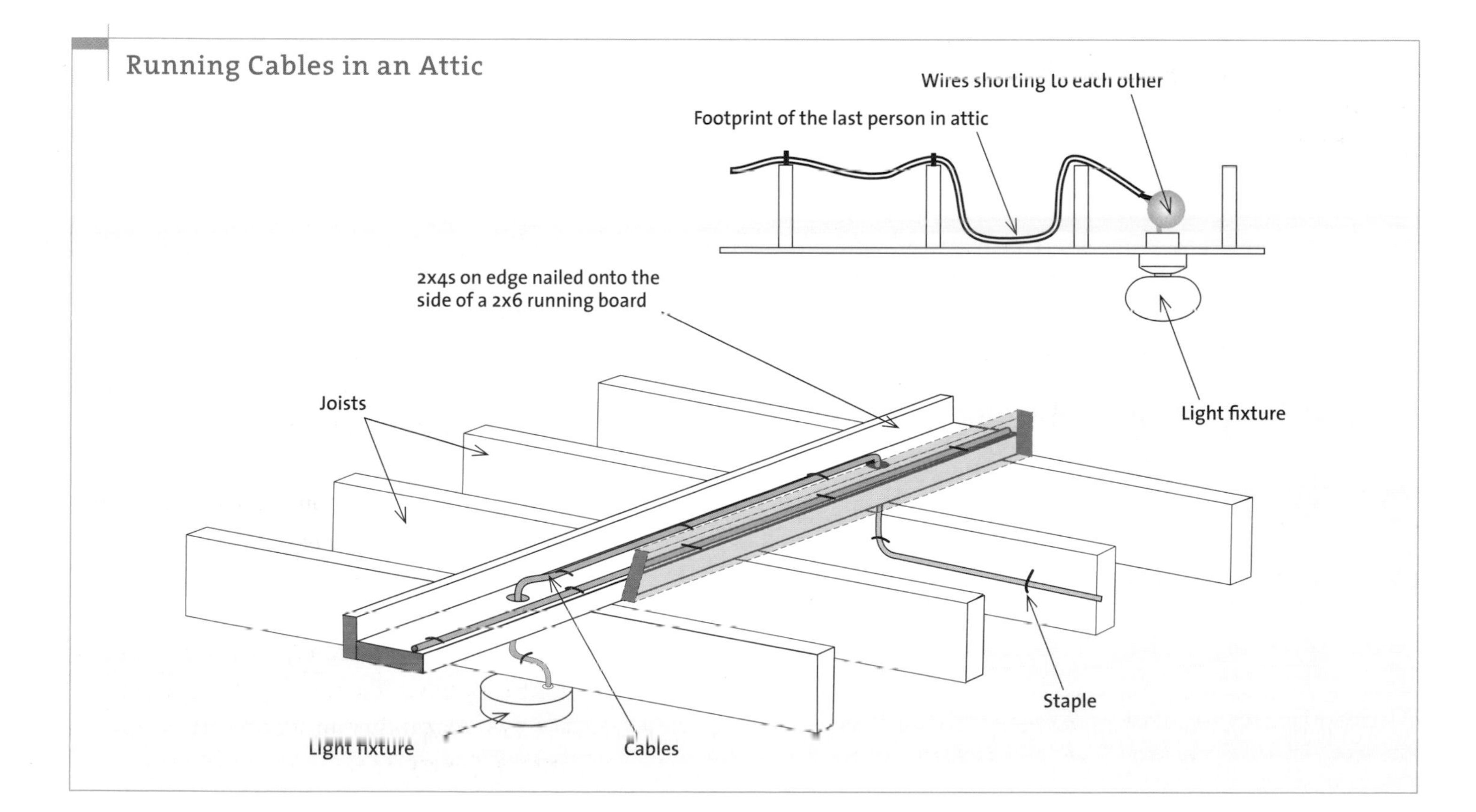

Work Safe • When routing cable through an attic, try to keep all cable flush against the sides of the joists and rafters. If possible, keep home runs close to an exterior wall where there is little headroom. When running cable over the edge of joists, use a 2x6 runner with 2x4s nailed to the side as kick plates. Any cable 7 ft. or more above the joist floor can be run over the top edges of the joists or rafters.

Wiring over and through Studs

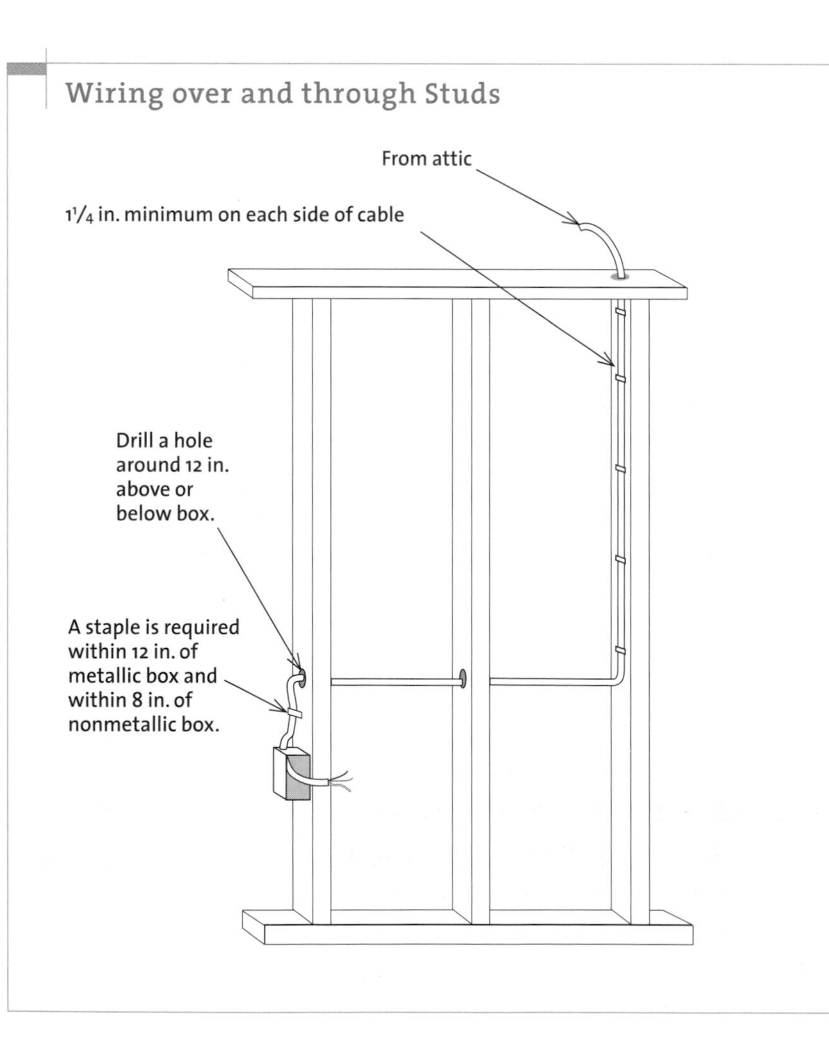

3. Once you reach the correct wall cavity, drill through the top plate and pull enough slack to get to the intended receptacle or switch.

4. Insert the cable through the hole in the top plate and pull the slack all the way through to its intended location. In the attic, staple the cable wherever needed (along the face or edge of the joists). Attach it every 3 ft.

5. At the main panel, cut the cable with enough slack to go from the large hole in the double top plate above to the bottom of the main panel.

6. Insert the cable from the attic through the double plate and into the main panel.

7. Mark the cable to indicate where it goes.

Along and through studs

When wiring vertically along a stud, keep the cable to the center of the stud (maintain 1¼ in. of space on each side of the wire) and staple it every 2 ft. to 3 ft. If there will be several cables run in a very narrow location, most inspectors will allow two cables under one staple. But if you need to stack more cables, or if there isn't enough room under the staple, stack the cables through special plastic brackets found at electrical supply houses. In addition, be sure to drill at least 6 in. to 8 in. above or below the receptacle or switch location, because code requires that the cable be stapled to the stud within 12 in. of a metal box and within 8 in. of a nonmetallic box.

Using a height gauge helps you drill all the holes at the same height for the cable runs. This gauge is a bamboo stick with a notch in one end in which the spade bit shaft rests. Remove the stick after the bit punches into the wood.

Make sure that you drill the studs at the same height on horizontal runs. Having cables zigzag up and down in the stud wall is ugly, wastes cable, and damages the jacket during the pull. There are many ways to keep the run straight. Experience is the first; most longtime pros can drill holes consistently in a straight line. Most beginners can't. The easiest way I've discovered is to use a little cane stick with a notch cut in one end. Just lay the spade bit in the notch and drill all the studs close to the same spot each time.

Wiring through Trusses

If trusses are used in construction, be wary. Do not drill or cut trusses to run cable (unless approved by the truss manufacturer and building inspector), because it could ruin the engineered specs and void the manufacturer warranty.

A floor truss has an open area around 12 in. to 18 in. tall. The space is perfect for running cable across without drilling through the trusses or installing a running board (it's okay to staple wire to the edge of the truss members). Attic trusses, however, have larger open spaces and may be used in accessible areas. If you need to run wire across the trusses, install a running board on top near the truss edge and staple the cables on top of the board. (The truss edge is located above the top plate along the outside wall.)

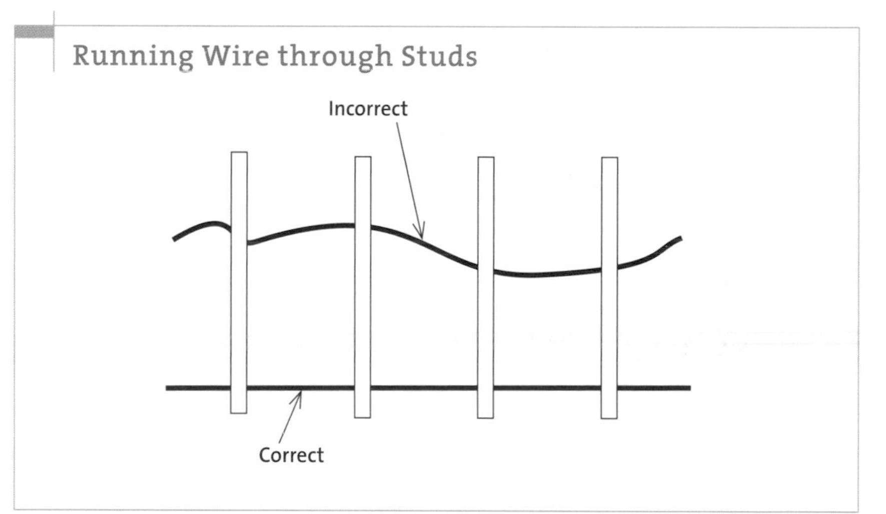

Around windows

Running cables around windows takes some forethought. Your first idea may be simply to drill every stud that gets in your way, including the triple stud on both sides of the window. But this method requires drilling six to eight times. Normally, it's better to take the cable up through the double top plate, through the attic along the exterior wall, and then down again

Wiring around a Window

To get cable around a window, run it up through the attic, through the studs below the window, or through the basement or crawl space.

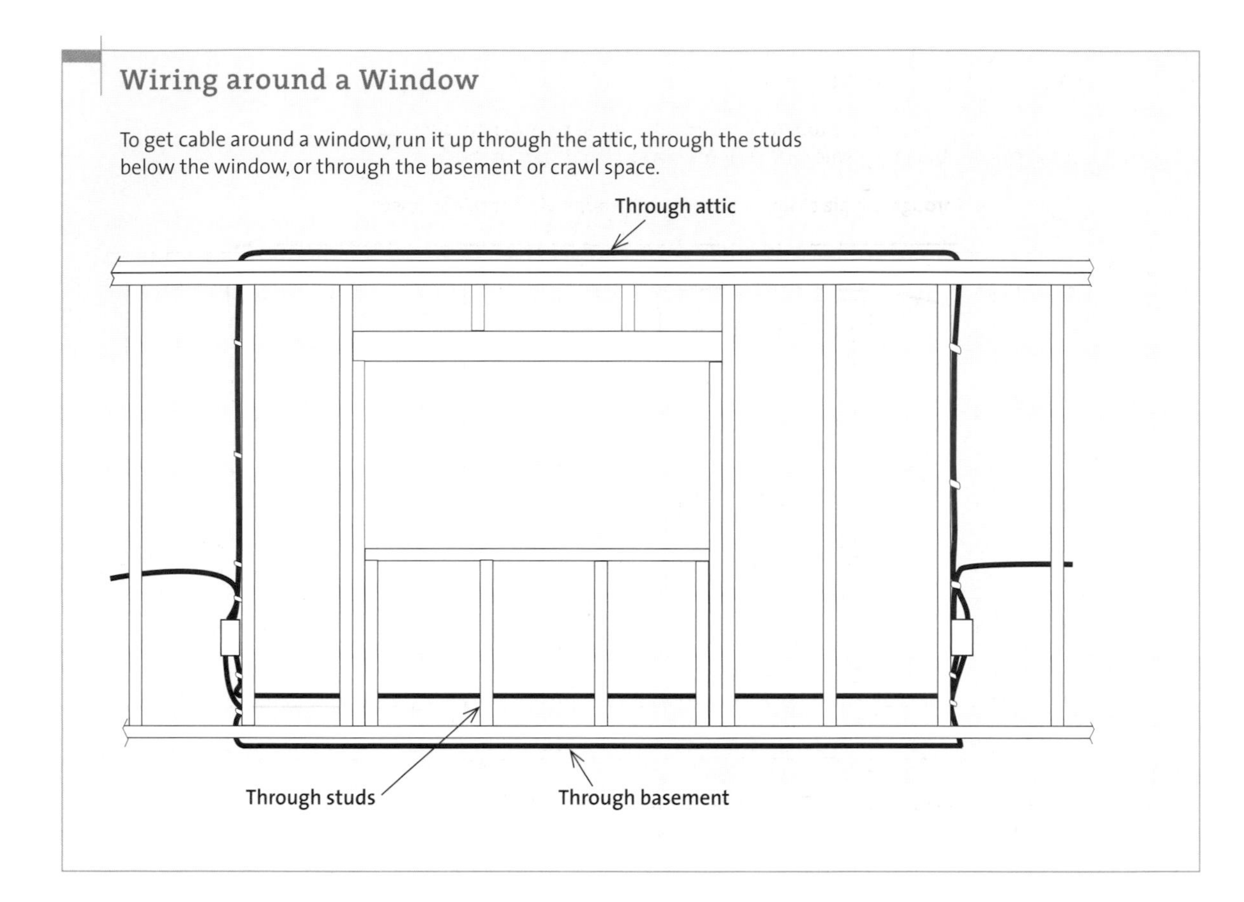

through the top plate to the wall cavity on the opposite side of the window. If you prefer, run the wiring through the basement. With the latter two methods, you only need to drill twice (see the drawing above).

Once in a while you may encounter a window that runs from top to bottom with no place to run the cable overhead due to an exposed-beam design. If there is no crawl space, your first option is to run the cable through the shim area between the header and the window frame. If that is nonexistent, you'll need to make way for the cable using a router or a circular saw and create a channel across the header to lay the cable. Once the cable is in, cover it all the way across with a 1⁄16-in. steel plate to protect it from nails and screws.

Around doors

Getting cable around doors also takes some forethought, but the logic is the same as getting around windows. The easiest route is to go up the stud and through the attic or ceiling joists, or you can go through the basement or crawl space. In these situations, you only have to drill twice. Another option is to run the cable through the cripple studs above the header, but this method requires a lot of drilling, and sometimes the door header is so high that there's not enough room over the door. Yet another option is to run the cable through the shim area between the door frame and the header and protect the cable with a 1⁄16-in. steel plate.

Wiring around a Door

To get wire around a door, run it through the attic, through the basement or crawl space, through the cripple studs above the header, or through the shim area of the door.

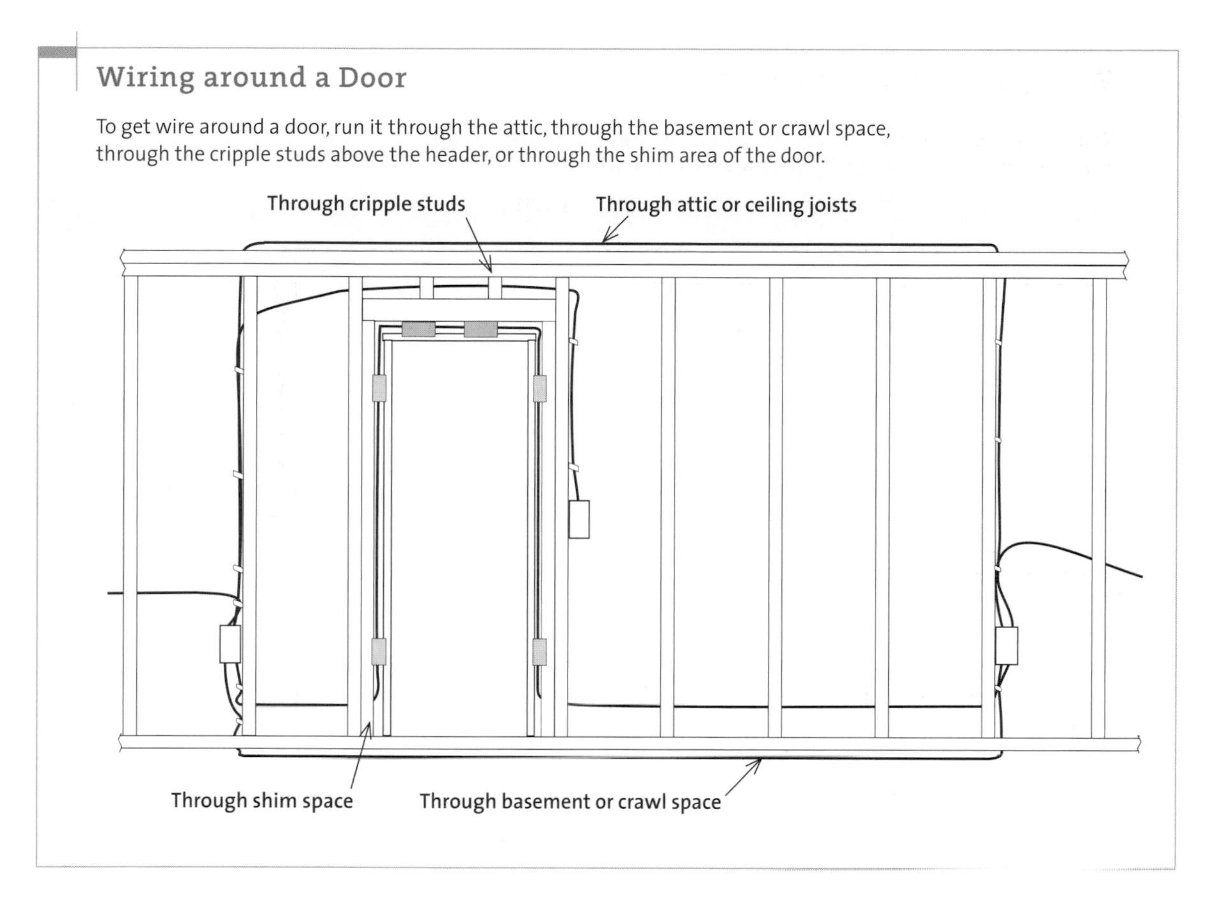

Once in a great while you'll encounter a house with an exposed-beam design and doors with solid headers with no shim room. Whenever there is absolutely no place to run a cable, use a router or a circular saw to create a channel in the header. Don't forget the 1⁄16-in. steel plate.

Boxed corners

Boxed corners are always problematic. If the boxed corner looks like a major problem, try to run the cable over and around it. If you cannot go around, which is rare, note whether the box is hollow or solid.

Hollow boxed corners A hollow boxed corner is easy to get through. Using a 1-in. spade bit, drill from both sides at the same level. Then it's simply a matter of fishing the cable through the holes. If a stud is in the way and you can't get the drill into the wall cavity (and if the carpenters don't mind), remove it temporarily. Carefully cut the nails at the top and bottom of the stud using a reciprocating saw with a fine-toothed blade, then lift the stud out. Cut parallel to the top and bottom plates, and don't ruin the stud by bashing it out with a hammer—the carpenters may massacre you. When you're finished, nail the stud back in place.

Fishing Cable through a Solid Corner

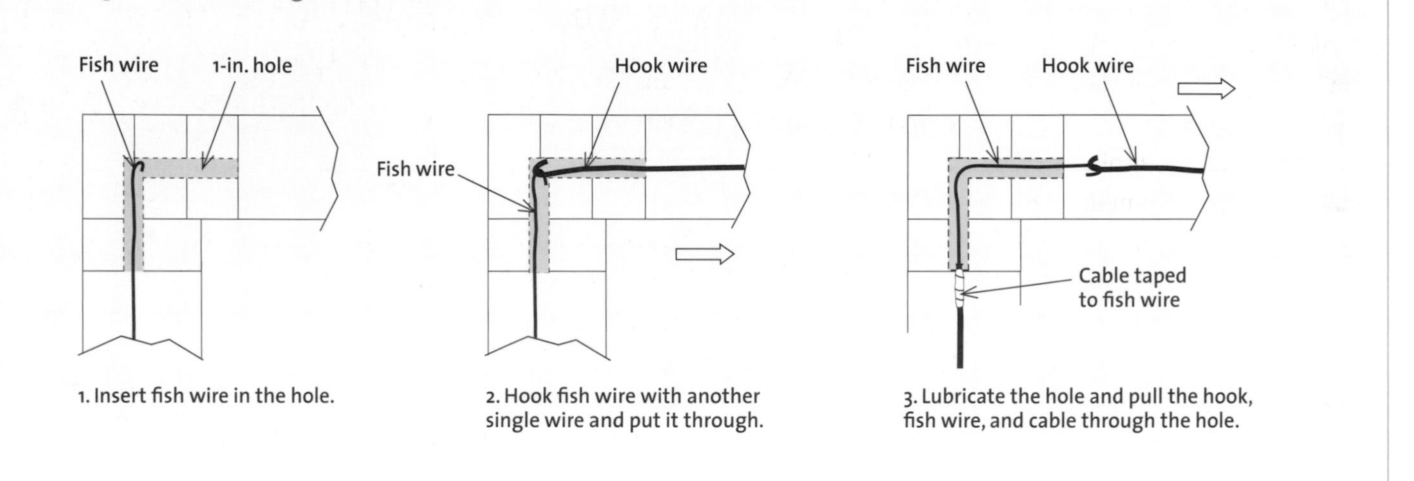

1. Insert fish wire in the hole.

2. Hook fish wire with another single wire and put it through.

3. Lubricate the hole and pull the hook, fish wire, and cable through the hole.

Solid boxed corners To get through a solid boxed corner, first remove any studs that are blocking your access with the drill. Next, drill a 1-in. hole a little over halfway into the corner. Be careful while drilling. The drill is working hard and could buck or stall, and there are a lot of nails in a solid corner. You can use a ½-in. drill or even a battery-operated one, but it will take longer and put a lot of wear and tear on the tool. Do not use a pistol-grip ½-in. drill because if it stalls in the wood or hits a nail the drill could break your wrist. If you use a spade bit for this operation, make sure it is very sharp. If you use an auger bit, use one that can go through nails, such as a Greenlee Nail Eater II. Once the hole on one side is complete, drill from the other side until the holes meet.

If the entire corner is so thick with studs that it isn't feasible to drill, then you have two choices. You can chisel or use a router to open a pathway along the stud face. Or you can open the pathway partway, drill through the corner at a 45-degree angle, and then continue the pathway. I prefer the latter method, because it doesn't put the cable in a 90-degree bend around the outside corner. If it's an inside corner, you have no choice but to have the cable in the sharp bend. Remember to attach steel guards to protect the cable.

Once the drilling is done, fish the cable through, as shown in the drawing above. This is easier said than done. First, be sure to clean all the wood shavings out of the hole before you start, and put some cable lubricant into the hole to make the cables slide easier. Insert a single wire through one hole to the very end. The single wire serves as a fish wire. Then, using a stiff wire with its end formed into a hook, reach from the hole on the other side and hook the fish wire. At the same time, pull and feed the fish wire until it comes out the other side of the corner. If it falls off the hook, reach in with some needle-nose pliers, grab it, and pull it the rest of the way out.

Drill a hole (at the same height as the holes in the boxed corner) through each stud you had to remove earlier and nail them back in place. Pull the cable through the boxed corner and all appropriate studs. Be careful not to damage the cable as you pull it through the 90-degree bend. It's wise to insert a second fish wire through the hole as the cable is being pulled through, just in case you decide later that a second cable needs to go around that corner, too.

Concrete walls and floors

Getting cables through concrete without the proper tools is very difficult. I use a rotary hammer with expensive carbide-tipped drill bits to drive through solid concrete. However, a simple star drill, or even just a hammer, will punch a hole through hollow concrete block.

There are a few methods for running cables through basement walls of cinder block or solid concrete. The least desirable way is to run them through the inside of the walls, with the boxes installed in cavities broken into the blocks. This, of course, cannot even be considered with poured walls. The most common method is simply to nail and glue pressure-treated boards where the conduit and boxes will be installed. Then attach the conduit, cables, and boxes to the wood. Otherwise, you'll wind up drilling and inserting some type of anchor for every box and anchor point for the conduit—and that's a lot of drilling and anchors to install. If the basement will have framed walls flush with the concrete walls, you can run the wiring through the framed walls.

Work Safe • If you can, predrill the studs that will make up a blocked corner before they're installed by the carpenter. Then all you have to do is pull the cable.

Chiseling and Drilling through a Diagonal Corner

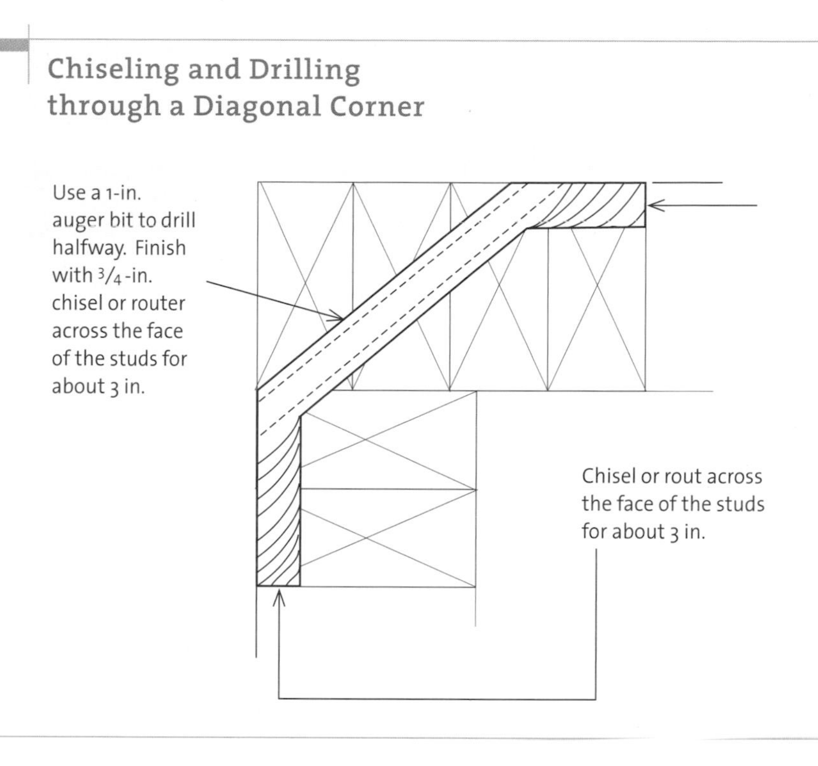

Never pour concrete around a wire or cable; always install conduit for the wiring. If outlets need to be in the middle of a concrete floor, which may be the case in a garage or shop, run the conductors through conduit before the pour and leave the conduit sticking out above the floor. Avoid installing boxes flush with the floor because water could pour into the outlet or trash could collect there.

Wiring outside

Under the 2008 code, you must use UF cable anytime such NM-style cable (even in conduit) exits the side of the house (to be exposed to rain). That is, you are no longer allowed to put NMB cable in watertight conduit outside of the house in wet locations. The code is vague here—one assumes that you can still run NMB cable under the house in a crawl space or between the joists under a deck.

How to Cut through 15½ in. of Wood with a Short-Shank Spade Bit

I prefer to use a short-shank spade bit to make a deep cut, because a long spade bit tends to snap in tight holes, it's hard to get the cuttings out of the hole with large bits, and an auger bit tends to jam in the wood during deep drilling. Although this example uses a spade bit, you can use the same trick with auger bits; just start with a large bit and taper to a small one.

A common ¾-in. spade bit can cut through 4½ in. of solid wood. A ¾-in. spade bit and a ⅝-in. spade bit with a 6-in. extension can cut through 9½ in. of solid wood. Drill to the full depth with the ¾-in. bit and then finish the hole with the ⅝-in. bit and extension.

Use a ½-in. spade bit with a 12-in. extension to drill through the last of the hole.

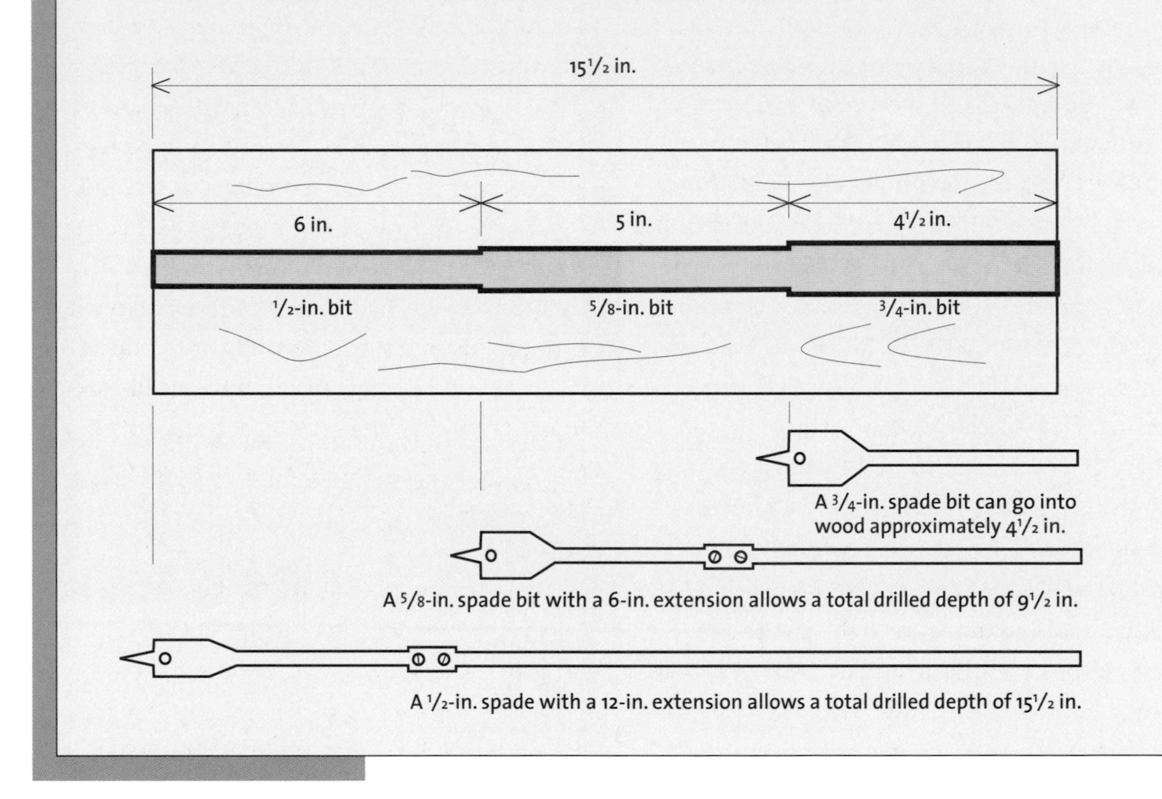

As a rule, bury electric cable at least 24 in. deep. This not only keeps you within code, but also prevents you from cutting the cable when you dig in your flower bed. If you know you are in an area that won't be dug up and are burying a GFCI-protected 120-volt line (for example, from the house to a pole light), you can raise the depth to 12 in. Tables within the NEC give depths for various situations.

Remember that the conductors, even UF, must be protected from physical damage. That means that you can run the conductor under the deck without physical protection, but as soon as it rises above the deck it must be in conduit to a high-enough point that it is no longer in danger of being damaged (normally around 8 ft.).

Work Safe • When using an auger bit to drill deep holes, wax the sides of the bit. The wax helps keep the bit from jamming in the hole.

Work Safe • Be very careful when pulling several wires through conduit, because the individual wires can get tangled easily. As you assemble the conduit, run a pull wire through each section. Once the conduit is assembled, tape all wires to the pull wire, smear cable lubricant all over the wires, and pull them through.

Routing Wires in Renovation Work

Renovation is not easy. Rewiring a house is very labor-intensive, sometimes taking half a day just to replace or add a single receptacle or switch. I try to run wires so that I make as few holes as possible in the walls. The easiest and cleanest routes are through the attic and basement, but these routes are not always available.

One of the first things you need to learn is how to find studs behind a finished wall. I normally use a stud sensor. This may or may not work on thick plaster walls. If all else fails, you can try drilling 1/32-in. or 1/16-in. holes or drive in a finish nail until you feel solid wood. Another option is to drill a small hole, insert a wire bent to a 90-degree angle, and then rotate the wire until you bump the stud. If the trim can be removed, cut into the wall behind the trim to locate the studs. Another trick is to remove a switch or an outlet cover and see which side of the box is nailed to the stud, then go every 16 in.

Opening a finished wall

There are several ways to open a finished wall. You may be tempted to use a jigsaw or reciprocating saw to open a wall, but I don't recommend it because of the high probability of cutting the existing wiring or even the plumbing in the wall. If you know for a fact that there are no wires or plumbing within a particular wall, then use a jigsaw to open the wall. But make sure the tool is equipped with a blade with many small teeth to reduce pulling, cracking, and splintering the finished wall. To keep the saw's bottom plate from scuffing the wall, cover the plate with tape. Try to cut through just the wallboard; adjust the depth of cut so that it just cuts through the wallboard. In many cases, I prefer using a sharp utility knife or a keyhole saw. A slow, manual cut is much safer and neater than a fast cut with a power tool.

Plaster on lath presents special problems. No matter how careful you are, sawing or drilling holes frequently splits the plaster or produces horizontal cracks in the wall. Old molding has the same problems. If the molding is unique, you may not be able to find a replacement piece if it gets damaged.

When drilling into walls, any drill will do, but it's better to use a cordless drill than an AC-powered one. If the drill bit of a cordless drill cuts into a hot wire, the drill, being mostly plastic, won't electrocute you. Never use an old, ungrounded metal-handled drill, unless you have a death wish. If you use a corded drill, make sure it is powered from a GFCI circuit or a GFCI-protected extension cord.

Wiring through the basement or crawl space

To run new cables through the crawl space or basement, first locate where they will run to the new receptacle or switch. The simplest way to do that is by first drilling a small pilot hole in the floor in front of the receptacle or switch in the wall. (If there is carpet on the floor, use a utility knife to cut a small X where you'll be drilling. This keeps carpet strands from getting caught in the drill bit.) Once you've drilled through, either leave the bit in the hole or pull the bit out and insert a piece of wire through the hole, which will give you a reference point in the crawl space. Move about 2 in. to 3 in. from the reference point, find the bottom plate, and drill through the subfloor and plate into the wall cavity. To help you find the bottom plate, look for nails extending through the floor that attach the plate to the subfloor. If you're replacing or rewiring an existing switch or receptacle, you should be able to see where the old cables go through the subfloor and into the wall cavity. Be careful not to hit your head on the nails, and wear safety glasses to protect your eyes from falling debris while you drill.

Creating a Reference Point in a Crawl Space

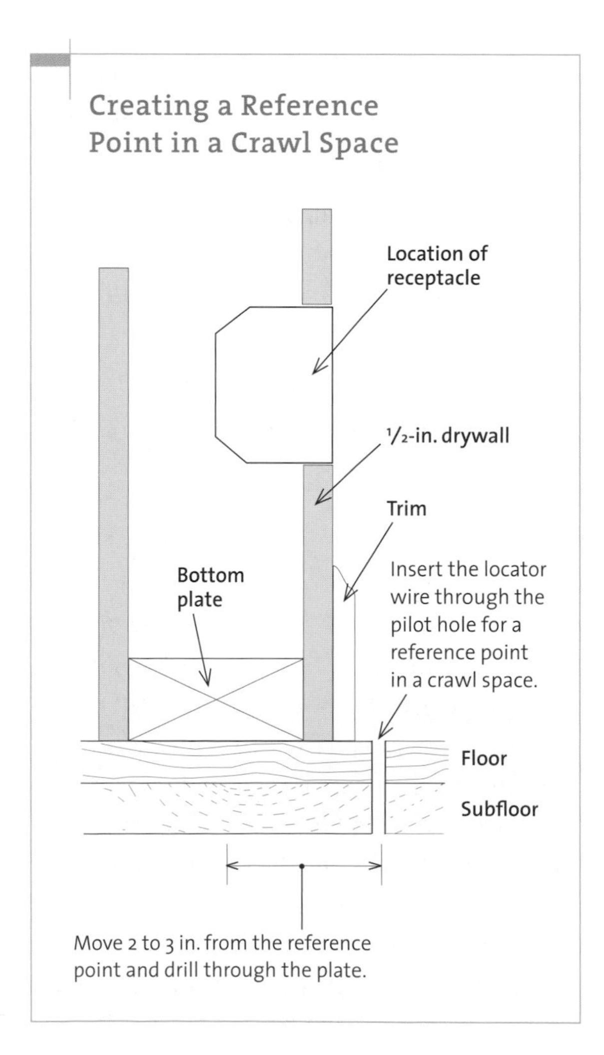

Wiring through a finished wall

Whenever a switch or receptacle is being relocated or added, you may have to route the new wiring through a finished wall. The first step is to de-energize all circuits within the wall. Bring in additional light, because de-energizing the circuit may also remove power to the lights in the room. Then locate the new box, trace around it on the wall, and cut out the hole for it. The best choice for remodeling work is a cut-in box because it has flaps that easily attach to the drywall. (If you choose another type of box, it may have to be nailed or screwed to the stud, which means removing a lot of the drywall.) If you're starting from an existing box that needs to be removed, another option is to use a mini-reciprocating saw to cut the nails of the existing in-wall box and remove it. If the drywall is being opened for a new box, the box should be located about 1 in. from the stud toward the center of the wall cavity. Do not cut the hole for the box so close to the center of the wall cavity that you won't be able to reach the adjacent stud with your drill.

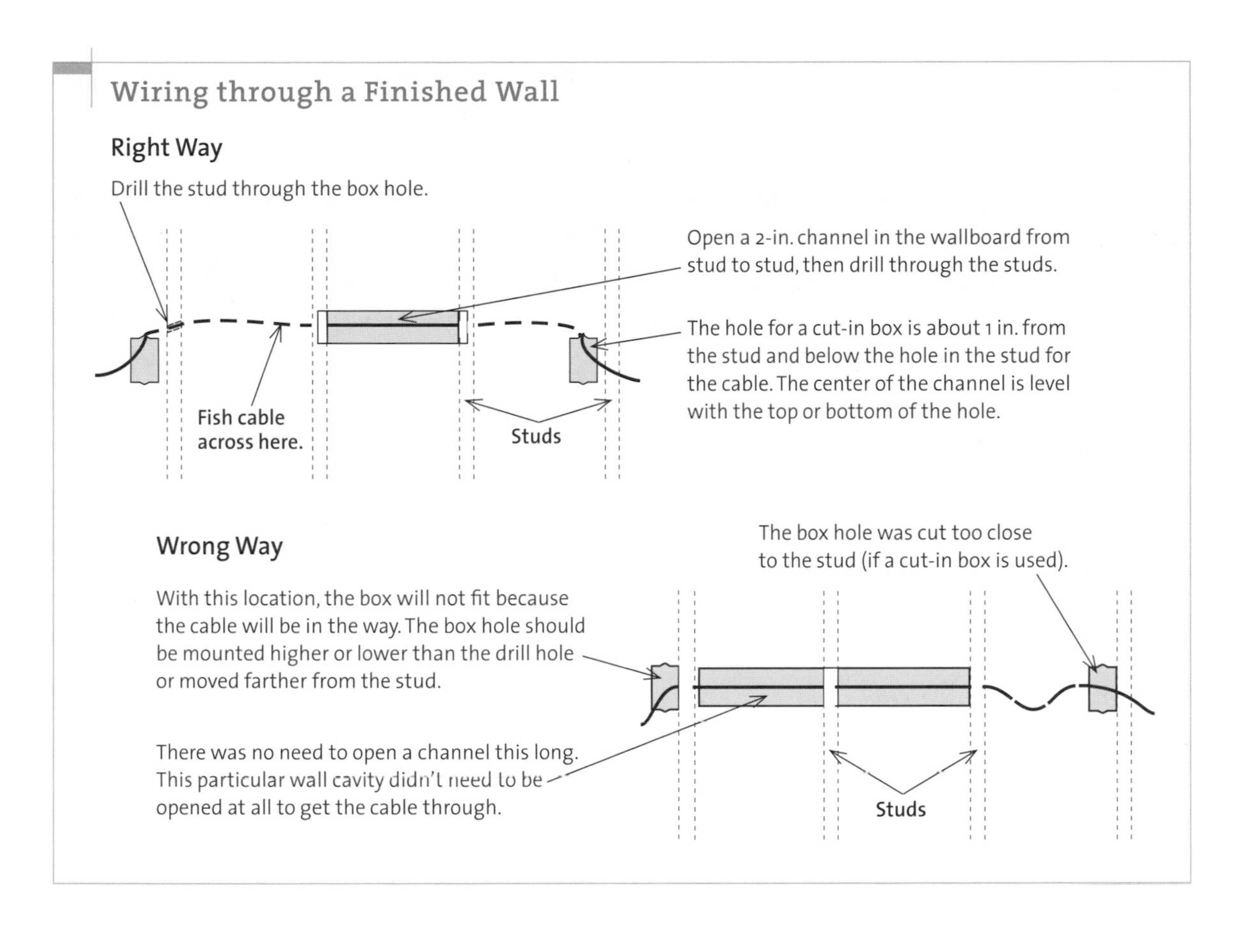

To run cable horizontally through several studs to a new box, use a utility knife or a keyhole saw to open a 2-in. channel through the wallboard from box to box. Drill through the studs and pull the cable through to the new box. If you're relocating or adding a box in the wall cavity adjacent to an old box, simply remove the old box from the wall and drill at an angle through the stud to the adjoining wall cavity. Make sure the hole is above or below the location of the old box. If the old box is staying, it won't fit back into the hole because the new wiring will be in the way.

When you install a new receptacle in a room and there is one in an adjoining room sharing the same wall cavity, power the new receptacle through the existing one. Turn off power to all circuits in that wall. Mark the location for the new box and open up the wall. Be aware that there are wires inside the wall that you don't want to hit. Do not place the boxes back to back, because the wall probably won't be wide enough. Instead, move the proposed box slightly to one side of the existing one.

Remove the existing receptacle from its box and find which side of the box is nailed to the wall stud. You have 16 in. center to center from that stud to the next one in which to locate the new box. If the old box is plastic, open up the cable slot with a screwdriver so that the new cable will go through smoothly. If the old box is metal, loosen the cable clamps and open up one of the knockouts or cable openings with a screwdriver. If you remove a knockout, install a connector to protect the new wire from chafing on the metal. Push a new cable into the old box, giving yourself 6 in. of cable out the box front. Cut the cable to its approximate length and feed it through the hole for the new box, leaving 6 in. Then install the new box.

Behind baseboard molding

You can route cable through the wall behind existing baseboard molding, but only when the molding can be easily replaced if it gets damaged. Do not try this method on old baseboard that is irreplaceable; it may split or otherwise become damaged as you remove it. This method is most commonly used in situations when the original electrician left out a feed wire to a switch or when an owner wants to add a receptacle or fixture after the finished walls are up.

The first step is to de-energize any circuits within the wall. Make a pencil mark along the top of the wall right above the trim. Pop the molding off as carefully as possible. Use a utility knife to cut any paint holding the molding to the wall and use a nail set to drive the trim nails through the wood, which will free the trim board. If you prefer, you can use a wide chisel or prybar to pull the trim off.

With the trim removed, open up a channel in the wallboard using a keyhole saw or a utility knife. Do not remove any finished wall above the mark. Using a long bit, drill ½-in. or ⅞-in. holes in the studs and pull the horizontal section of cable first. Wear safety glasses and be careful not to hit any nails. Run a fish wire from the proposed box hole, tape the new cable to it, and pull both into and out of the hole. Remove the fish wire, run the new cable

Routing Cable behind Baseboard Trim

To run cable through the studs behind existing wallboard, remove the baseboard section. Cut out the wallboard from the center of the stud nearest the existing box to the center of the stud nearest the new box.

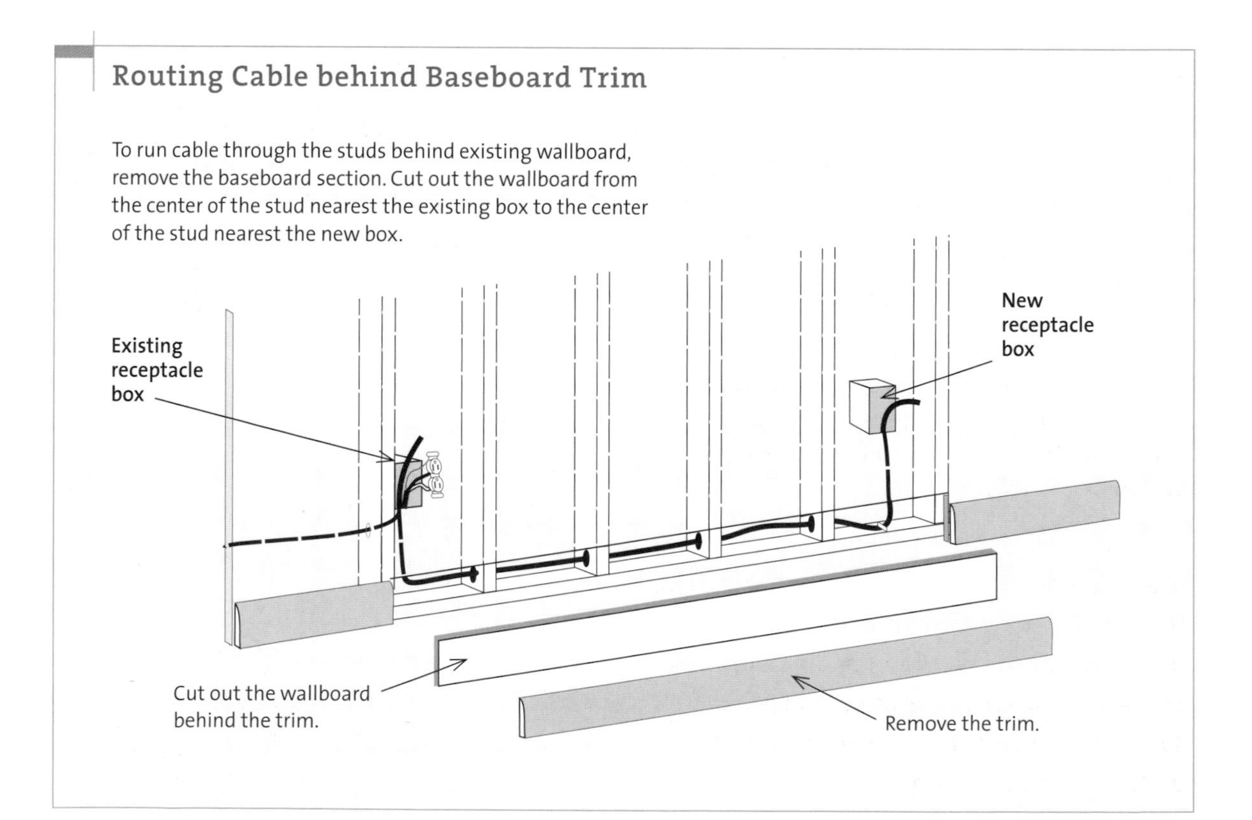

Never use a screwdriver to remove a staple or you'll damage the cable. If you can't get the staple out with electrician's pliers or end nippers, try an old-fashioned nail puller. Simply put the staple in its teeth and pull back on the handle.

into the box, and insert the box into the wall. If you're bringing cable into an existing receptacle box, run a fish wire from the bottom of the box to the opened area behind the trim. Tape the new cable to the fish wire and pull both up into the old box.

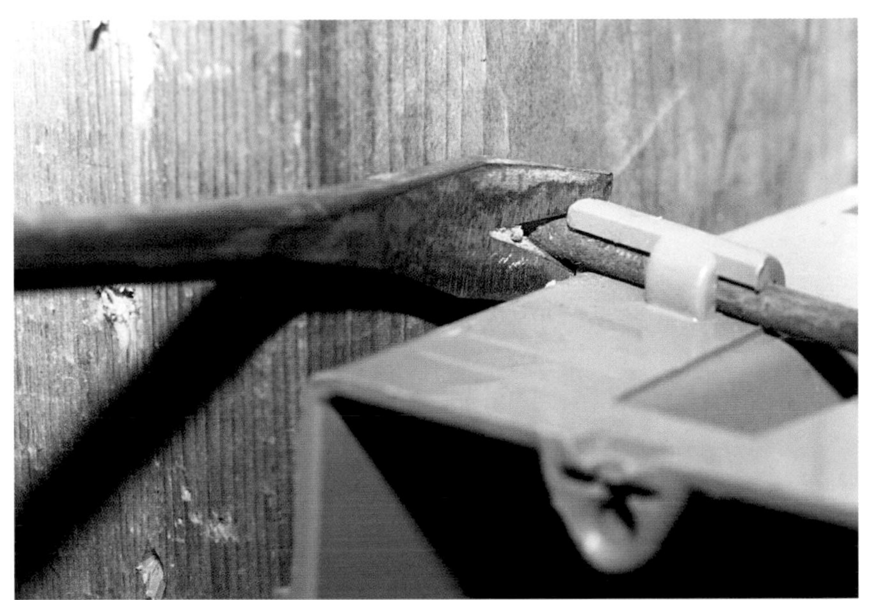

A trim nail puller may look like just a bent screwdriver with a forked tip, but it is the perfect tool for removing an integral nail box from a stud without damaging the box.

My trim pull kit consists of a 6-in. prybar (with the edges ground down), a piece of cardboard, and an outlet box with one side cut out. The box serves as a pivot point and keeps the wall from getting dented. The cardboard keeps the box from marring the wall. Once this prybar pulls out the trim 1/4 in., I replace it with the larger prybar.

Around doors

Routing wiring around a door is normally done when a wall switch was installed on the wrong side of the door frame or when a new switch is being added and power is coming from the receptacle on the opposite side of the door. The procedure also works for windows. As always, first de-energize the circuit you're working on and any that may be within the wall. Remove the trim. Then, using a sharp chisel, notch the shims and anything else that may prevent the cable from going around the door. Route the cable from the receptacle behind the baseboard, if possible, then up and around the door, squeezing it in around the notched shims. Staple the cable to the shims and protect it where necessary with steel plates. If large sections of the cable are exposed, run it through steel conduit. If a switch will be mounted adjacent to the door, open a hole for the switch box about 6 in. to 8 in. from the jamb. To get the cable from the shim area to the switch box, drill through the door frame from the switch-box side. Insert the cable through the frame and into the box, then install the box.

Through the ceiling or attic

Wiring through the ceiling or attic is normally done when the wiring to a ceiling light is being replaced or when a new ceiling light is being installed. As always, de-energize the circuit to the light and any other circuit in that wall. If the old light is present, remove it and the old

Routing Cable around a Door

If there is no other route, run the cable around the existing doorway. Remove the trim around the door and cut away some of the drywall to expose the shim area. Chisel a little from each shim, lay the cable between the jamb and the frame, and staple it to the shims. Cover each exposed section with a 1⁄16-in. steel plate.

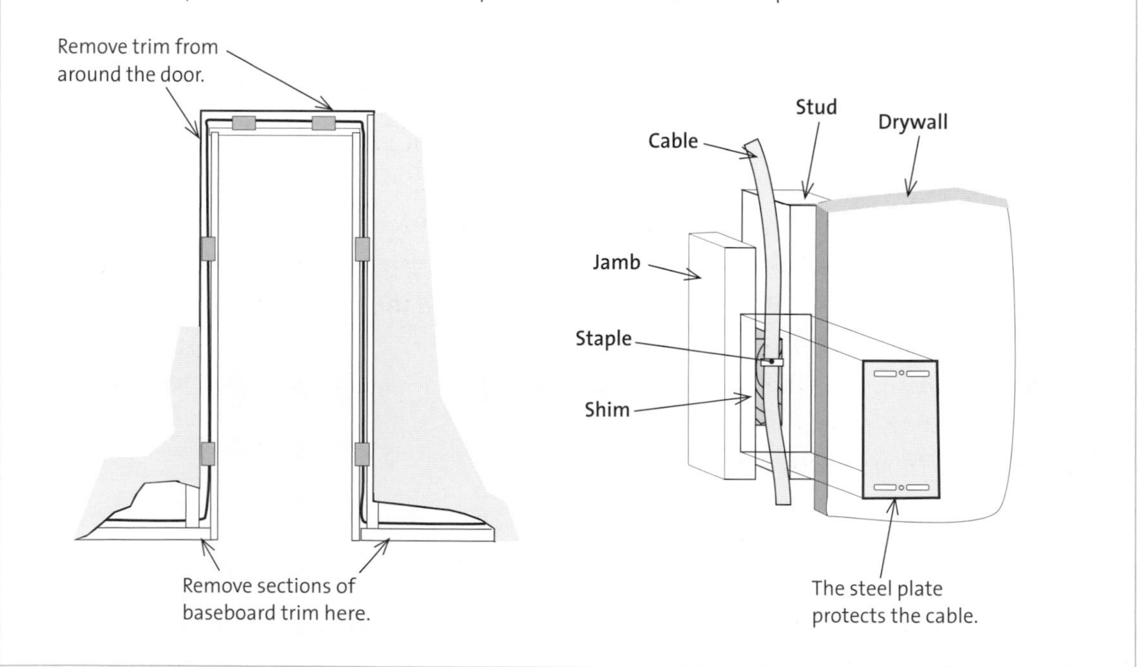

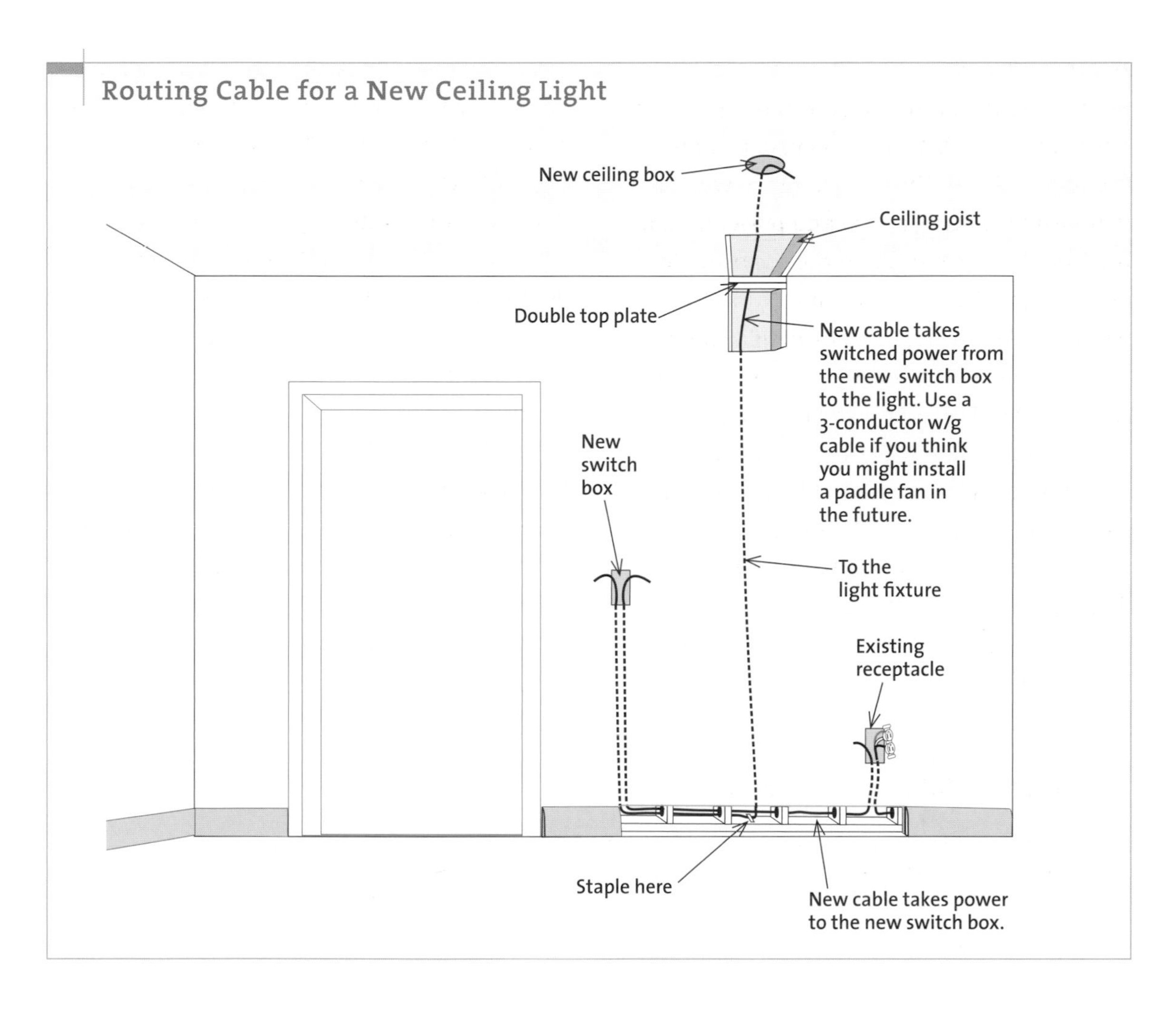

box. If the light is a new installation, cut out the hole for the fixture box. Be sure to have the box in hand so that you know what size hole to cut. In this situation, it's also important to run the wiring parallel to the ceiling joists so that you don't have to open up the entire ceiling to run the cable through the joists.

Next, remove the baseboard trim, cut a channel in the wallboard, and drill through the studs. Cut holes in the wall for the switch box and two slots in the wallboard, one at the top of the wall and the other in the ceiling. The holes must be big enough for a drill to get into them. Drill a 1-in. hole through the top plate with a right-angle drill (remember to wear safety glasses). Run the cable from the receptacle to the new switch box to the ceiling fixture using fish tape or another method, and then install the new boxes.

Fuses and Circuit Breakers

Officially, fuses and circuit breakers are called overcurrent devices because they provide overcurrent and short-circuit protection. That is, if the circuit pulls more current than the device is designed for, the fuse or breaker will automatically open the circuit and it will have to be manually reset. Located in panels and subpanels and sometimes at the point of use (some appliances have built-in fuses), overcurrent devices are designed to protect the house wiring from overloads, short circuits, and ground faults.

An overload occurs when a load placed on a circuit is greater than the rating of the overcurrent device, such as when a washing machine's motor goes bad and starts pulling 30 amps instead of 20. A direct short means that the hot and neutral wires are either directly or indirectly touching, such as when a hot wire comes loose in an outlet box or when a nail is driven into a wire. A ground fault is a short in which the hot wire touches something that is grounded. For example, a ground fault occurs when the hot wire of a drill touches the drill's grounded metal case. If overcurrent protection is not in place in any of these instances, the excessive current flowing through the wire produces enough heat to damage the wire's insulation or jacket, and it may start a fire.

There are many types and classifications of fuses and circuit breakers. Entire books have been written on this subject alone. In this chapter, I focus on the most common fuses and breakers that are used in the home.

Fuses

Although today's fuses provide an excellent source of overcurrent protection, it wasn't always that way. They had to go through years of development. Early fuses were simply pieces of small-diameter copper wire connected in series in the circuit. Because the wires were small, they had a high resistance, causing them to overheat and eventually melt or explode when exposed to excessive current. Although the system worked, it was both a fire hazard and a hazard to workers. The copper-wire fuse evolved into a lead-alloy fuse, but it still had the same exploding problems. The next invention was a zinc element inserted in a fiber tube filled with fire-resistive powder.

The problem with old fuses—those prior to 1940—was that they took too long to open. This delay caused massive damage to the circuit as a result of the huge current buildup during a fault. Today's fuses are much more sophisticated than those of yesteryear. Modern fuses blow quickly, which cuts the current off in an instant—this is much faster than a circuit breaker operates. The two common fuses we use today are cartridge fuses and plug (sometimes called glass) fuses.

Cartridge fuses

Cartridge fuses (designated Class H by UL) are the oldest type of fuses still in use today. These are one-time-use items. In the old days, you could unscrew the ends of a fuse and replace the element. Today, once the element is blown (melted), you have to replace the entire cartridge.

The two most common cartridge fuses are the ferrule and the knife blade. A ferrule-type

Cartridge fuses (left) come in two common sizes for use around the home. The larger one is used in circuits for mains (normally 60 amp) and stoves (normally 40 amp) in old-style fuse boxes. The smaller one is normally for point-of-use locations, such as water heaters and old electric stoves. Inside a cartridge fuse (above) you'll find a fire-resistive material around a metal bar element. The size of the element determines the amperage of the fuse.

A knife-blade cartridge fuse allows a better connection for higher amperages (60 amps and greater).

fuse protects up to 60 amps. It consists of a fiber tube with two metal end caps that make contact with the circuit. Inside is a zinc element surrounded by fire-resistive powder. A knife-blade cartridge fuse is a ferrule-type fuse with flat metal tabs sticking out from the end caps. It protects circuits above 60 amps and was used for mains in old 100-amp fuse boxes.

The common problems encountered with cartridge fuses are not in the fuse itself but in the fuse holder or fuse box. The clips that hold the cartridge fuses sometimes get pulled apart over years of changing fuses, and the rivets that hold the fuse clips onto the fuse box loosen over time. Both of these conditions result in the fuse—and sometimes the entire fuse box—arcing, burning, and overheating. Eventually, the heat destroys the fuse by disintegrating the fiber shell. If you can't figure out why a fuse keeps blowing, the fault may lie with the panel. If that's the case, you must replace the entire panel; it cannot be fixed.

Burned areas normally signify a loose connection on the fuse holder or within the panel.

Plug fuses

A typical plug fuse has what is called an Edison, or standard, screw base. They are available in ½, 1, 2, 3, 5, 6, 8, 10, 15, 20, 25, and 30 amps, all of which are rated at 125 AC volts. Plug fuses are further subdivided into general-purpose and time-delay fuses. General purpose is the common, fast-acting fuse. Time delay is used for circuits that protect motors when they start (motors have a high start-up current). Standard-base fuses can be used only as replacements in existing installations that show no evidence of tampering or overfusing.

These are Edison-type plug fuses. **No matter what the amperage is, they all share the same thread pattern.**

The problem with standard fuses is that they all have the same type of threaded base. It's easy to substitute one amperage fuse for another and thus create a fire hazard. This design flaw led to the development of type-S fuses, which are basically the same as standard fuses but with slightly different threaded bases for the various amperages. Type-S fuses can also be identified by their nonmetallic threads. These fuses are allowed in new construction; standard fuses are not.

For older boxes, install S-type Edison-based adaptors. These adaptors, which screw into the fuse box, prevent someone from inserting the wrong-size fuse into the base.

Type-S fuses have specific thread patterns for various amperage fuses. The 30-amp fuse (right) cannot be screwed into the fuse socket of the 15-amp fuse (left).

Old-style main fuse boxes

Old-style 60-amp and 100-amp fuse boxes continue to provide overcurrent protection in many older homes throughout the country. A 100-amp panel may still offer satisfactory service as long as it hasn't been hit by lightning or been overloaded and damaged (look for melted areas around the box and for any areas that give off excessive heat). However, 60-amp boxes should be updated to provide higher amperage, because most of those panels were designed with only four plug-in fuses. You can't power a modern house with only four circuits.

Overfused wiring is quite common in old fuse boxes. Instead of 15-amp fuses protecting the panel's 14-gauge wire, someone has installed two 30-amp fuses (green) and one 20-amp fuse (red). By the time the fuse opens the circuit, the 14-gauge wire's insulation will be damaged and could potentially start a fire. This typically happens when the loads are maxed out, blowing the low-amperage fuses, and the homeowner increases the fuse size, not realizing the danger potential to the wiring.

At the time, 60-amp boxes were adequate for the loads that were placed on them. But as the years went by, circuits and appliances were added, overloading the circuits and eventually blowing the fuses. Instead of upgrading the service, many people simply replaced the proper amperage plug-in fuse with a higher-amperage one (usually 30 amps). Thus, wiring that needed 15- and 20-amp protection became overfused, allowing excessive current to heat the wiring and kill the fuse box or start a fire. If you have a fuse box in your house, check the plug-in fuses for any 30 amps (they are usually green). If you find any that do not go to a dryer or water heater, replace them immediately with the proper size fuse. If in doubt, install 15-amp fuses. Even better—upgrade the service. Modifying a fuse to make it work when it is already blown is a disaster waiting to happen.

Circuit Breakers

Circuit breakers (officially called molded case circuit breakers, or MCCBs) are the most common type of overcurrent protection used today. Simply put, a circuit breaker senses and measures overcurrent in a circuit and opens the circuit. A breaker is designed to be ambient compensated. That is, as air around the breaker heats up, the breaker senses this heat and derates itself. Thus, if a 15-amp breaker is sandwiched between two breakers generating a lot of heat, the 15-amp breaker may kick off at 12 amps.

Most breakers use a two-part system to protect the circuit: one system for mild overloads and another for severe short circuits. In a mild overload, a bimetal strip, called a thermal trip, heats up and bends backward, eventually tripping the breaker. For severe short circuits and

Checking Fuses

Sometimes you can inspect a fuse just by looking at it. A cartridge fuse may be blackened and distorted, the contact terminals may be arced and burned, or the body may have disintegrated. With plug-type fuses, look at the window to see if the element is broken or blackened.

Many times a fuse element looks good when it isn't. As a double check, use a multimeter or a similar device.

A multimeter uses two probes to check a fuse. It sends a low-amperage current into one end of the fuse (via one probe); if the current comes out the other end of the fuse and enters the second probe, the fuse is considered good. The multimeter indicates this by a tone, light, or numbered reading, depending on the device.

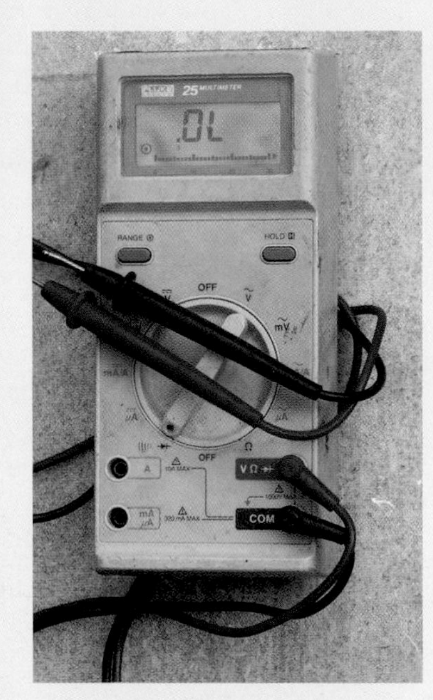

The fuse multimeter above has two types of testers: a continuity tester (where it is set now) and an ohms, or resistance, tester (two positions to the right). To use the device, place a probe on each fuse contact (it doesn't matter which probe goes on which contact). The continuity tester sounds a tone and the resistance tester provides a resistance reading (close to zero). If you do not receive an immediate continuity indication, scratch the probe into the contact's metal to bypass any corrosion.

To do a continuity check on a cartridge fuse, place a probe on each of the fuse's metal end caps. Sometimes the caps come loose as the fuse body distorts.

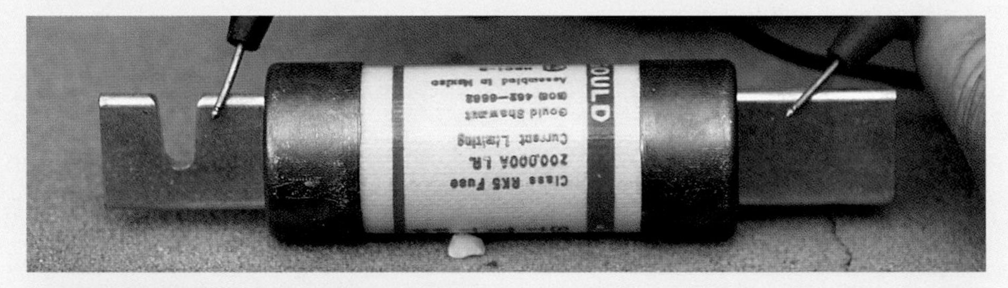

To check a knife-blade fuse, place a probe on each blade.

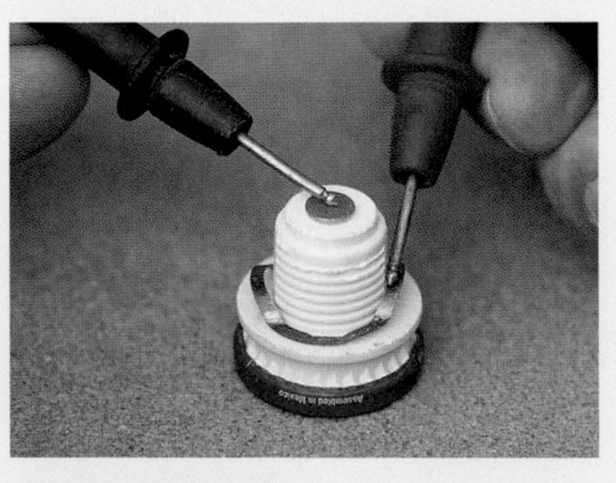

A type-S fuse needs a probe on the center terminal and a probe on the metal band below the threads.

A plug fuse's terminals are the screw threads and the center terminal. Place a probe on each one.

Above Code • Although all these breakers work in basically the same way, I generally use full-size breakers. Full-size breakers are easier to turn on and off and absorb heat better than smaller breakers do. They also fit in any panel; half-size and dual breakers fit only in panels specially designed for them. In addition, installing too many half-size and dual breakers makes it much too easy to get into "panel fill" violations, and the panel invariably winds up looking like a bird's nest with no room to work.

ground faults, an electromagnet bolsters the bimetal strip, pulling it back even faster for an almost instantaneous trip. When the breaker trips, it opens to a position halfway between on and off. After you have identified the problem on the circuit and remedied it, you can turn the breaker to full off and then back on again. Never hold a breaker on when it is trying to trip off.

Most breakers for the home are designed for copper and aluminum conductors and for wire-insulation maximum temperatures of 60°C and 75°C. The temperature rating and CU/AL should be printed on the breaker. Most are also labeled as HACR, meaning they can be used for heating and air conditioning equipment.

In addition to other specs, every circuit breaker has an ampere, voltage, and short-circuit interrupting rating. The ampere rating is on the breaker handle; it indicates the maximum current the breaker can carry without tripping. Standard residential breakers are rated at both 120 and 240 volts, because those are the common voltages within a panel. The short-circuit interrupting rating is based on the maximum amount of fault current available from the utility. Thus, if the utility transformer generates 10,000 amps of current, the breakers in the residence should be rated to at least that amount of current. Residential breakers have ratings ranging from 10,000 amps to 65,000 amps. However, for a single-family dwelling, the available fault current rarely exceeds 10,000 amps.

Single-pole breakers

Single-pole breakers control current on loads that use only one leg of the incoming 240 volts. Standard single-pole breakers come in 15-amp, 20-amp, 25-amp, 30-amp, 40-amp, and 50-amp ratings. Most people rarely use

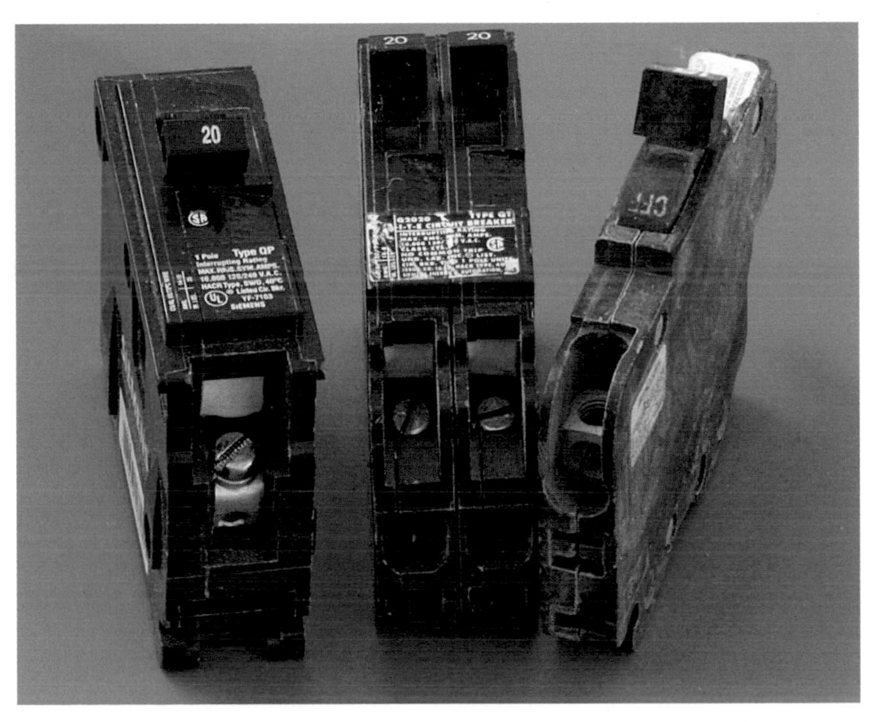

Compare the different types of single-pole breakers. From left to right: a common 1-in.-wide breaker, a dual breaker that only fits onto slotted buses (also 1 in. wide), and a half-size breaker. Not all manufacturers make half-size breakers, which only fit into panels designed for them.

Installing and Wiring a Circuit Breaker

Hooking up a circuit branch to a breaker is fairly simple and basically the same for both copper and aluminum wire. First, strip off just enough insulation to slip the wire into the breaker. Don't strip off too much, though—if more than 1/4 in. of bare wire is showing outside the circuit breaker, the wire may short out on another wire. Make sure the screw is on tight and install only one wire per breaker. (The Square D QO® line breaker is the exception; it can have two wires on its 15-amp, 20-amp, and 30-amp breakers.)

When connecting aluminum wire to a breaker, follow these instructions. Without nicking or cutting the wire, completely strip the insulation from the end. Wire-brush the exposed conductor strands to remove any oxidation (the wire should be shiny). Apply a coat of an approved antioxidant and insert the conductor into the breaker, making sure that all strands are contained. Then tighten the screw to the specified torque indicated on the breaker.

Different manufacturers attach their breakers to the panel in different ways, so it's important to follow the maker's directions. Turn the power off first, then hook the wiring side of the breaker under a panel lip, or notch. Push the breaker down onto the bus tab. You should feel a solid stopping point as the breaker seats. Turn the power back on and check to be sure everything is working properly.

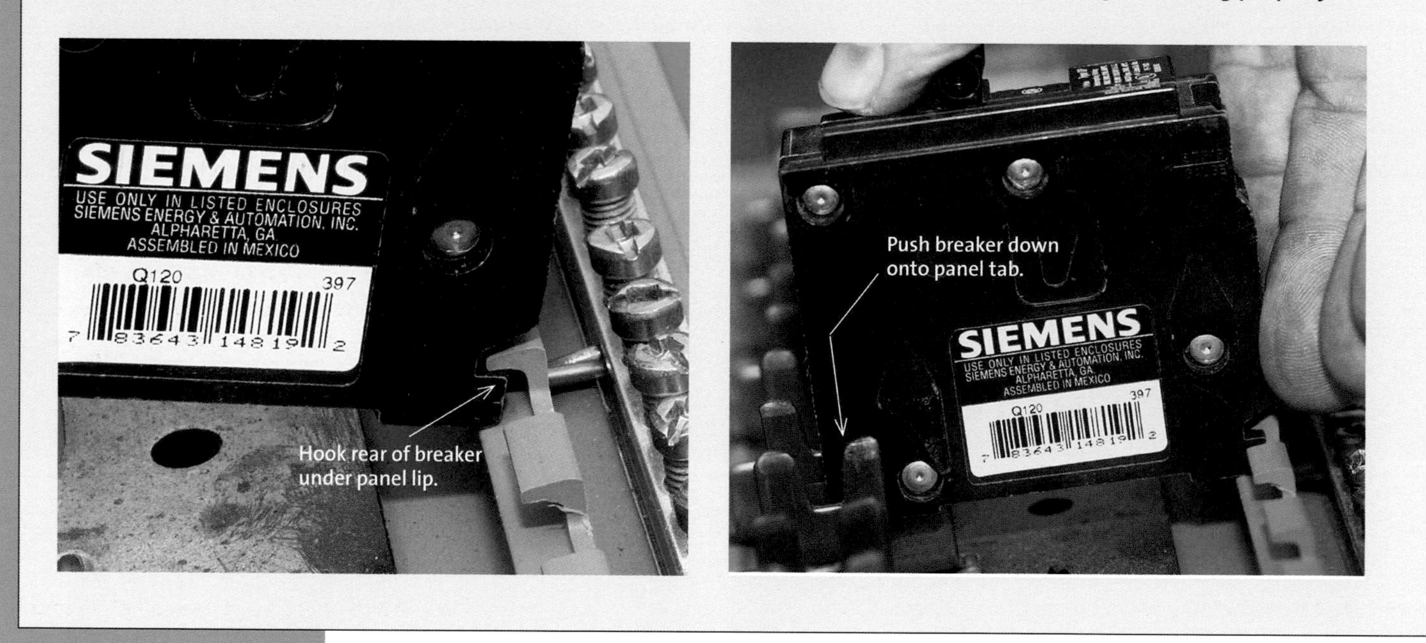

anything other than 15 and 20. Single-pole breakers are also available in three varieties: half size, full size, and dual. The half-size breaker is as thin as a pancake and hard to turn on and off due to its tiny handle. Both full-size and dual breakers are 1 in. thick; however, dual breakers control two circuits.

Double-pole breakers

Double-pole breakers, rated at 120/240 volts, control current on loads that use both legs of the available 240 volts at the panel. They normally consist of two single-pole breakers with one handle and a shared internal trip mechanism. To install them, simply hook the side that faces the gutter under the panel notch

Panel-Fill Violations

A panel-fill violation occurs when dual breakers are installed in solid-tab panels; the panel then houses more breakers than it was designed to hold. To prevent this, some panels have solid tabs that allow only full-size breakers. Dual breakers have a metal clip on the back that prevents them from fitting into a solid tab.

When a cost-conscious contractor installs a cheap, small-volume panel that holds only a few breakers, the electrician may eventually run out of room. To get more breakers into the panel, some electricians remove the metal clip so that the dual breaker can fit on a solid tab. This is not only illegal, but also dangerous: Stuffing a panel like this can create a fire hazard.

A former electrical inspector for Roanoke County, Virginia, told me of an inspection where clips from dual breakers were littered all over the floor in front of the service panel. Looking into the panel, he saw an overabundance of dual breakers. Needless to say, the panel failed inspection. The bottom line: Install a full-size 40/40 panel. The cost difference is insignificant.

Work Safe • Do not swap one manufacturer's breakers for another, even if they fit. Unless they are listed for use in other panels, assume that a manufacturer's breakers can be used only in its own panels. If you do install another manufacturer's breakers, it voids the panel's warranty and the inspector has a right to fail your work. In addition, if there is a panel failure, the panel's manufacturer will put the blame on the "foreign" breaker and the person who put it there.

Two breakers are inserted onto a panel bus. The dual breaker (left) cannot slide down onto a solid tab like the common single-pole breaker (right) can. A metal band inside the back of the dual breaker prevents it from being inserted into a panel for which it is not designed.

A 2-in.-wide double-pole breaker stabs onto both hot tabs of the panel's bus to power 240-volt and 120/240-volt loads. It consists of two single-pole breakers sharing a common trip.

and push the other end onto the panel bus (the same way you install single-pole breakers). Double-pole breakers are available in sizes ranging from 15, 20, and 25 amps all the way up to 200 amps. The 15-amp and 20-amp breakers are normally used for low-wattage water heaters, baseboard heaters, and pumps; 30 amps are used for water heaters, dryers, and heat pumps; 40-amp and 50-amp breakers are used for stoves; and breakers larger than 50 amps are used for electric heating systems.

Like single-pole breakers, double-pole breakers come in two-in-one units, called quads. When there are only two slots available in a panel for two double-pole circuits, you may have to use a quad, but only do so if the panel has slotted tabs.

Be careful when switching quad breakers on and off. The two inner breakers are tied together to power one double-pole device. The two outer breakers are tied together to power the second device. The long handle connecting the two outer breakers can be tricky to operate. To turn the device on, push up on the center of the handle; to turn it off, pull back on the center of the handle. However, if you force one side of the breaker up by pushing on one end of the handle, you could wind up having one side of the 240-volt circuit on and the other side off.

This quad breaker comes as a dual double pole. There is one 30-amp double pole on the two inside terminals. The two outside terminals are ganged together to make a second 30-amp double pole.

This quad breaker has a 30-amp double pole and two 20-amp single poles.

Finding Fuses and Breakers

You need to know which breaker or fuse goes with what receptacle. To determine this, plug a radio into a receptacle and listen for it to go off as you throw the breaker. Or a faster method is to use a breaker/fuse locating device. This is a two-part device: one plugs into the receptacle (it sends a signal through the wire) the other, a receiver, picks up the signal at the breaker. Be sure to follow the tester's instructions thoroughly. For example, you may need to turn it on away from the panel, then drag the device down the breakers (on one side of the panel) and then re-drag again listening for the tone. I find these work but can give an occasional false reading. So do not assume a particular breaker is the one you are hunting until you throw the breaker and verify loss of power at the outlet. Once the outlet is linked to its breaker, remove the outlet cover and write the breaker number on the back of the outlet plate.

Warning: These transmitters can damage the electronics in some of the more sophisticated GFCI receptacles (I've done it). The GFCI continued to give power but the life protection was fried. I'm not sure if the same damage potential exists for AFCIs. Only time will tell.

To find the breaker or fuse that controls a certain receptacle outlet, you need a plug-in transmitter (bottom) and a receiver (top).

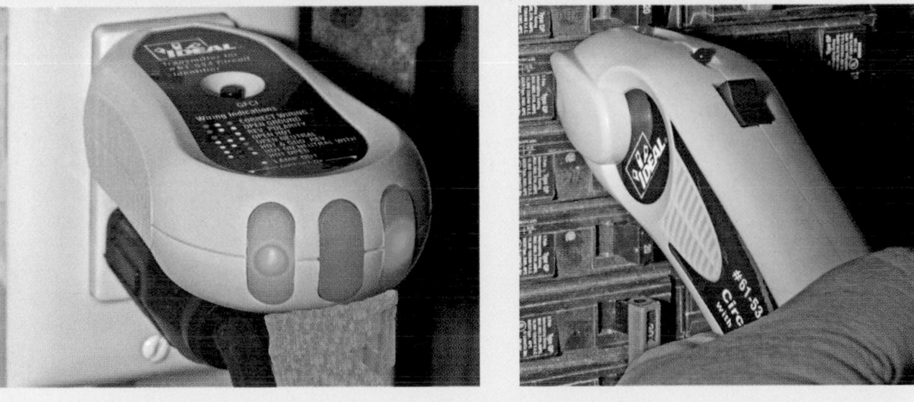

To locate the breaker that controls this GFCI receptacle, plug the transmitter into the outlet and drag the receiver down each breaker until you hear a tone (calibrate the receiver as per instructions).

After determining the controlling breaker, write the breaker number on the back of the cover.

CHAPTER 8

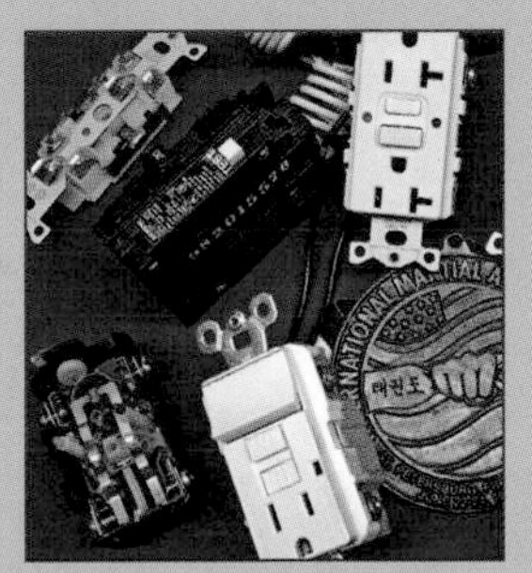

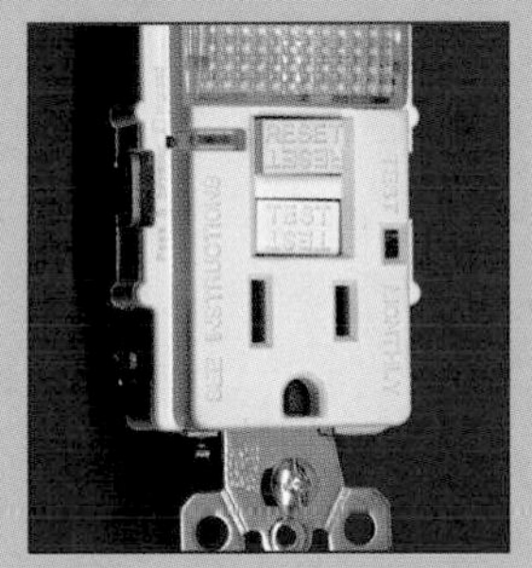

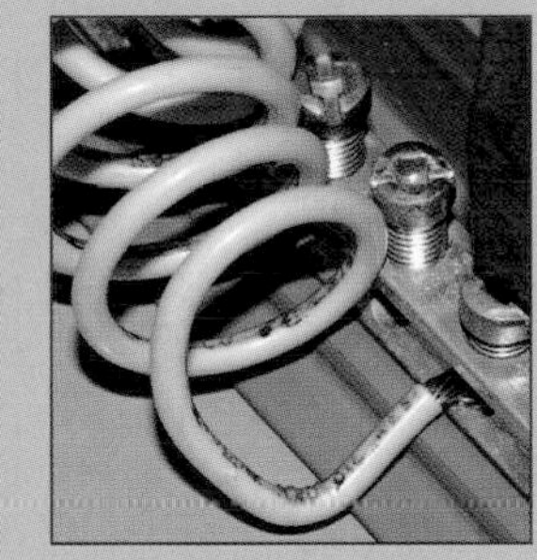

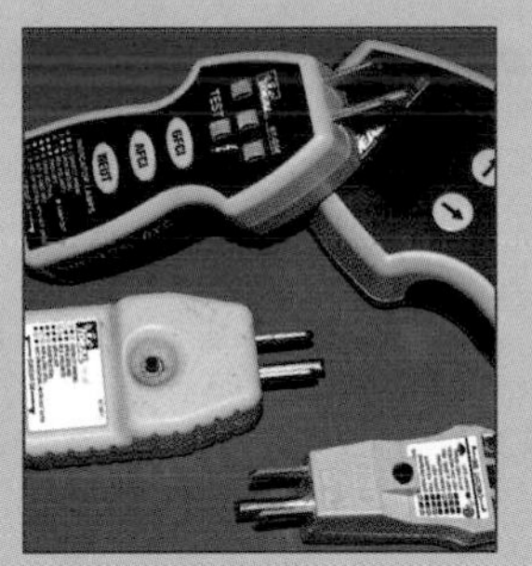

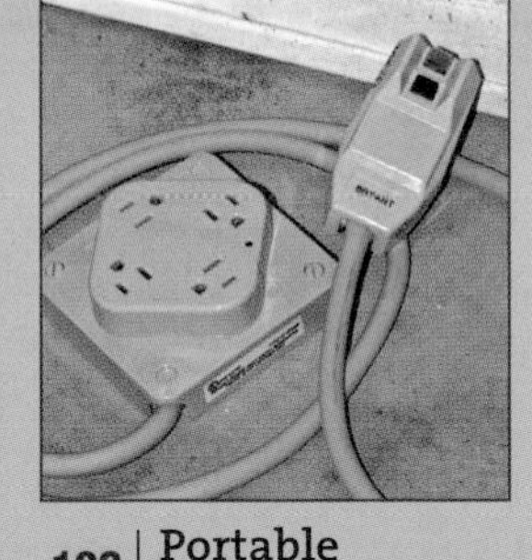

Ground-Fault Circuit Interrupters

If a thing or device could get a gold medal for saving lives, my vote would be for the GFCI, or ground-fault circuit interrupter. This simple-looking device has saved the lives of countless thousands.

Ground-fault protective devices come in many designs, each with a four-letter code (IDCI, ELCI, ALCI, etc.) that seems to have come straight out of a Pentagon dictionary. But regardless of the different four-letter acronyms, all have one thing in common: protection of either people or equipment. A class-A device opens the circuit if the ground-fault current exceeds 5-6 milliamps (0.005-0.006 amp). A class-B device opens the circuit when the ground-fault current exceeds 20 milliamps (0.020 amp). Class-A devices are what we commonly install around the house. Class-B devices are typically used on electrical equipment for lighting pre-1965 pools, which had higher current leakage than more modern lights.

The old movie scene of a radio dropping into the tub and electrocuting the ill-fated bather could not happen in a bathroom properly wired with today's GFCI protection. This chapter will illustrate how GFCIs work, how to test them, and more important, where and how to install them properly.

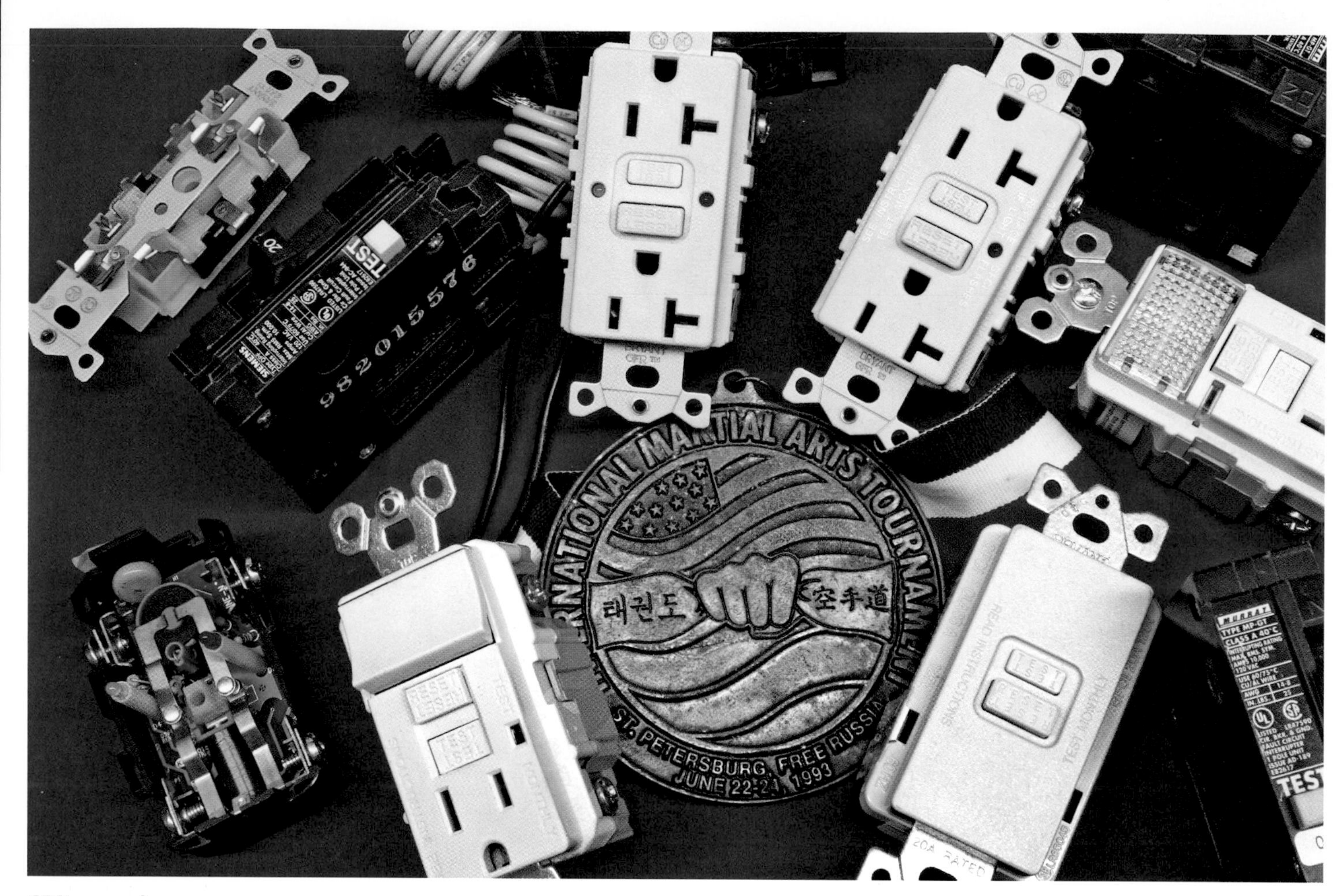

GFCIs come in many guises and they all deserve a gold medal for lifesaving. Hundreds of lives, including mine, have been saved due to the use of ground-fault circuit interrupters.

How a GFCI Works

General-purpose, 120-volt household circuits have current flowing to and from the load on two insulated wires. Power is brought to the load on the black wire, flows through the load, and then returns via the white wire. A GFCI compares the current flowing to the load with the current coming from the load. The current, or amperage, should be equal. If there's a difference, the electrons must be flowing somewhere other than the load (such as through you to ground), and the GFCI will open the circuit. Such a current can be as low as 0.006 amp and doesn't require a grounding wire to work.

The GFCI opens the circuit in $1/25$ to $1/30$ of a second, which is 25 to 30 times faster than a heartbeat. You'll still receive a mild shock, which will feel like a pinprick, but don't panic. If you receive a shock while on a ladder, freeze. Jerking your arms or legs wildly could make you lose your balance and fall off the ladder, causing more harm than the shock itself.

As good as GFCIs are, they can be fooled. As long as the current flowing through the black and white wires is equal, the GFCI operates as though everything were okay. Thus, if you're standing on an insulator or a nonconductive surface, such as a dry board, and you place your body between the black and white wires, the electricity will flow out the black wire, through your body, and back into the white wire. No current will leak through you to ground, so a ground fault won't exist yet you would be electrocuted. This happens because

How a GFCI Receptacle Works

As long as the current on the black wire remains equal to the current on the white (return current), the GFCI operates like a normal receptacle. But if the return current becomes less than the input current, the GFCI will immediately shut off the power to the load.

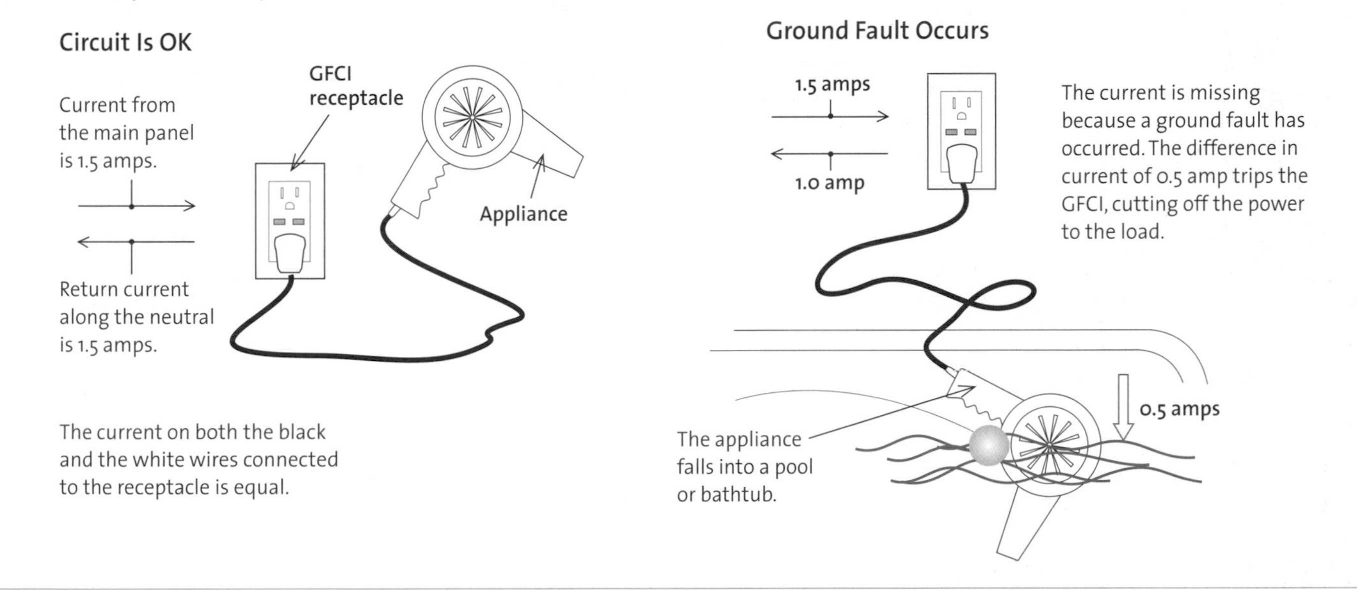

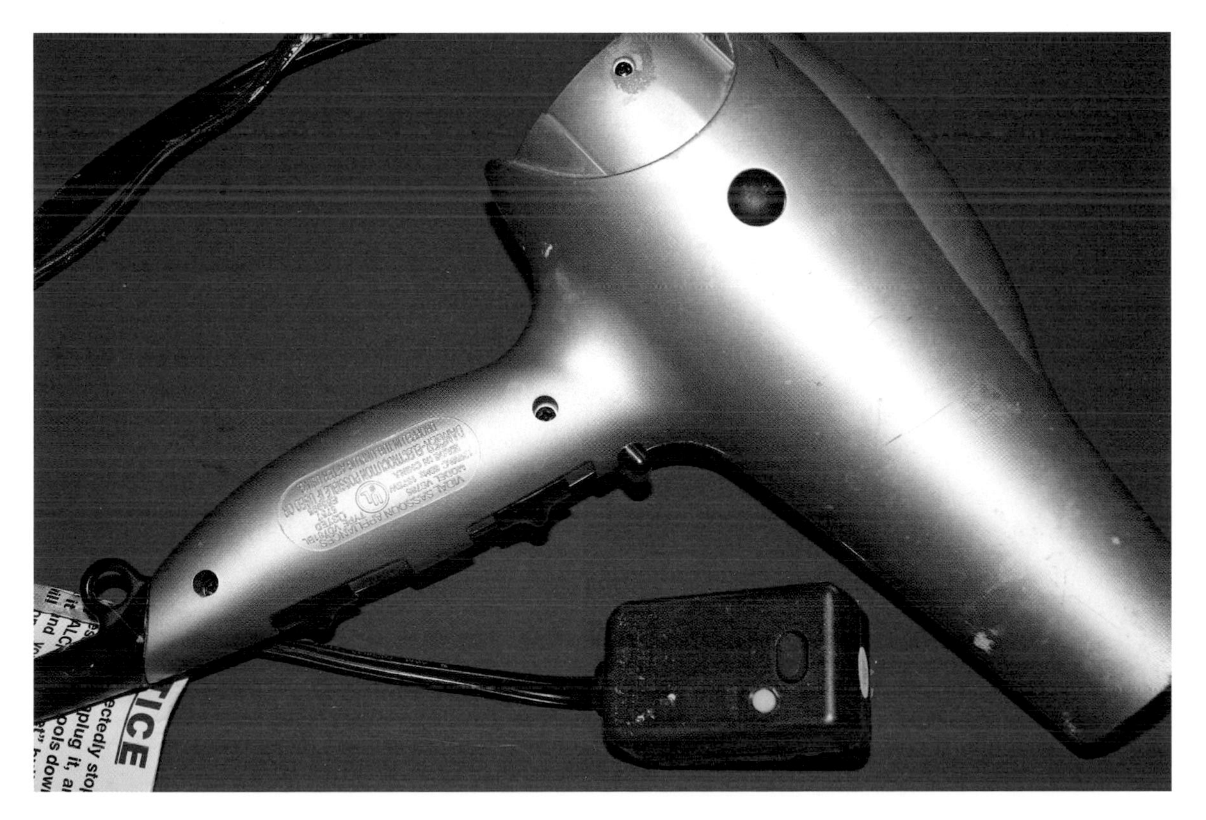

Blowdryers now come with their own fault protection (an immersion detector circuit). This does not exclude the requirement for the blowdryer to be plugged into a fault-protected outlet.

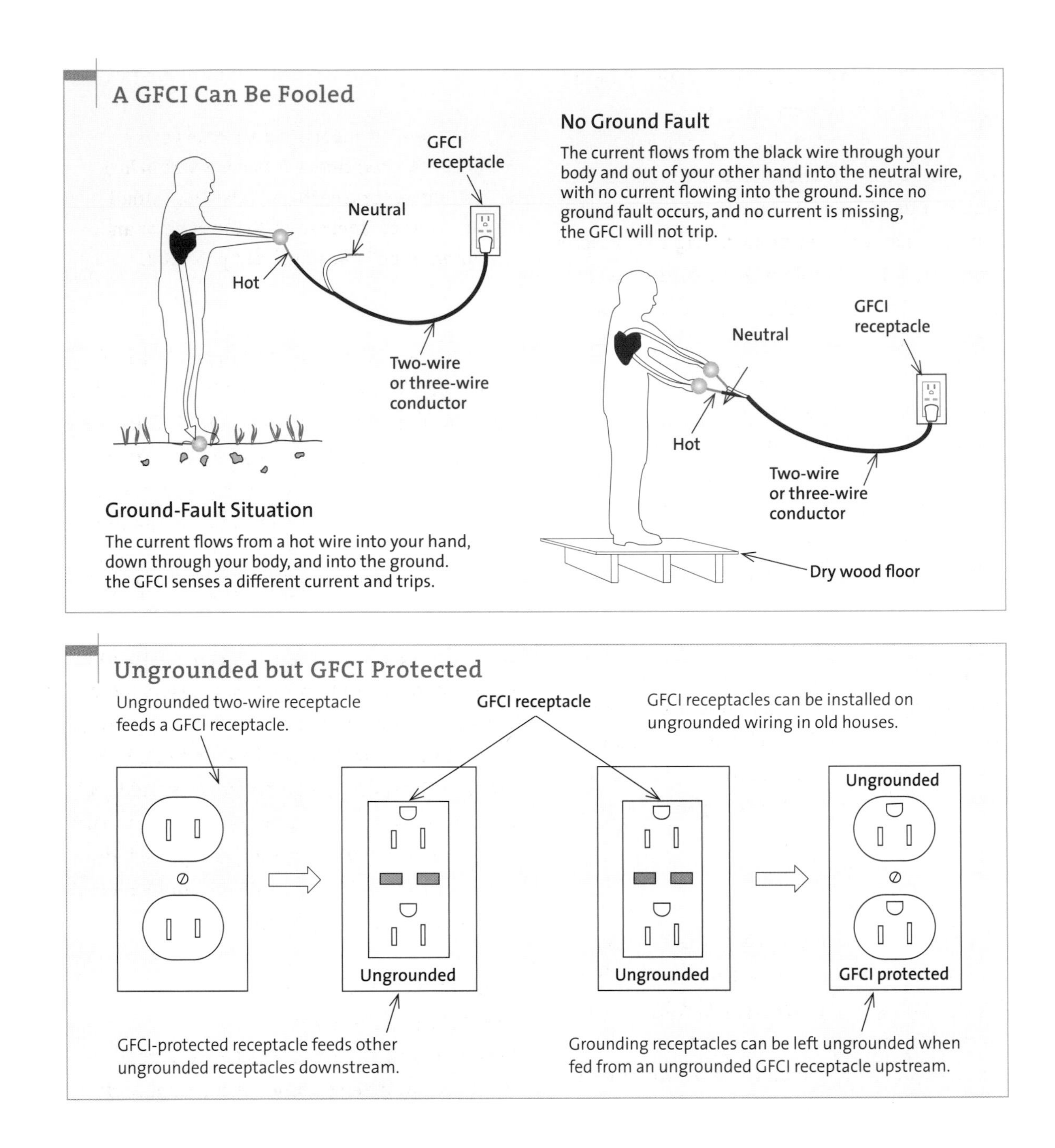

a GFCI can't tell the difference between you and a toaster, sending your heart into ventricular fibrillation, a wildly erratic rhythm, and you'll be dead in a few minutes. It would be as though there were no GFCI protection on the circuit at all.

In that situation, a circuit breaker would not have helped. Breakers for general-purpose receptacles trip only when the current exceeds 15 or 20 amps—2,500 times more than it takes to produce death. Breakers, and even fuses, are designed to protect the wiring within the household against excessive current. They are not designed to protect your life.

GFCIs can be used in a few unusual situations. For instance, they can be used ungrounded if there is no ground on the circuit (such as an older house using 2-conductor ungrounded NM cable or knob-and-tube conductors). The logic is that you are safer with an ungrounded GFCI than with an ungrounded receptacle. Thus, you can replace any ungrounded 2-prong receptacle with a GFCI receptacle. Just be sure to mark the GFCI as being ungrounded. In this case, it's important to note that the GFCI may protect you from a ground fault, but it will not ground your three-prong appliance, nor will it protect against surges and spikes.

A GFCI can also feed ungrounded two- or three-prong receptacles. Those downstream plugs will have the same protection. The GFCI will trip if there's a ground fault in any receptacle in the circuit. As before, if a three-prong receptacle is ungrounded, you must label it as such.

Types of GFCIs

The two most common types of GFCIs used around the home are the receptacle and circuit-breaker types. Receptacle-type GFCIs, or GFCI receptacles, fit into a standard receptacle box and will protect any load working out of that receptacle and any downstream load from its output terminals. Circuit-breaker types, or GFCI breakers, fit into the main control panel just like a standard breaker and, if wired correctly, will protect all the receptacles on the circuit.

GFCI circuit breakers are more expensive and less convenient than GFCI receptacles. GFCI receptacles can be reset at the point of use; GFCI breakers require you to go to the main panel. However, I've found that GFCI breakers tend to last a lot longer than GFCI receptacles, and they are a good choice if you forget to regularly test GFCIs to make sure they're working properly.

Work Safe • Electrical accidents can arise not just from contact between a live conductor and earth, but also from simultaneous contact with a live conductor and another body at a different potential.

GFCI receptacles are available in 15-amp and 20-amp configurations and both 120 and 240 volts. GFCI breakers are available in configurations up to 30 amps from most manufacturers, and a few manufacturers make them in configurations up to 60 amps. Breakers rated above 60 amps (which are used for spas) are hard to find, but they can be ordered. GFCIs are sized just like regular receptacles and breakers. For instance, a 20-amp GFCI circuit breaker goes with a circuit containing 12-gauge wire. Do not install a 20-amp unit on a branch circuit that uses 14-gauge wire.

New receptacle GFCIs are better than ever

Ever evolving, the newest GFCI receptacles need a lot more electronics inside than their early cousins because they now are required to do a lot more than protect life. Modern GFCIs are less error-prone and also help solve problems that arise from improper wiring. They also are required to function better in moist conditions, give fewer false trips, and survive surges and overvoltages. In other words, they work better and last longer than earlier generations.

Breaker or Receptacle

I always recommend using GF breakers, as opposed to GFCI receptacles, to protect outside outlets, simply because the life protection lasts longer. However, there are times that such logic fails—essentially when the physical distance to the breaker involves excessive inconvenience, or you simply don't mind changing out the GFCI.

For example, if the route from the outlet to the service panel is 100 steps up a steep climb, you're better off using a receptacle ground fault with its reset button within easy reach. Or if the trip back is an excessive physical distance, uphill or down, that takes a long time to walk. Or, no matter what the distance or climb, it hurts to walk because you have bad knees or use a walker. Or if you are affluent enough that the convenience of having the reset button close by outweighs the extra cost of changing out receptacle GFCIs every few years. It's all cost versus convenience. Hopefully the new longer-lasting GFCI receptacle designs may be longer-lived, making such debates moot.

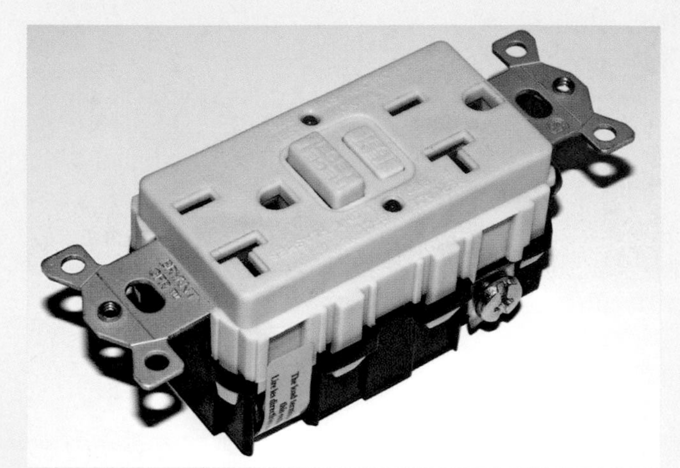

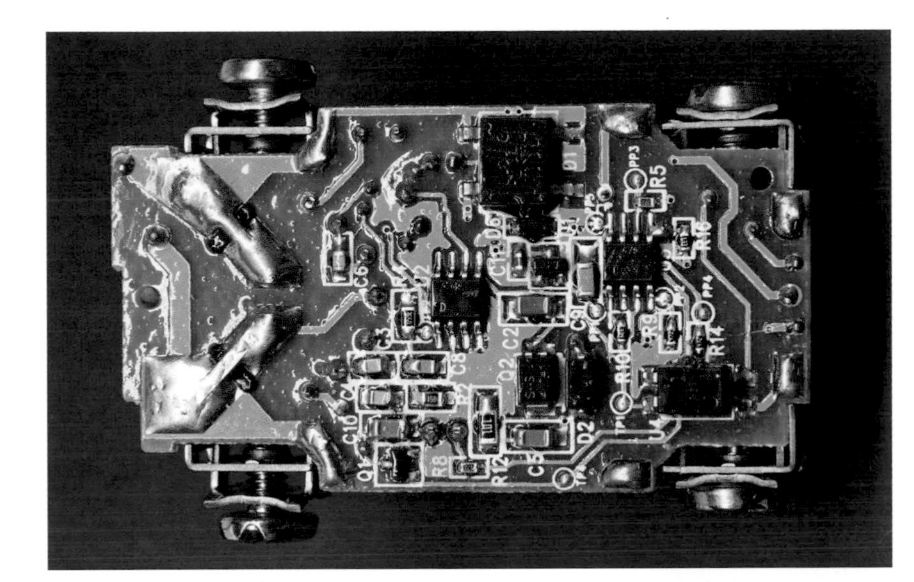

The guts of a modern receptacle GFCI. **It is much more sophisticated than one of just a few years ago.**

To lower the incident of accidental cable reversal that occasionally occurred in the old style, GFCI receptacles now come with the two load screws taped over. This removes the question of which two screws to use for the incoming power cable because you only see the two line terminals. The installer should never remove the tape unless a load terminates there. There is one additional terminal, the ground terminal, but few people miswire it because it is painted green and labeled "ground." However, if the installer still manages to reverse the cables, there should be no power to the face of the GFCI for the newest designed GFCI receptacles.

Besides simply complying with the newest performance requirements, GFCI manufacturers have started one-upping each other, going beyond minimum requirements, and trying to bring to the consumer the best quality GFCI available. For example, both Bryant® and Hubble® took the monthly push-button test of the GFCI out of the hands of the consumer and put it in the electronics of the GFCI. That is, the GFCI tests itself every 60 seconds.

An additional feature you find in some GFCIs is the ability to deliver power even after the reset button has kicked. Once the reset button is pushed, the red light starts to blink and power is restored—albeit without life protection. This feature expands the useful areas of GFCIs. Many areas now can have ground-fault

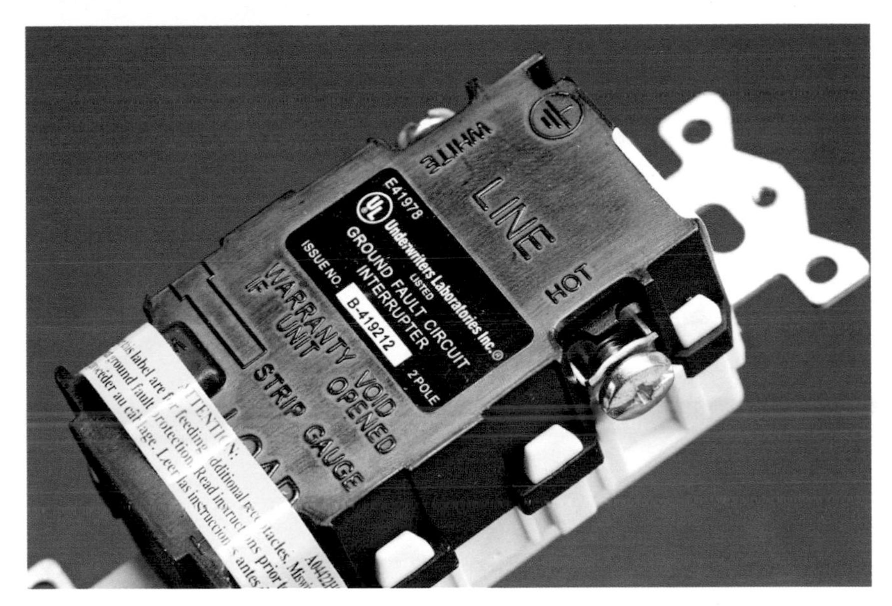

To keep from confusing load and line, **load now comes with tape over the screws. Remove the tape only when you want to protect receptacles downstream.**

Picking the Right Box

The inherent problem with receptacle GFCIs is their mass: They take up a lot of room in the outlet box. This creates a major problem when you are required to install a GFCI in the box of an old, always-too-small receptacle outlet. There just isn't enough room. I'm sure the code people didn't think of that when they made the requirement. This means replacement of the box, which is no easy chore in an older home, or putting the entire circuit on a GFCI breaker.

This lack of room also applies to new boxes that are of small volume, such as a handy box. Under no circumstances do you want to install a receptacle GFCI in a 1½-in.-deep handy box. This is a "push and pray" situation. The wires, when crushed into the back of the box, will cut into the metal and short out with a bang.

The box you do want to use is a deep plastic box, which gives you lots of room for both the GFCI and the splices that are behind it.

protection that didn't have it before because of the worry that a false trip would permanently remove power. Currently only Hubble and Bryant are making such receptacles. This is the design you need when loss of power would result in property damage as in a sump pump installation. The obvious disadvantage of having power without life protection is that children might reset the bathroom GFCI receptacle and jump back into the tub with the faulty appliance. Because of that potential, I don't recommend installing this type in the bathroom. The same logic, and problem, exists for adults using outside receptacles. If the adult is lazy and doesn't replace the GFCI when life protection is lost and plugs in the electrical device anyway, he is at risk. Bottom line: Be selective.

Ground-Fault Protection

GFCIs are extremely sensitive to changes in electrical current, cutting off power when they sense one. When installing a GFCI on a freezer, refrigerator, sump pump, and the like, consider installing an alarm on the equipment that will alert you if there's a problem. Lights—unless you have a good reason to do so—should never be on a GFCI-protected circuit. If the GFCI trips, you might be left in the dark, which could be dangerous, especially if you're trying to find your way out of a wet, unlit bathroom. Also, GFCIs cannot be used for ranges, ovens, cooking appliances, and clothes dryers whose grounded neutrals are connected to the frame of the appliance.

For life protection, the NEC requires GFCIs in a few locations in a house: kitchens, bathrooms, garages, crawl spaces, unfinished basements, decks, porches, and outbuildings (and don't forget the job site, as mentioned earlier). In wet locations, such as on kitchen countertops, all receptacles must be GFCI protected. Most appliance cords are 6 ft. long, so any appliance within 6 ft. of a sink can fall into the water. Because bathrooms are almost always wet, GFCI protection is required for all receptacles, whether they are located in a light, in a medicine cabinet, or in a wall. The overhead bathroom fan or light—or combination fan/light—needs GFCI protection when it's mounted in a bath or shower enclosure. In addition, the unit must be approved for damp or wet locations.

GFCI protection is required in any general-purpose receptacle in a garage; in a crawl space at or below grade level; in an unfinished basement; in a boathouse, dock, or seawall; and in areas around swimming pools, spas, and

Light Indicators (Bryant and Hubble) of GFCI Function

Condition	Green LED	Red LED	Receptacle Power
Normal Condition	On	Off	On
Ground Fault Condition	On	On	Off
Test Button Pressed	On	On	Off
Reset Button Pressed	On	Off	On
Reversed Line/Load Connection	Off	Off	Off
Auto-Test Detects Circuit Failure	Off	Flashing	Off
Reset Button Pressed	On	Flashing	On

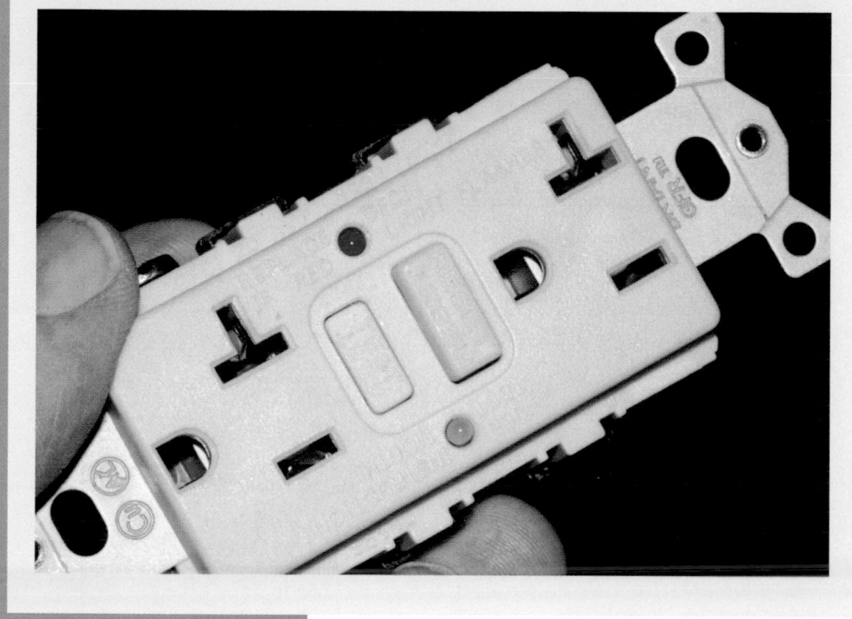

Some GFCIs come with computers that test the outlet every 60 seconds and have a series of lights for indication and troubleshooting.

hot tubs. When it comes to GFCI protection, swimming pools, spas, and hot tubs, because of the obvious danger involved, are complicated and should not be handled by an amateur. Those areas should be looked at and wired by a licensed master electrician.

All outside receptacles, including those in outbuildings, need GFCI protection with a weatherproof-while-in-use cover. Typical examples are Christmas-tree lights, electric-fence chargers, and low-voltage transformers. The receptacle cover must be suitable for wet locations while the appliance is in use. Such covers sometimes look like bubbles on the front of receptacles. However, TayMac makes several models that are flush with the exterior siding or inside wall. You cannot use a cover with a metal, spring-latched door. That type of door is used simply to cover the outlet when it's not in use. (For more on watertight covers, see the next chapter.)

Light on the Load Side

It is not against code to put a room's overhead light on the load side of a GFCI receptacle. The NEC does not dictate common sense nor does it deal in physical safety, other than to prevent shocks. For example, it is not against code to be standing on a metal ladder when working on wiring, it's just stupid. Thus the NEC is silent on such issues. When the GFCI trips, you do not want to be left in a dark bathroom with the shower running, sitting in a tub full of water, and stepping onto a wet floor—or a dozen other scary scenarios. Following that same logic, do not put the entire bath on a ground-fault breaker if it includes the overhead light. Bottom line: If a house has the bath light on a ground fault it should be changed immediately before someone gets hurt.

Silicone-Filled Wire Connectors

When wiring in a GFCI or a GFCI-protected outlet around docks and piers, use silicone-filled wire connectors. If you use common wire connectors they will soon start corroding. And if you leave the skirt pointing up, they will fill with water. Before they sold silicon-filled wire connectors I made my own by injecting silicone into the wire connector before I inserted it over the twisted wire splice. For more information on both silicone and common wire connectors, see the chapter on receptacles and boxes.

New codes require the use of tamper-resistant receptacles for household 125-volt 15- and 20-amp receptacles. Look for TR on the front.

Wiring GFCI receptacles

In order to save lives, GFCI receptacles and circuit breakers must be wired correctly. And with so many wires involved, it's easy to connect one incorrectly. GFCIs have both load and line connections—four screws and wires to figure out, plus the ground wire. Sometimes the load connections are on the right, sometimes on the left, and other times at the top or the bottom. There is no way to tell any of the connections apart without reading the print on the back of the GFCI. And the print can sometimes be difficult to read.

Unusual GFCIs

Blank GFCIs, with no receptacle outlets on the face, are also available. These are used for protection in areas where you obviously don't want or need a receptacle, as in hard-wired installations such as you occasionally find in fountains, pools, and hot tubs. GFCIs with switches and nightlights are also making their way onto the market. The switch is independent of the GFCI. That is, it can be wired to send unprotected power as well as life-protection power. Decorator GFCIs (not shown), of course, could not be far behind once GFCIs were introduced and are available at most large distributors that carry a decorator line. These work exactly the same way as ordinary GFCIs but allow you to spend more money, if that is what you want.

Unusual GFCIs. A plain face front (left) is used for spas and whirlpools. The switch (center) GFCI can turn on the lights at the lav, and the nightlight GFCI (right) gives a little illumination when needed.

Some GFCI receptacles provide power even if they are wired backward (power from the service panel is wired to the load side instead of to the line side). If the downstream receptacles are wired from the line side of the GFCI, they may not work correctly: The GFCI may not be able to sense a current imbalance. Worse, if the GFCI is wired incorrectly, it may still test normally. (Some GFCIs now have built-in sensors that light up if they are wired improperly. Others, may not work at all.) To wire a GFCI receptacle, follow these steps:

Wiring a GFCI Receptacle with Both Protected and Unprotected Cables

A GFCI receptacle can provide ground-fault protection to regular receptacles downstream. Wires from the main panel should be attached to the line side; wires feeding protected receptacles downstream should be attached to the load side.

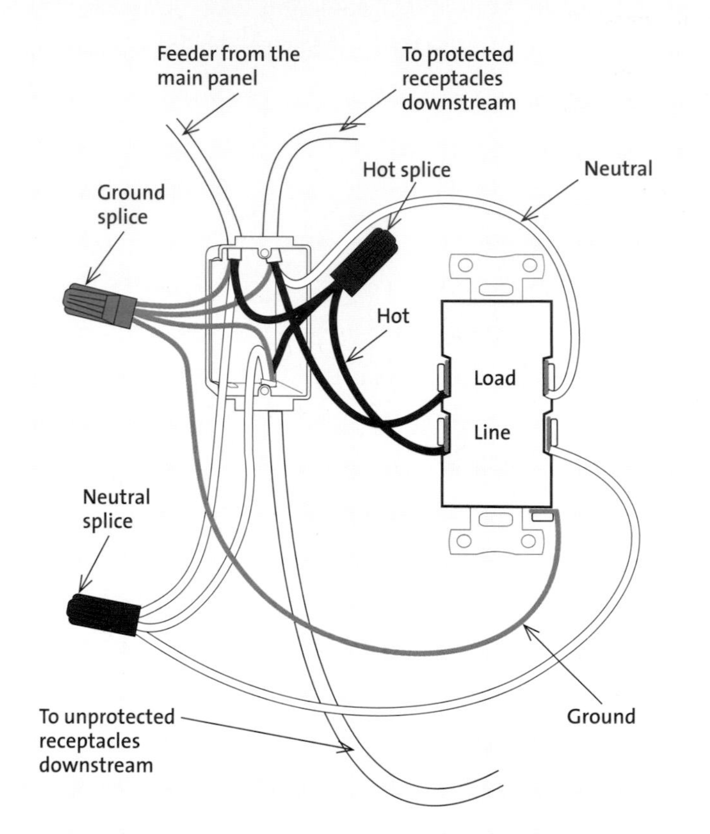

Wiring a Single GFCI Receptacle

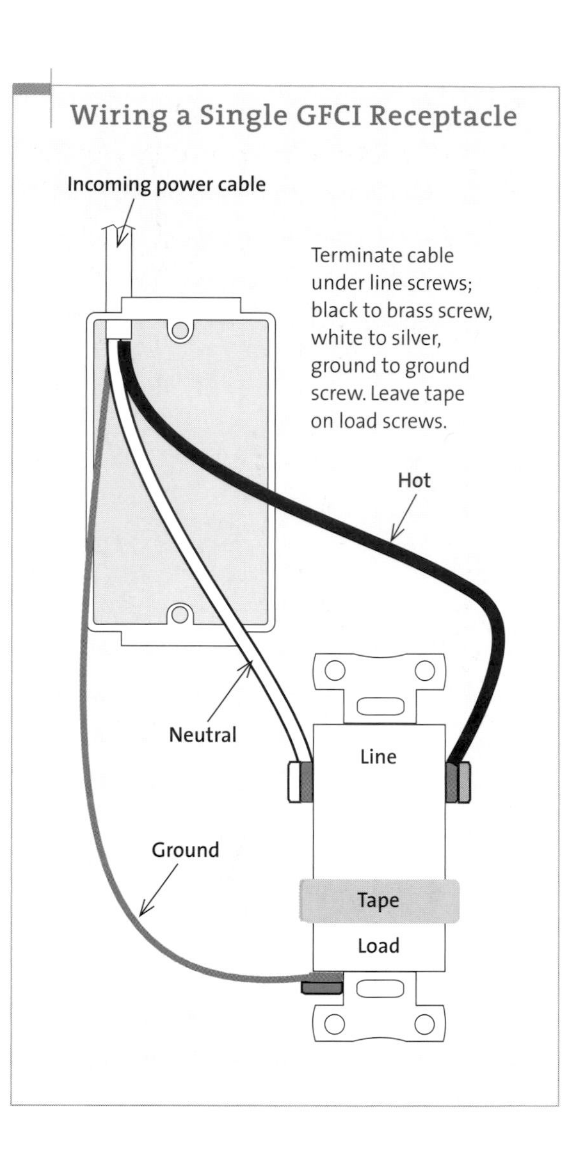

1. Locate the terminals for LINE HOT and LINE NEUTRAL. Connect them to the feeder wires from the main panel (black to hot, white to neutral) to bring power into the GFCI. Once connected, protected power will be available at the GFCI and at the load terminals. Unprotected power is taken off the two feeder wires before they attach to the receptacle line terminals. Never put two wires under any screw; instead, use wire nuts and pigtail off the splice to the GFCI.

2. Identify the terminals for LOAD HOT and LOAD NEUTRAL (these may be covered with tape). These provide outgoing protected power, if you want it. If you have no need for any more protected receptacles in the circuit, nothing will connect to these two terminals. However, any receptacle that uses these two terminals for its feeder will have ground-fault protection provided by the GFCI receptacle. If used, all protected receptacles should be marked on the

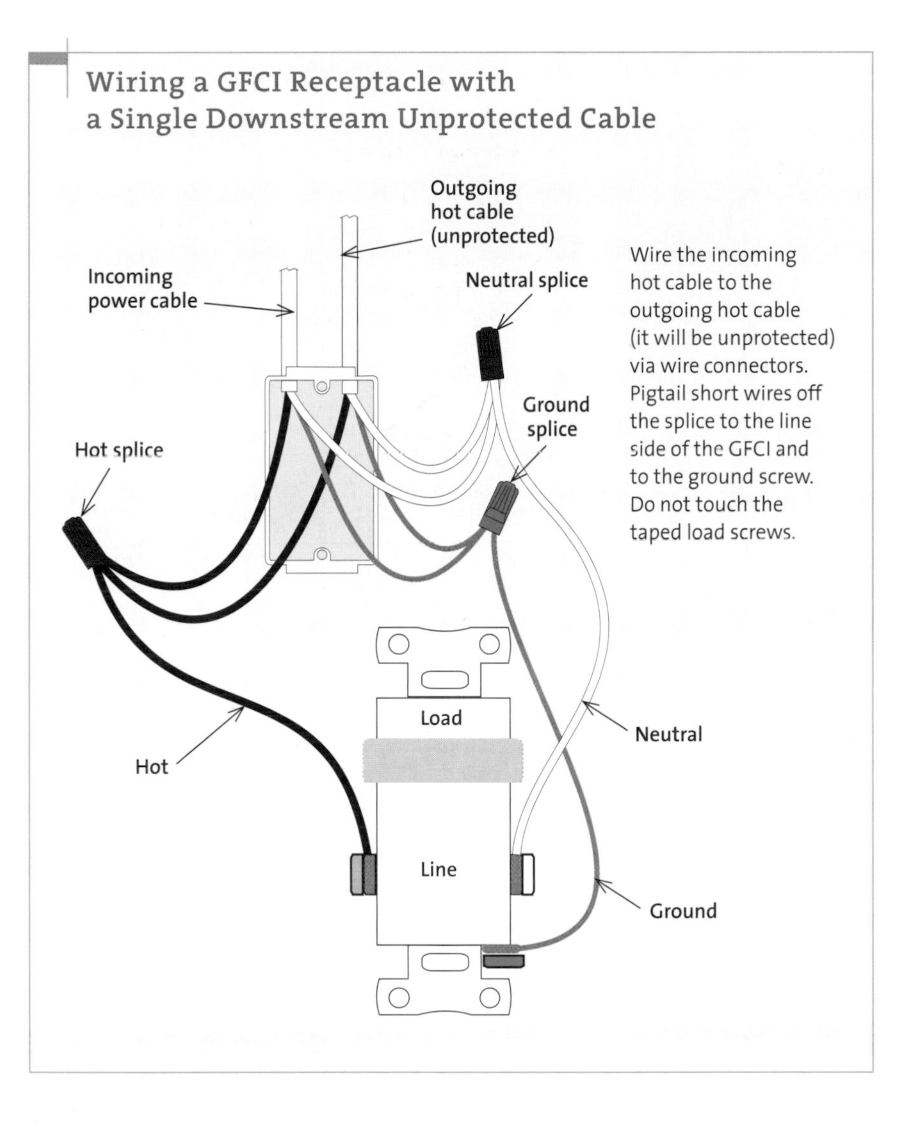

receptacle cover as "GFCI protected." Little stickers are normally provided with the GFCI receptacle for this purpose.

3. Connect the grounding wire. Though the grounding wire should always be connected if a grounding circuit exists, a GFCI doesn't need a ground wire to work. All it needs to cut off the circuit is a current imbalance from hot to neutral. For this reason, GFCIs can replace old two-wire receptacles.

Wiring GFCI circuit breakers

A GFCI circuit breaker protects wiring just like a standard breaker; that is, by tripping when the current exceeds the breaker's rating. It also provides GFCI protection, tripping when current is missing in the circuit. Before installing the breaker, make sure it matches the gauge of the branch's wiring: a 15-amp unit with 14-gauge wire, a 20-amp unit with 12-gauge wire.

Locate the breaker lettering to determine which side the black or white wire goes on before you install the breaker. Once installed, the lettering will be lost behind another breaker.

Can I Substitute?

Many times I have wanted a 15-amp GFCI receptacle, only to settle for a 20-amp because that is all they had. And sometimes vice versa. The question is, "Can I substitute one for the other?" The answer is, "Sometimes yes and sometimes no." If you have a 20-amp circuit with 12-gauge cable, you can use either. Reason being, a 15-amp load coming through a 15-amp receptacle can certainly be held by a 20-amp breaker. However, if you have a 15-amp breaker with 14-gauge cable, you certainly cannot put a 20-amp receptacle on that line. Reason being, the minute a 20-amp receptacle is installed on said circuit, it allows a 20-amp load, such as a window air conditioner, to be plugged in. And certainly a 15-amp breaker will not be able to hold it.

Wiring a GFCI breaker is fairly simple but still requires care, because the wiring can easily be reversed. A standard GFCI breaker has a white wire attached to it. This wire, normally coiled like a pig's tail, connects to the neutral bus and is the breaker's 0-volt reference.

However, some breakers simply have screw terminals for the neutral and hot wires. The circuit neutral conductor and the circuit hot conductor will each have a screw terminal on the breaker. Take a close look at the two screw terminals—the writing is hard to read. One reads neutral and receives the circuit's white wire; the other reads hot and receives the circuit's hot wire. Do not reverse these two wires! If reversed, the circuit may test normally using the GFCI's test button. It may trip using GFCI push-button testers on the downstream receptacles, but it may *not* provide ground-fault protection. The only way you will know there's a problem is by using a plug-in tester on one of the downstream receptacles to verify polarity.

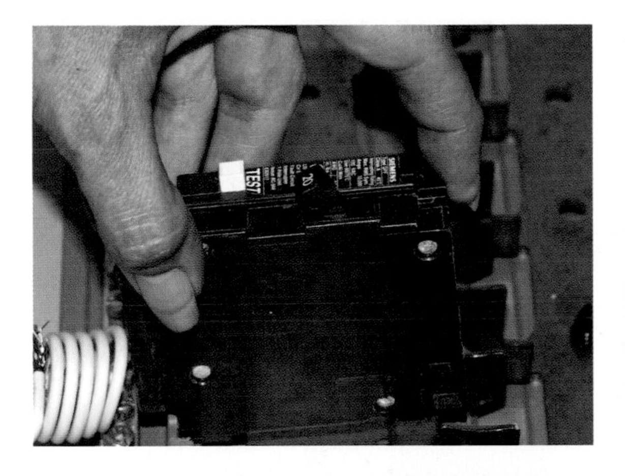

To install a GFCI breaker, turn off power to the panel, snap the breaker into the panel as any other breaker, attach the white coiled pigtail wire to the neutral bus (it must be by itself), and attach the white and black load wires to their respective screws on the breaker.

Testing GFCIs

All GFCIs should be tested at least once per month with their own test buttons. The test button places a current imbalance on the circuit to see if the GFCI will open the circuit. When a test button opens a circuit, a GFCI receptacle opens both the neutral and the line terminals. A GFCI breaker differs in that when it trips, it opens only the hot line—the neutral stays intact. If a breaker has been wired backward, the intact line is now the hot line and current can still go to the load—and to you. After a test, press the reset button on the receptacle to ready the GFCI. To reset a GFCI breaker, turn the breaker to full off and then on again.

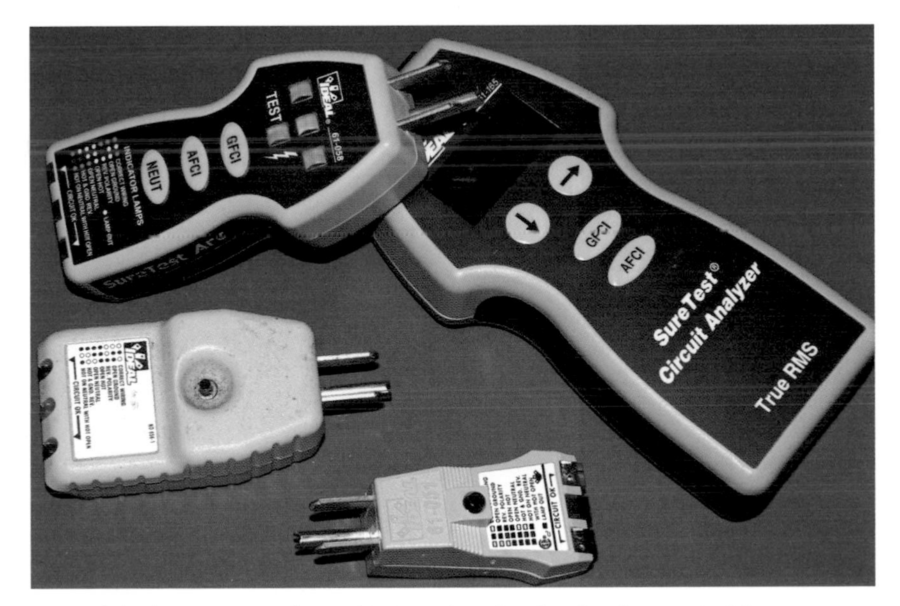

Ground-fault testers can be sophisticated or simple. Simple ones test for proper wiring, whereas sophisticated ones perform myriad tests and have digital readouts.

Take Note • GFCIs are not toys. I mention this because one of my customers had a child who loved to press the test and reset buttons on a GFCI receptacle I had installed. She liked to hear the click as the GFCI tripped off. She kept doing this time after time, day after day, click after click, until finally the button broke off and flew across the room.

GFCI testers

Many people used to (or still) rely on plug-in testers with a push button to test GFCI receptacles and breakers. In the early days, many GFCI testers weren't listed. It wasn't until 1993 that testing criteria for GFCI testers finally caught up with their design. Today almost all testers are listed to the 1993 standard. Thus if you are an old timer and have had your tester since day one, I suggest you purchase a new one.

Plug-in testers are fine, but for a more exact test of the absence of voltage you might want to use a voltage-measuring device. If your life depends on the absence of voltage, use a multimeter. A multimeter (not a plug-in tester or a touch-glow voltage tester) is also the only way to verify full presence or full absence of voltage on a circuit. If the voltage is only 70 volts due to a faulty device (caused by, say, internal corrosion), the plug-in tester's lights will still glow as though it had full voltage. The reverse is also true: It takes around 53 volts to light those lamps. Thus the checker could indicate "no voltage." However, as the prong of the plug-in tester is removed it could loosen the corrosion inside the outlet and the power could come back to full voltage and zap you.

If there isn't a test button or if you are testing receptacles at the distant end of a GFCI-breaker-protected circuit, you may need to use a plug-in tester. However, if the GFCI breaker doesn't trip, don't automatically condemn the breaker. Retest the unit, using the test button on the breaker itself. If it doesn't trip, then it's the breaker's fault; if it does trip, then the fault may be due to a wiring problem with that specific receptacle or with the in-wall wiring.

You may also use plug-in testers to test receptacles working off the load side of a GFCI receptacle. If the test button on the GFCI receptacle works but you cannot trip a distant receptacle, write it up as a potential wiring problem, but realize that the problem may lie with your tester. Low-cost checkers use the incoming voltage as a reference. If the voltage is low, the checker may be unable to leak enough current to trip the circuit. Better-quality plug-in checkers use a device that compensates for lower voltage.

Portable GFCIs

Portable GFCIs are necessary on job sites or when working in older houses that have no existing life-circuit protection. In other words, if the outlet you plug into is an ordinary outlet protected by an ordinary breaker that won't trip until it exceeds 20 amps, you are playing a dangerous game. It only takes a few thousandths of an amp to produce death.

The NEC recognizes this and made it a code requirement that contractors use ground-fault protection on job sites. And because your life at home is no less valuable than when you're at work, you need to run all your power tools off a protected outlet there as well. This doesn't always happen, however, because it's too much trouble to wire them in, or else you don't feel

Four different types of portable GFCIs. The inline unit (bottom left) I use as an extension of the tool's cord. The triplex has a rubberized end, (top right) whereas the square quad (top left) has a hard-to-break plastic end. Both the triplex and the quad have test and reset buttons in the plug. The plug-in unit (bottom right) is quite useful where you want ground-fault protection but you don't want to rewire anything. Here it is used at the clothes washer.

comfortable working with wiring or working in a hot breaker box. The solution is to use portable GFCIs, the ones that are integrated into extension cords. With these protected extension cords, there's no working with hot wiring, no working in a crowded receptacle box. Just plug in the cord and you have life protection and considerable piece of mind.

Underwriters Laboratories recognized years ago that extension cords need added safety protection. The problem is that a GFCI built into an extension cord (as opposed to protected inside a wall) will likely take significant physical abuse. It will be stepped on, pulled at, driven over, have heavy items dropped on

Receptacle (125-volt, 15-amp and 20-amp) Locations That Commonly Need Ground-Fault Protection (for more information see Chapter 6)

Location	Comment
Kitchen countertop	A protected receptacle is required at each counter wall space 12 in. or wider. No place along the counter can you go more than 24 in. without an outlet. Areas behind stove and sink are considered dead areas. Outlet to be no more than 18 in. above counter and cannot be face up. Exceptions can be made for the handicapped. Ask inspector.
Kitchen island or peninsula	GFCI outlet required if long distance is 24 in. or greater and the island short distance is 12 in. or greater. Outlet to be no more than 18 in. above counter and cannot be face up. Exceptions can be made for the handicapped. Ask inspector.
Bath	At least one protected receptacle needed within 3 ft. of sink lip. All such receptacles within 3 ft. of sink can be in cabinet (side or face but not facing up), provided they are not more than 12 in. below the countertop. Receptacles shall not be installed within or directly over the footprint of a tub or shower.
Utility, wet bar, and laundry sinks	Within 6 ft. of sink rim.
Outside the dwelling	Typically a dwelling needs a GFCI-protected outlet in the front and back no more than 6½ ft. above grade. Exception for a single dedicated receptacle close to the roof for deicing or snow melting.
Unfinished basement	All receptacles, no exception (2008 code change). See Crawl Space for possible problems.
Residential garage	No exceptions (2008 code change).
Outbuilding	At grade level
Crawl space	No exceptions (2008 code change). This one may cause significant problems in that many sump pumps have enough internal leakage to trip a GFCI the minute it comes out of the box.
Boat hoist in dwellings	Both outlet for and hardwired boat hoist. Common sense dictates that any 125-volt outlet at a pier or dock have ground-fault protection.
Vending machines	If you are affluent enough to have a vending machine outside, ground fault it and have the cover watertight while in use.
Farms	GFCI protection is required for outlets in damp or wet locations and in dirt confinement areas. Even trickle current has been shown to adversely affect farm animals—even to lower the amount of milk cows produce.
Storable pool	All electrical equipment and power supply cords used with storable pools and all 125-volt receptacles within 20 feet of the pool walls. Receptacles cannot be located within 10 feet of a storable pool.
Fountain	Receptacles within 20 ft. of a fountain. Common sense dictates the fountain's pump/motor have ground-fault protection, and any lights around it.
Balconies, decks, and porches over 20 sq. ft.	At least one 15- or 20-amp and 125-volt receptacle is required at no more than 6½ ft. overhead. Add GFCI protection if these receptacles may be used via extension cords to power outside power tools.

Note: **Some locales may differ as states adopt codes at different times and some delete certain sections.**

it, and will itself be dropped, often from great heights. With such wear and tear, it would be rather easy to have an open neutral, a hot wire touching the metal outlet box, and a broken GFCI—all on a long exposed cord waiting to shock someone. The GFCI cannot trip because, with the neutral open, the electrical power has been cut off. Bottom line: This is why you should not make your own GFCI extension cord by simply installing a cable and GFCI receptacle into a metal handy outlet box that you lug around outside. It's against code and, more important, very dangerous.

Protection within a corded GFCI is done via a relay that senses an open neutral and cuts all power immediately—both hot and neutral. This system works well. What is rarely discussed, though, is the "interruption" problem. As you plug and unplug cords or perhaps even bump the male plug at the relay, the GFCI will unlatch and cut the power. Of course, you don't know this until you are up the ladder and pull the trigger on the drill. To prevent this, use an "auto reset" portable GFCI. Many corded GFCIs come with both manual (you have to push the reset button if tripped) and automatic reset. The auto reset works after a voltage interruption (plugging and unplugging cords) or if the GFCI just gets bumped on the floor. However, you still have to manually reset the GFCI if it trips via a ground fault. Of course, if the fault is still on the line, such as a grounded tool lying on wet grass, it will not reset.

Being portable doesn't mean the GFCI always has to be on a cord. Several manufacturers make a plug-in GFCI for places like damp, unfinished basements. One of the easiest ways to get GFCI protection (say for a clothes washer on a damp floor) without having to work in a hot panel or receptacle outlet is to use a plug-in multiple-outlet assembly that incorporates GFCI protection (it will need a three-prong outlet). Beware, do not cut the ground lug off the portable GFCI just to plug it into an ungrounded receptacle, or use a cheater plug. Any grounded cord means the manufacturer requires that its appliance be grounded, which means it needs a grounded receptacle.

In summary, every pro and do-it-yourselfer needs a portable GFCI, preferably with an auto reset. Store it where you store your corded tools. When you pick up the tool, pick up the GFCI as well. Think of it as an integral part of the tool and use it every time you plug in the tool. Your family will appreciate it.

CHAPTER 9

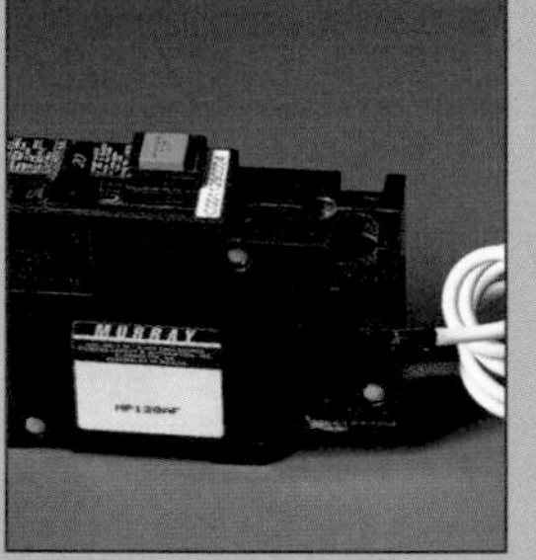

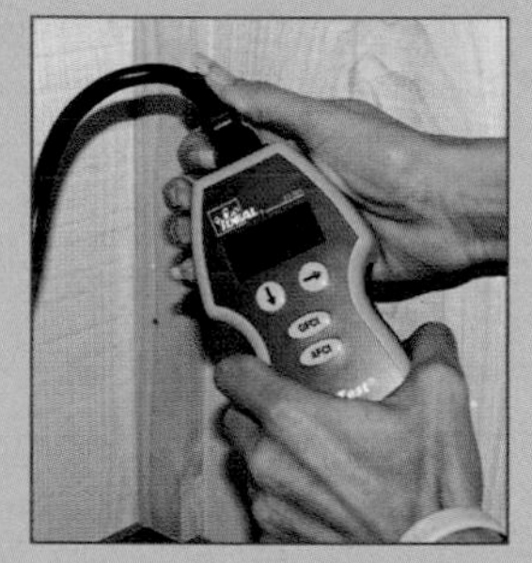

Arc-Fault Circuit Interrupters

Arc-fault circuit interrupters (AFCIs) are circuit-breaker-type devices that recently have come on the scene. They install in the service panel like a common GFCI breaker and are required for any common 120-volt 15- or 20-amp branch circuit outlets feeding any room in the household that isn't already being protected by GFCIs, typically living rooms, bedrooms, halls, sewing rooms, etc.

The purpose of an AFCI, other than functioning as a common breaker, is to detect series and parallel arcs that could start a fire in the house wiring and extension cords. A 30 mA to 50 mA (0.030-amp to 0.050-amp) ground-fault detection circuit was added to the breaker to make it more sensitive. Some AFCIs are dual listed as both AFCIs and GFCIs and will trip at around 5 mA. AFCIs trip faster than a common breaker, which hopefully will save lives (arcing and sparking are blamed for about 40,000 home fires and 350 deaths each year). In this chapter I will discuss how to install and wire AFCIs and the types of arcs they detect.

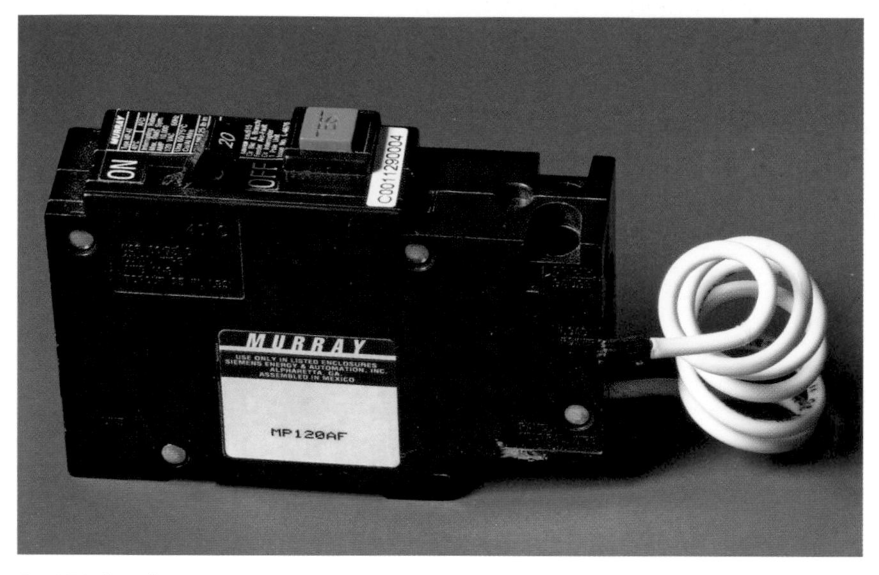

An AFCI breaker is required to protect non-GFCI-protected 125-volt 15- and 20-amp household receptacles on all new construction.

How It Started

In 1993, the Electronic Industries Alliance (EIA) proposed a change to the electrical code that would lower breaker trip values and therefore help prevent house fires. The proposal was rejected and the industry instead turned to arc-fault prevention. Later, several circuit breaker manufacturers approached Underwriters Laboratories with the idea of creating an electronic circuit breaker that could sense potentially damaging arcs within a wiring system.

At about the same time the Consumer Product Safety Commission contacted UL to study the new technology and better understand how it did its job of detecting arcs—good and bad. Both UL and the CPSC agreed that "arc fault technology, once fully developed, appeared to be very promising as a technology to significantly reduce the risk of fire beyond the scope of present conventional circuit breaker protection."

Then the competition began. Every manufacturer came out with their own design and AFCIs became a requirement of the National Electrical Code on January 1, 2002.

What Is an Arc Fault?

An arc fault is an unintentional electrical discharge characterized by low and erratic current, which can start a fire. There are three common types of arcs that may occur in your home:

- Hot to neutral: The most common example is a too-tight NM connector and a too-tight staple.
- Hot to ground: An arc between a wire and ground that isn't sustained long enough to open the circuit of a common breaker.
- Series: A wire that is cut but comes back in contact with itself through use or other movement. Extension and line cords are primary examples.

Regular circuit breakers don't always trip in these instances because it's not something they were designed to do. For example, most breakers rely on a certain sustained amount of heat in order to trip. If the surge is short enough, a tremendous current can flow through the overcurrent device (breaker) without tripping it. (This is why lightning surges of 10,000+ amps can flow through a breaker.) Arcs that are less than the breaker's overcurrent rating also occur. For example, an arc current of less than 20 amps may not trip the breaker, but it can start a fire.

Above Code • The first NEC code cycle to adopt AFCIs was in 2002, and only bedroom receptacles were required to be protected. The 2005 code cycle left out the word "receptacle."

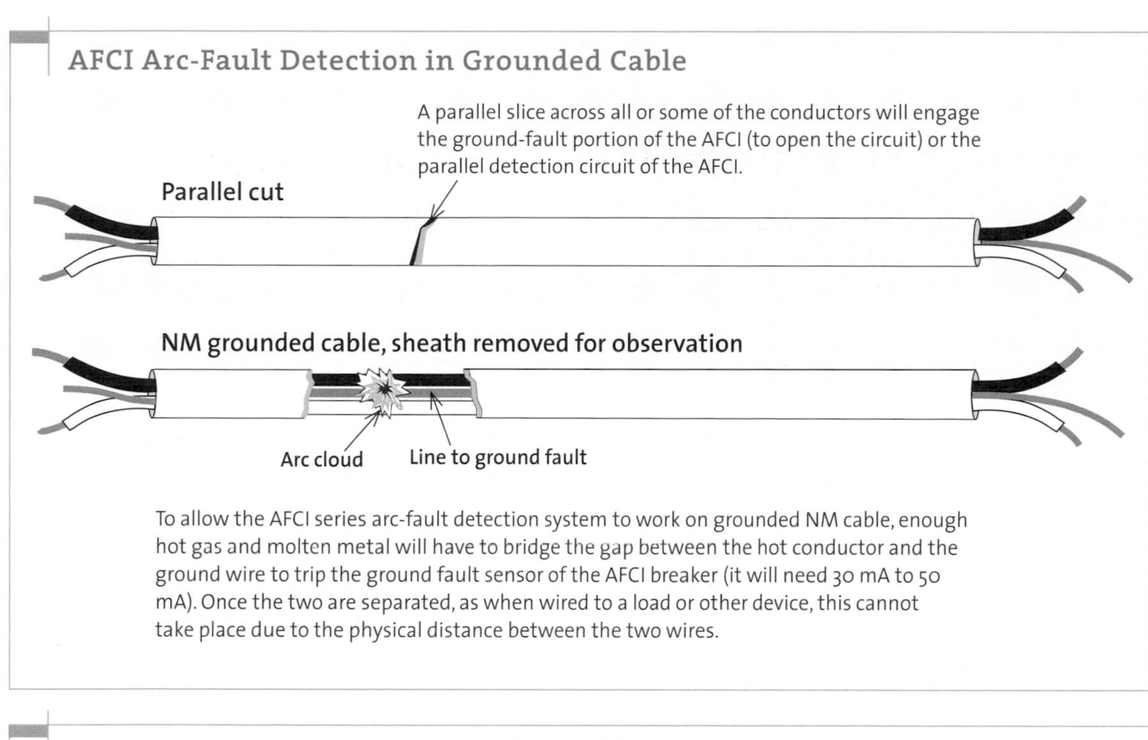
AFCI Arc-Fault Detection in Grounded Cable
A parallel slice across all or some of the conductors will engage the ground-fault portion of the AFCI (to open the circuit) or the parallel detection circuit of the AFCI.
Parallel cut
NM grounded cable, sheath removed for observation
Arc cloud
Line to ground fault
To allow the AFCI series arc-fault detection system to work on grounded NM cable, enough hot gas and molten metal will have to bridge the gap between the hot conductor and the ground wire to trip the ground fault sensor of the AFCI breaker (it will need 30 mA to 50 mA). Once the two are separated, as when wired to a load or other device, this cannot take place due to the physical distance between the two wires.

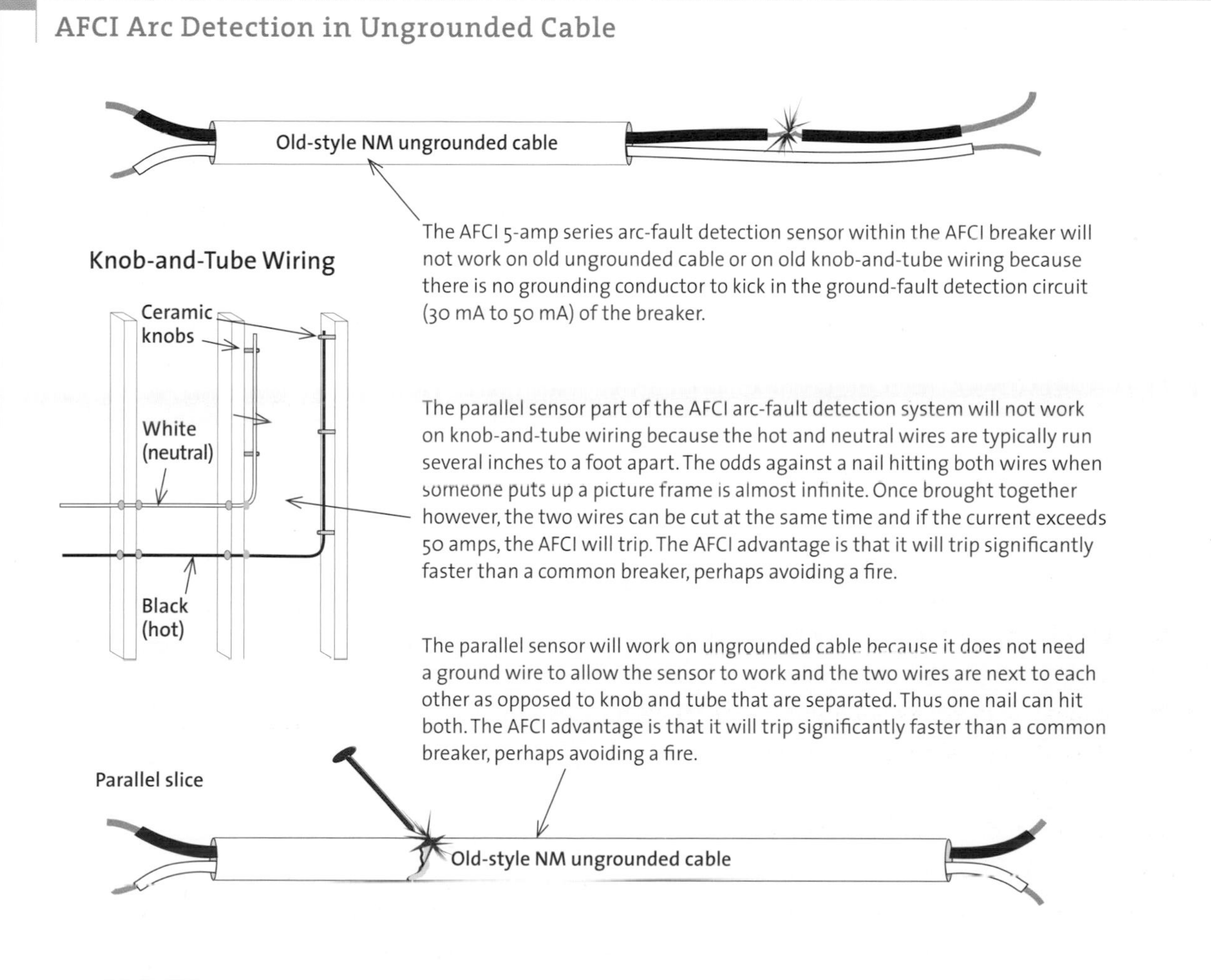
AFCI Arc Detection in Ungrounded Cable
Old-style NM ungrounded cable
The AFCI 5-amp series arc-fault detection sensor within the AFCI breaker will not work on old ungrounded cable or on old knob-and-tube wiring because there is no grounding conductor to kick in the ground-fault detection circuit (30 mA to 50 mA) of the breaker.
Knob-and-Tube Wiring
Ceramic knobs
White (neutral)
Black (hot)
The parallel sensor part of the AFCI arc-fault detection system will not work on knob-and-tube wiring because the hot and neutral wires are typically run several inches to a foot apart. The odds against a nail hitting both wires when someone puts up a picture frame is almost infinite. Once brought together however, the two wires can be cut at the same time and if the current exceeds 50 amps, the AFCI will trip. The AFCI advantage is that it will trip significantly faster than a common breaker, perhaps avoiding a fire.
The parallel sensor will work on ungrounded cable because it does not need a ground wire to allow the sensor to work and the two wires are next to each other as opposed to knob and tube that are separated. Thus one nail can hit both. The AFCI advantage is that it will trip significantly faster than a common breaker, perhaps avoiding a fire.
Parallel slice
Old-style NM ungrounded cable

WHY YOU NEED AFCIS

A wire connector has fallen off the neutral splice and the ground wire is pressing against bare neutral.

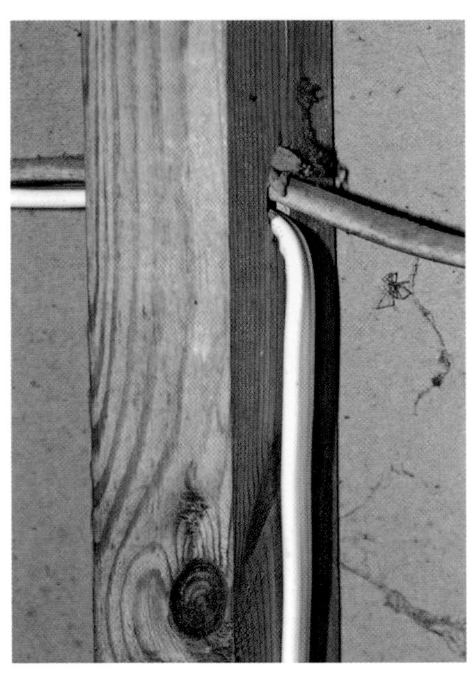

Pinch point. Never pull an immediate turn with a cable. It can heat up and arc.

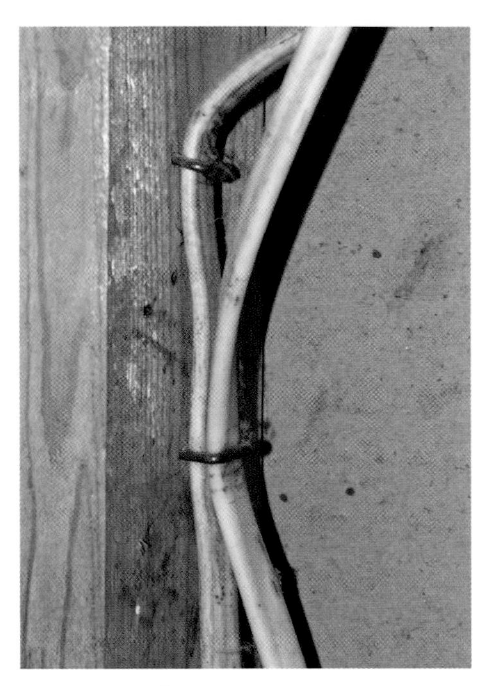

Hitting a staple one too many times can short out a single cable (top), and placing too many cables under a staple can do the same (bottom).

A panel entry without an NM clamp can cut or puncture a cable.

NM clamps tightening on the side of cable can damage the cable (top). A singular clamp if overtightened can damage cable (extremely common) (middle). Three cables (against code) in a NM clamp almost guarantees a damaged cable (bottom).

How AFCIs work

AFCIs sample the electrical current and use electronic circuitry to detect the waveforms produced by arc faults created by a nail hitting a cable inside a wall, a cable or cord overheating, a cable or cord being pinched, or a staple being sunk into a cable. A bad waveform is nonperiodic and nonrepetitive and has certain frequencies associated with it. The AFCI will sample all the data and open the circuit, if needed.

What makes AFCIs so controversial is that it is very hard to create a do-it-yourself field test to prove that they actually work. Because of this, some wonder if they will work as expected in actual practice. Just remember that AFCIs will not open the circuit for every arc in every situation. They will not eliminate all arc fires, as some advertise they will. They work for specific arcs in specific situations.

The series arc sensing is done indirectly through ground-fault technology. The series arc must send enough material across from the hot wire to the ground wire to create a 30 mA to 50 mA ground-fault current in order for the GFP sensor in the breaker to send the signal to open the breaker. Currently, without a ground wire, the series part of the sensing network does not work, though perhaps future generations of the AFCI will. The parallel cut is simply like a knife cutting the cable. Here, any circuit breaker will work, though the AFCI will act faster to hopefully minimize the chance of fire.

Thus, for ungrounded wiring in old houses, the series 30-mA ground-fault sensor (line-to-ground) is almost nonfunctional on the cable run because there is no equipment-grounding conductor running with the cable. However, it will work if the hot or neutral conductor gives a fault against a grounded armored cable or in wiring outside the wall whose conductors are adjacent to each other.

Take Note • Do not confuse AFCI combos with AFCI/GFCIs. The Combos are AFCIs with GF sensing at around 20-40 ma. True AFCI/GFCI units have ground fault sensing at around 5 ma and are quite expensive. Thus to stay out of trouble, use AFCI combos for branch circuits requiring AFCI (fire) protection and GFCIs for circuits requiring life protection.

Wiring AFCIs

AFCIs physically snap into the panel like any other breaker. In fact, they look like GFCIs (right down to the pigtail wire) and can be confused with them if you don't read the fine print on the breaker's sticker. To help prevent a mix-up, some manufacturers produce AFCIs with different color test buttons than GFCIs, which have white test buttons.

It's easy to get confused about where the black and white wires terminate on the breaker. Before you insert the breaker, look at the stamped plastic on both sides to determine which wire goes where. One side says "white" or "neutral," the other side says "black" or "hot" or perhaps "load." Once you insert the breaker between other breakers, you can no longer see these markings. To install, simply snap the breaker into the panel like any other breaker and connect its white pigtail to the grounded/neutral bus in the main service panel (just like a GFCI breaker). Connect the black wire and the white wire to their respective screws. Typically this takes a very narrow blade screwdriver.

How Many AFCIs Do You Need?

Codes now require almost all rooms in the house to use AFCIs except the garage, kitchen, laundry, and bath. Expect problems—including skyrocketing costs and occasional loss of power. The latter could crash computers and cut off people on any kind of life support, and, of course, all the gamers and TV watchers will lose power and start yelling. And there would be no lights and no hardwired smoke alarms.

Instead of using one AFCI breaker for all required rooms, consider having one AFCI feed only two required rooms. This would limit the inconvenience and aid troubleshooting in the event of a problem. Ideally, there should be one AFCI breaker per required room (Above Code), which would keep the problem contained to that specific room which would ease troubleshooting. The problem, of course, is the tremendous expense of the numerous AFCIs.

An AFCI works the same whether a white pigtail is left long and coiled (as it comes out of the box) or cut short.

Testing AFCIs

Admit it. All of us electrical guys have a fascination with testers. We test for this and we test for that. We just love to test. And here comes AFCIs and we start looking for a tester for them as well.

The AFCI breaker can be tested with its own test button, but it would be nice if we could test the AFCI by plugging a tester into the most distant receptacle. This would verify that we have a continuous circuit from the last outlet back to the AFCI breaker. In other words, we want to be able to test an AFCI branch circuit just as we do a GFCI branch circuit.

Circuit analyzers do more than a simple tester or multimeter because of what they can show about a circuit. The Ideal SureTest Circuit Analyzer (61-165), for example, can test an arc-fault breaker for a parallel arc (line to neutral) as well as perform a 30-mA ground-fault-leakage test. In addition it will tell you the amount of current it took to trip the breaker and the time it took to trip. If wanted, it will give you the ground resistance it sees on any of the three conductors back to the main panel. Add a common GFCI test as well as a variety of other tests and you have a circuit analyzer.

One very handy test a circuit analyzer can do (and a low-cost plug-in checker can't do) is to test for a bootleg ground. This is when there is no ground at the outlet but the installer ran a jumper from the neutral to the ground screw

Take Note • Terminate the white neutral wire of the AFCI on the bus by itself—one wire per screw.

GFCIs and AFCIs Are Not the Same

Although they may look similar, GFCIs and AFCIs are not the same. GFCIs are designed to sense small amounts of leaked current from the hot wire to ground (ground fault). They are not designed to sense arc faults. Warning! I do not advise installing AFCIs on circuits, even bedroom circuits, that power life-sustaining equipment. However, my advice goes against code so you might need to get an exemption.

behind the outlet to fake the reading on a common plug-in tester. A circuit analyzer will display "FG" for false ground. For AFCI breakers, its internal electronics will read the connection between ground and neutral and pop the breaker. If you read an FG at an AFCI protected outlet and the AFCI breaker has not opened the circuit, either the AFCI or the wiring to it is bad. Remember, however, that the short from neutral to ground could be accidental as well—such as a bare neutral wire touching the side of a metal box—and the AFCI breaker will kick the circuit open. The analyzer will also check for shared neutrals, which is when the neutral is hitting a grounded metal box or a bare neutral conductor, or is simply tied into the neutral of another circuit.

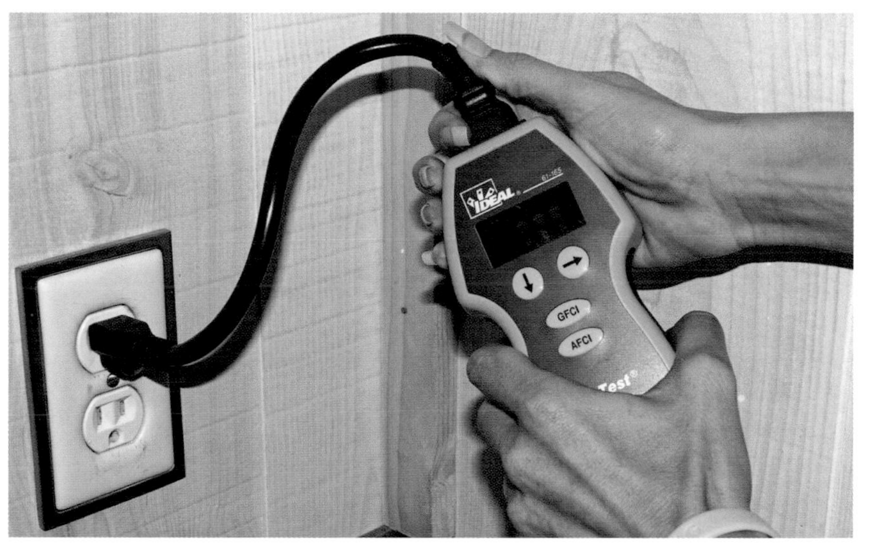

This circuit analyzer can do an outlet check, an AFCI parallel arc-fault test, a shared neutral test, an AFCI 30-mA ground-fault leakage test, a GFCI test, a false-ground check, and myriad other tests and measurements via a display.

Troubleshooting an AFCI Circuit

Well, it happened. You put the AFCI breaker on a bedroom circuit in accordance with code (and to feel more safe and secure) and the darn thing keeps tripping. The problem is that an unlimited number of rooms are allowed on an AFCI circuit, which means an unlimited number of receptacles, lights, smoke alarms, fans, window air conditioners, sconces, can lights, and so on. Troubleshooting an AFCI breaker can become a screaming nightmare.

But all is not lost. Following a logical troubleshooting procedure will allow you to find

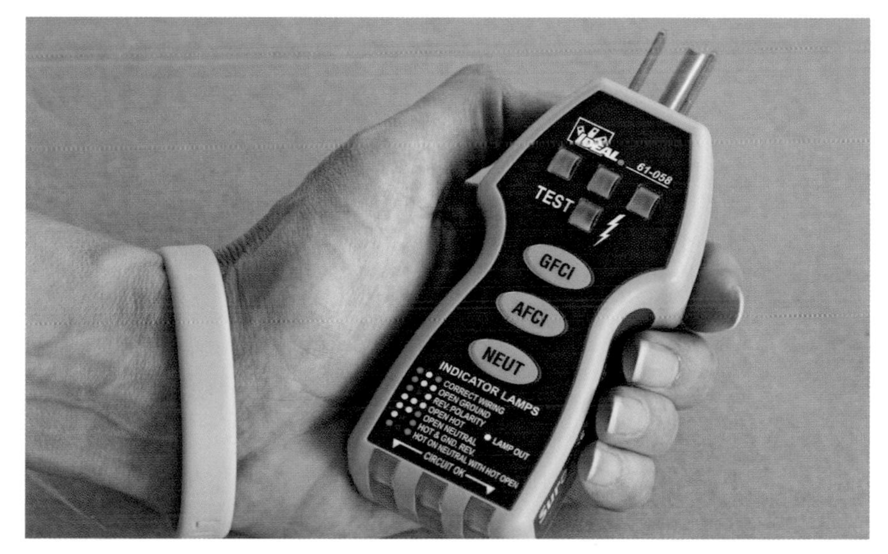

This plug-in tester can do an outlet check, an AFCI parallel arc-fault test, a shared neutral test, and a GFCI test, but it cannot do a 30 mA ground-fault leakage test for AFCIs and cannot check for false grounds (it has no display).

AFCIs and Smoke Alarms

The controversy around AFCIs doesn't stop with how they work or how they are tested. It continues on to what they protect. The problem here is that if there is a fire, say in an adjoining room, as soon as the AFCI wiring is involved and the AFCI kicks, power is removed from the smoke alarms.

Thus, if there is a fire in an adjacent room that has no smoke detector and it burns into the AFCI wiring, the AFCI will remove the power from the smoke detectors and there may be no alarm if its batteries are dead. This may be why some jurisdictions did not adopt the AFCI requirements, and others that did removed the requirement to include smoke detectors on the AFCI circuit. Bottom line: Verify that all your smoke detectors have working batteries.

the problem. The following is a list of things to do and to ask yourself as you attempt to narrow the location of the problem from anywhere to an exact spot.

- Check the breaker first. Many have been known to die an early death, or even to be bad right out of the box. There have already been AFCI recalls and at least one manufacturer seems to have quality-control problems. So turn off the breaker and remove the hot and neutral wires. Then turn the breaker back on. If it kicks back off, you have a defective AFCI breaker.
- Were you running anything when the AFCI tripped? If so, look to that appliance for the problem. It might be faulty or just noisy. Or the load might have been too large for the circuit. Exceeding 15 amps on a 15-amp breaker and 20 amps on a 20-amp breaker will open the circuit. For long-duration loads like a plug-in heater, you should not exceed 80 percent of the circuit trip value. If it isn't the appliance, look to the outlet it's plugged into.
- Have you recently drilled or nailed into the wall? For example, have you hung a picture, screwed in a shelf, or put up curtain rod supports? If so, you might have hit a wire with a nail or screw.
- Is there an extension cord plugged into an outlet? It could be pinched, overheated, or have a bad end. Unplug the cord and see if the breaker still trips. If there is a GFCI on the AFCI circuit, it could be tripping the breaker.
- Are there any outside outlets on the AFCI circuit? These collect moisture, corrode, and can easily trip an AFCI breaker.
- Do you have loose receptacle connections? That is, do you have receptacles where the plugs fall out? This is an indicator that the receptacles are old or just cheap. It is possible that the wire attachments to the receptacle in its back are loose and causing an arc that was received by the AFCI breaker, which in turn opened the circuit. Receptacles that have holes in the back for connections are known for loose connections.
- Have you recently installed anything new, such as dimmers? These have been known to produce enough noise

as to give a false trip to the AFCI. Staying at or below 1,000 watts of total dimmed lamps per AFCI circuit should result in nuisance-free performance. Additional loads can be on the circuit, as long as they're not dimmed loads.

- Is there a sump pump on the AFCI circuit? If the seals break down it will allow leakage current, which will trip the AFCI and remove the power on the pump circuit.
- Do you have a multiple-outlet assembly plugged into a receptacle? If so, unplug it.
- Unplug all appliances that are plugged into a receptacle and turn off all wall switches. Turn the AFCI breaker off and back on. If it does not trip, the problem is in an appliance or a light. Next, turn on all the wall switches one by one. If the breaker kicks, the problem is in the fixture wiring of that particular wall switch. If the breaker does not kick, look for a problem in an appliance. Plug them in one by one until the breaker trips to indicate the bad appliance.

Advanced troubleshooting

Sometimes the trouble is with the wiring. The AFCI can be fine, but it won't work if somebody damaged the cable. To troubleshoot this, first look closely at the NM connector that contains the cable at the service panel. If it's too tight it may be cutting into the cable and tripping the breaker, which is a common occurrence.

Another common problem I have is with somebody new on the job stapling the NM cable so tightly that it cuts across the wire and shorts out the cable. To solve this I suggest using insulated staples and giving the new installer a short course beforehand.

Something else you may find is that either the neutral is touching the ground wire in one of the electrical boxes or the neutrals are mixed. With two different circuits in the box, you should not have combined neutrals. Either will make the return neutral current different into that circuit than the hot current that left, resulting in an AFCI trip.

Don't use single-pole AFCIs on shared neutral circuits, because the current flowing out and the current flowing back is monitored and must be equal. The AFCI won't know the difference between sharing the neutral current and a high-resistance ground fault that is leaking away the current. You can use double-pole, shared-trip (the little arm that connects two side-by-side breakers) AFCIs. However, if there is any installation mistake and a neutral is lost, accidental or otherwise, major problems, including a fire, can ensue.

Let's say a neutral is dropped and some 120-volt appliances receive 240 volts. You'll hear explosions like firecrackers going off one after another. Some appliances could even catch fire. It's not the AFCIs that create the problem but the use of dual-pole shared neutral circuits and a dropped neutral. Bottom line: Do not use shared neutral circuits on AFCI wiring (or any other wiring for that matter). If the breaker trips, it will be time consuming and very expensive to figure out where the problem is.

CHAPTER

10

Receptacles and Boxes

You're about to learn all the top-secret information about receptacles, boxes, and wiring that contractors rarely discuss. You're putting your ear to the keyhole and listening as we discuss what to buy and what to avoid, what looks good and what really is good, and what works and what doesn't—before you install a receptacle, not after. In this chapter, I discuss standard receptacles seen throughout the home.

Receptacle Boxes

One problem many electricians face is not thinking ahead and having the right box for the job. When buying an outlet box, consider whether it is deep or shallow, metallic or nonmetallic, and a nail-on (also called captive or integral) or a bracket-mount type. Each one has advantages and disadvantages. Not knowing which receptacle box you need leads to many problems, and one of the most common is using a box that's too small.

Outlet boxes come in different depths (higher cubic in. capacity). Always go for the deepest, as they will allow more room for wires, connectors, and receptacle body.

Volume problems

Most receptacle boxes have one common problem: They don't have enough room or volume inside the box to contain both the receptacle and the splicing for the incoming and outgoing cables. If the box is too small and doesn't have enough volume, the wires become overcrowded, resulting in broken and shorted wires and possible damage to the box's receptacle-holding threads because everything you shoved in is pushing to get out. To prevent this from happening, the NEC regulates the amount of wires and devices that can go into a receptacle box.

The regulation assigns a limit to the number of conductors each box can carry (if you stuff in too many cables, it is called a cable-fill violation). Nonmetallic boxes come in 16 cu. in., 18 cu. in., 20.3 cu. in., and 22.5 cu. in., and the volume is stamped inside the box. To determine the volume of a metal box, check the manufacturer's catalog or use an NEC chart.

Above Code • Rather than go through meticulous calculations each time you want to put a cable in a box, use the deepest boxes available. A single-gang box (the box used almost exclusively to house receptacles) should be 3½ in. deep (22.5 cu. in.), the same depth as a common stud wall. Do not run more than three 12-gauge cables or four 14-gauge cables in this box.

Sometimes it is physically impossible to install a deep box. A typical example is when a carpenter furs out a basement wall by laying the studs flat against the concrete or block wall. This makes the wall cavity only 1½ in. deep. Because you can't use deep boxes unless you knock holes in the wall for every receptacle, you must go sideways to get the volume. Don't be tempted to use the commonly available metal handy boxes just because they fit; they don't have the volume for in-and-out cabling. In this case, use a nonmetallic 1½-in.-deep, 4-in.-square box, which has an additional front plate that's raised to hold the receptacle. Install the cover plate before the finished wall goes up so that the finished wall can be cut around the raised area.

Metal boxes

Once the standard of residential construction, metal boxes are now used less frequently. Nonmetallic boxes are more popular because they are nonconductive, less expensive, and faster to install. However, metal boxes do have three significant advantages over nonmetallic boxes: They are strong, have more available designs, and can often be stacked. Stacking allows the rectangular switch or receptacle box to increase in size to single, double, or triple. In addition, metal boxes are still needed where

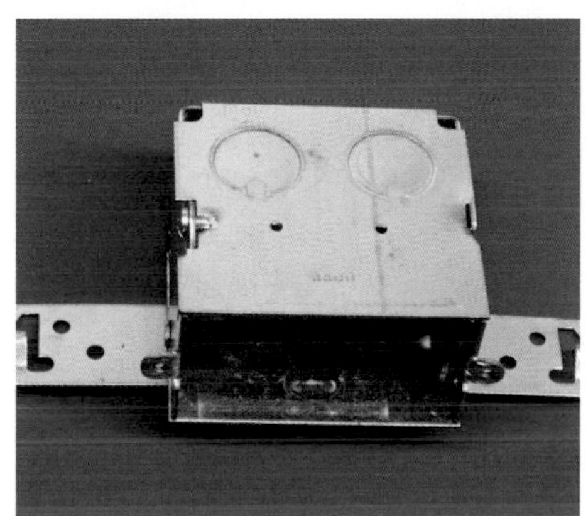

Metal boxes come in myriad sizes and types—more so than nonmetallic. They are stronger than plastic but are conductive and will require grounding. The latter means that if you short a hot screw onto the box, the breaker will kick—not so with plastic. As a precaution, when dealing with metal boxes I tend to cover the hot and neutral screws with electrical tape (see pp. 210-211).

Old-work boxes are available in metal as well as plastic. However, metal boxes need to be grounded (and thus require more labor to install) and can cut insulation if they are overcrowded in the box.

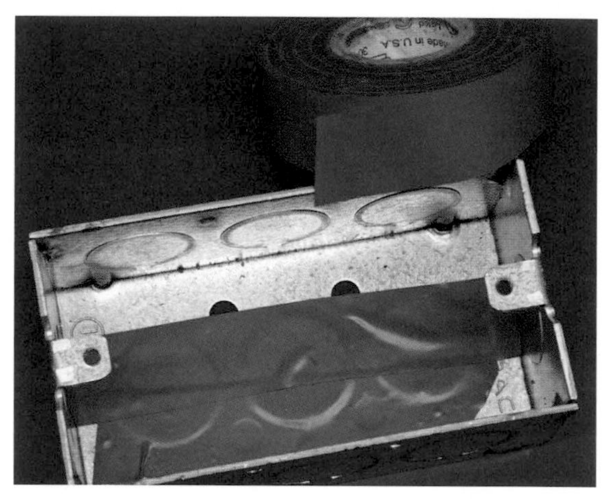

For any metal handy box or small-volume metal outlet box it is wise to put electrical tape against the back of the box to keep the insulated wires from being cut by the metal of the box as they are shoved in.

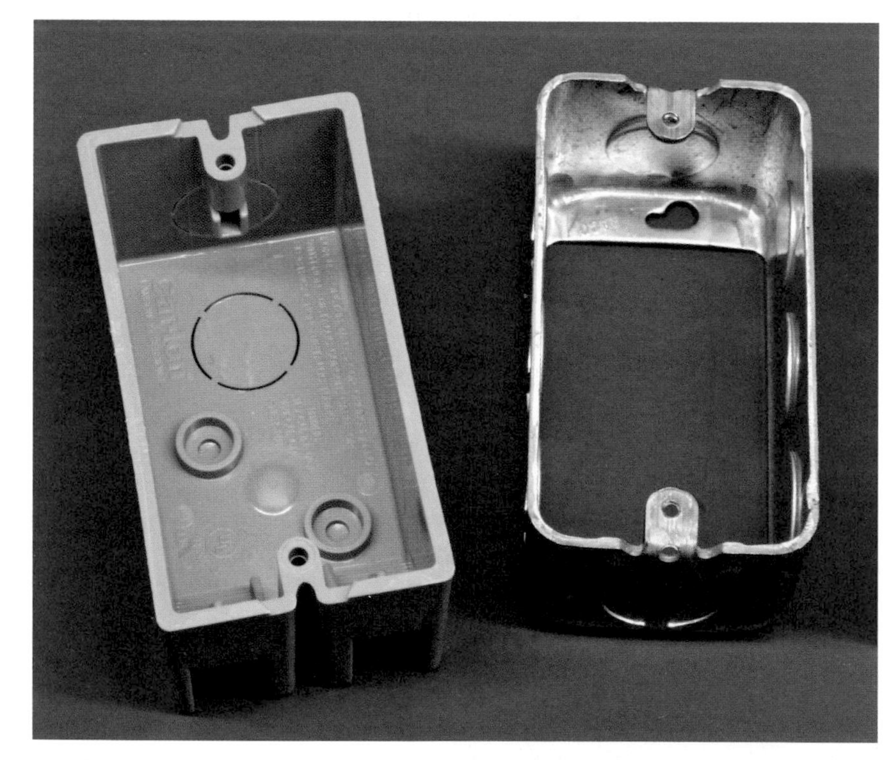

Handy boxes, metal or plastic, are the most misused of all outlet boxes. Because of their small volume you are only allowed a single cable within the box. A box with a hollow back can be bolted to a handy box with a solid back. This doubles the amount of splicing room in the box and allows plenty of room for a second cable.

the entire box is exposed and open to damage, such as along a basement or garage block wall, or in a situation where the required design is only available in metal.

Handy, or utility, boxes are the most misused of all the metal boxes. The common rounded-corner, 1½-in.-deep box is only allowed one 14-gauge cable with switch or receptacle—anything more is a cable-fill violation. However, I don't know of any electrician, including me, who hasn't installed one or two 12-gauge cables with receptacle in a handy box. And when the cables were pushed in and power applied—bang. When volume is needed in metal boxes, 4-in.-square boxes or larger are the best to have around. To keep the insulated wires from being cut by the back of the box, consider insulating the back of the box with electrical tape.

When using a pancake box, make sure that you do not create a cable-fill violation by putting anything in it other than one cable. Some

Metal Boxes Are Conductive

Metal boxes were once the mainstay of the trade. There are those who still refuse to use anything else. But let's face it, when using a metal box, there is always a risk of shorting the receptacle screws against the sides or having insulated conductors cut by the metal as the receptacle is pressed back into the box. This is extremely common when a GFCI replaces a smaller, standard receptacle. Many electricians cross their fingers and look the other way as they shove in the GFCI. (One of my helpers blew out an entire circuit doing this.) And you can no longer get by having the neutral short against the metal box; that will now kick an AFCI. To prevent the screws from shorting onto the sides of the metal, always wrap a few layers of electrical tape (Scotch Super 88) around the receptacle to cover the hot and neutral screws. Even better, use nonmetal boxes when possible.

There is very little gap between the receptacle screws and the side of a metal box. Covering the receptacle screws with a couple layers of electrical tape prevents them from shorting onto the sides of the metal box.

An alternative to taping over the screws is this Cooper® receptacle with the flip cover over the screws. Great idea—long overdue.

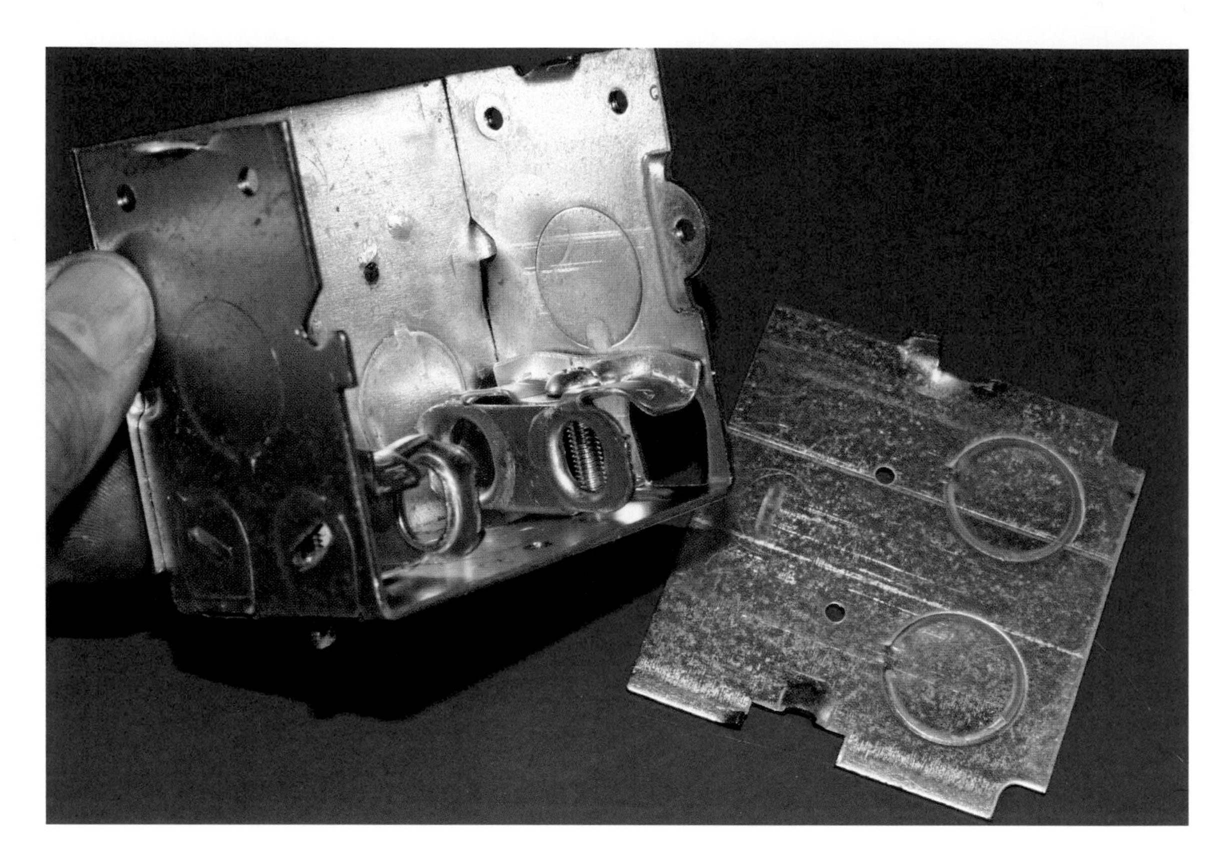

One advantage some metal boxes have over nonmetallic is that their sides come off and they can be ganged together to increase their size (via boxes the same size and type). This box will terminate metallic clad (MC) cable.

lights have round bases with such little depth that they won't fit over the ½-in. pancake; in that case, you must recess the box in the wall.

One significant advantage metal boxes have over plastic is that some can convert from a single gang box to an unlimited number of gangs by the simple removal of a screw.

Above Code • Unless you have a specific need that requires a metal box, stick with plastic ones. The one exception to this rule is overhead paddle fans. Although some nonmetallic boxes are approved for paddle fans, I don't trust them yet. In my opinion, there is simply no way a plastic box can be as secure as a metal one. I'm always afraid the movement will strip out the plastic female threads and the fan will fall.

Nonmetallic boxes

Nonmetallic boxes are the most common receptacle boxes used in residential construction. They are both inexpensive and nonconductive. Although materials vary, the most common are flexible PVC plastic, rigid plastic, and fiberglass. These boxes are normally used throughout the house; however, the common rectangular box, such as the integral-nail box, should not be used if the entire box has to be surface-mounted in a habitable area. In addition, the common rectangular, nonmetallic box should not be used for hanging lights. Boxes designed for hanging lights are listed as such and have a larger attachment screw than standard boxes do.

Receptacle boxes differ from switch boxes in that they rarely have more than a double gang (two duplex receptacles in a box). Switches commonly use three- and four-gang boxes for all the switching required at entryways.

One too many hammer hits will warp a thin plastic box.

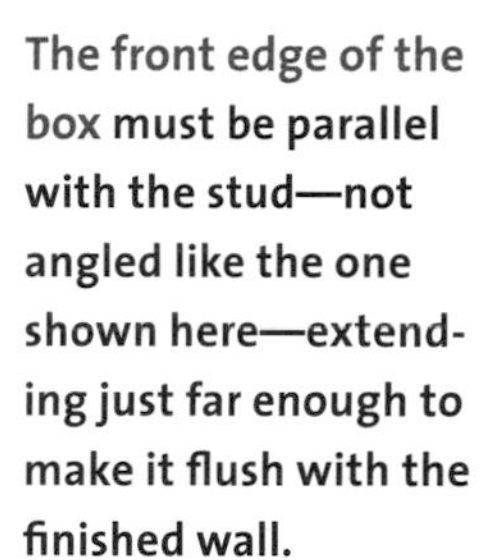

The front edge of the box must be parallel with the stud—not angled like the one shown here—extending just far enough to make it flush with the finished wall.

Fiberglass boxes are sturdy but can easily be destroyed by one misplaced hammer blow.

The most commonly used receptacle is a single-gang, nail-on box. For this type of box, simply nail the box to the stud, leaving just enough sticking out to be flush with the finished wall. Three challenges occur here: getting the distance out from the stud exactly right, keeping the box parallel to the stud, and keeping the box from being distorted or broken as it is nailed onto the stud. To prevent the first two, take your time. Some boxes even have raised edges or ribs to help you judge the distance out from the stud. In addition, you can use a piece of ½-in. plywood as a spacer to help get the distance right. Hold the plywood to the stud and align the edge of the box with the plywood.

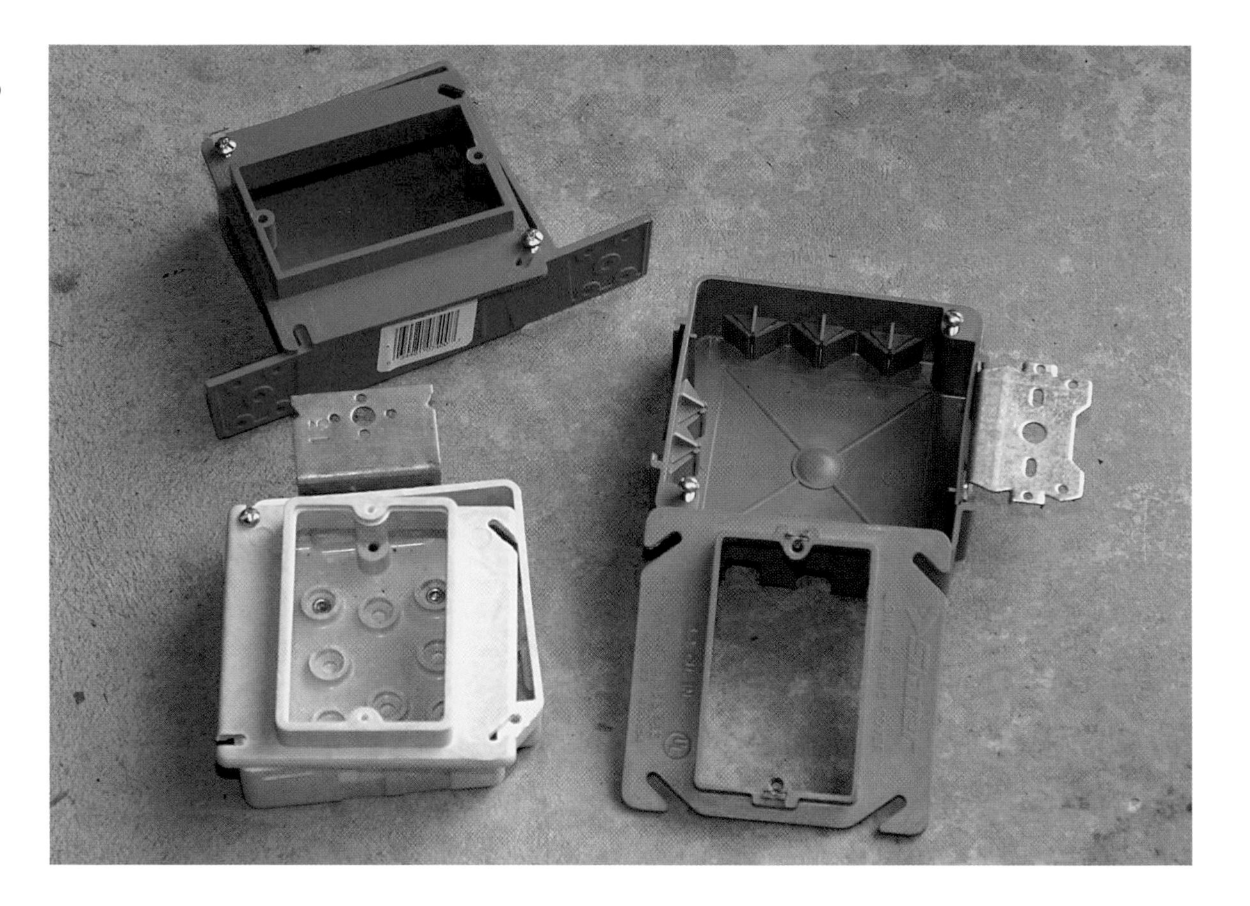

Use a 4-in.-square box for shallow, 1½-in.-deep walls. The extra width provides the volume needed to hold all the wires. The receptacle mounts to the cover plate, which in turn screws to the box.

To prevent breaking the box as you nail it in, simply tap the nails—don't drive them hard as the nail head comes close to the box. The more precisely you get the box mounted on the stud, the less trouble you'll have making the receptacle fit against the finished wall. If you don't want to be bothered by constantly solving these problems, install the slightly more expensive side-mount box. This box has a bracket mounted to its side that nails to the front of the stud, placing the box at a prescribed distance (normally ½ in.) out from the stud and keeping its face parallel to the stud. This box will not work with thin wood-panel walls.

To finish off a basement, it is common to frame out by placing 2×4s flat against the wall. Installing them flat may increase the usable square footage of the room, but it does not allow much room for ordinary outlet boxes. Instead, you have to use large 4-in.-square by 1½-in.-deep nonmetallic boxes. Though shallow, the large footprint is big enough to accommodate both outlets and wires. Be sure to attach the box with drywall nails so the drywall can fit flush to the wall adjacent to the box.

Cut-in boxes

To install receptacle boxes in existing hollow walls, special boxes called old-work or cut-in boxes are used. These boxes do not mount to studs; instead, they attach directly to the finished wall, using plastic ears in front and

Cut-In Receptacle Boxes

Cut-in boxes are used mostly in remodeling. They attach to the face of the finished wall with drywall ears on the front. The support mechanism on the back sandwiches the box to the wall.

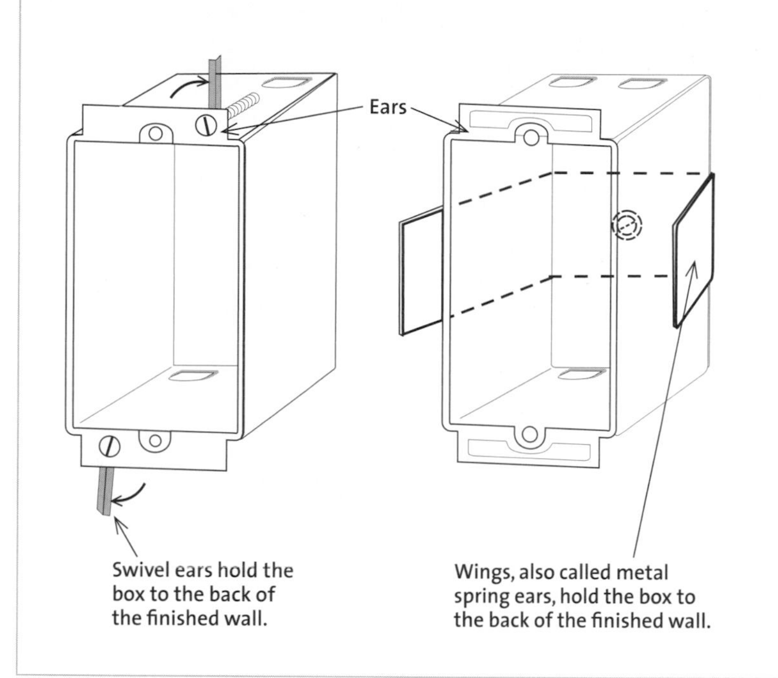

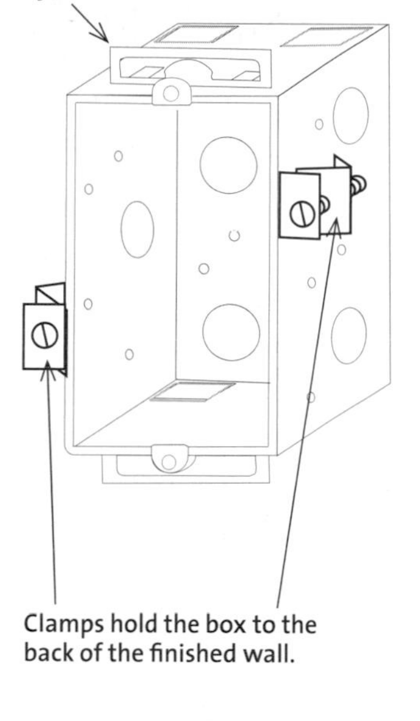

supports (either swivel ears or wings, also called spring ears or clamps) in back. The plastic ears hold to the drywall surface and the supports in back sandwich the drywall to hold the box in place. In addition, you can make any box with drywall ears work like a cut-in box, as long as you have an adapter to do so. Remember, if you're installing a cut-in box in a wall with a receptacle directly on the other side, offset the new box a bit so that the boxes are not back-to-back (usually there's not enough room in the wall cavity to do this). Be wary if you install a cut-in box in a wall that separates the house from the garage. Once you cut into the wall you have opened a firewall—meaning that you have just created an opening for a garage fire to enter the house.

Cut-In Adapter

Any box with drywall ears can be used as a cut-in box with this adapter, which is simply two metal straps that fold around the box to support it from behind the finished wall.

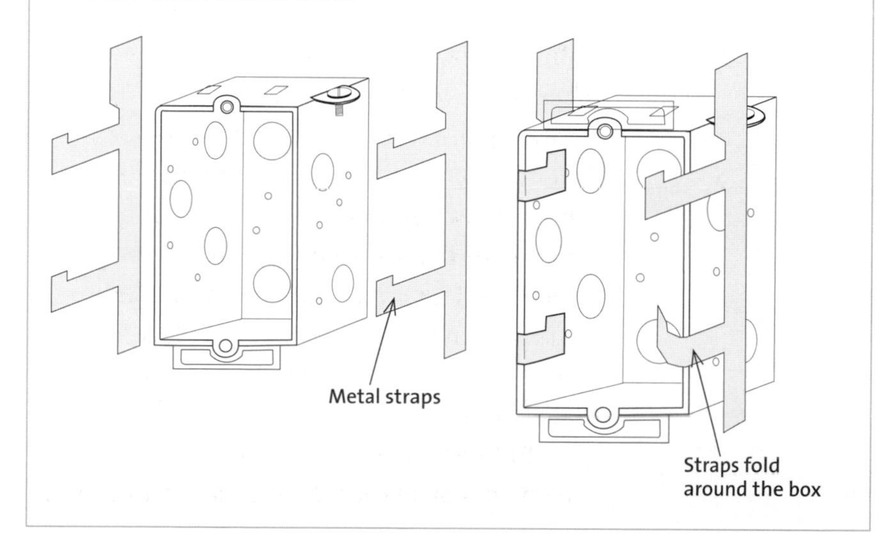

Cut-in boxes allow for few errors during installation. The attachment system on the cut-in box will only work if the hole is small enough for the box's plaster ears to press onto the finished wall front. If the electrician is sloppy and cuts the box hole too big, the box will simply fall through the hole. A good electrician will cut the hole exactly to fit the box. Some manufacturers furnish templates to aid in cutting out the hole. However, with other manufacturers, you're on your own.

If you accidentally cut a hole that's too big for the box, cut a thin cedar shim (the kind used to shim doors and windows) 2 in. to 3 in. wider than the hole. Put some glue on the shim, slip it into the hole, and pull it back against the inside of the finished wall. Once this is dry and secure, cut a piece of drywall the exact size needed to make the opening the correct size and glue it on top of the stick.

A typical spring-loaded watertight cover is no longer code for outside use. While in use, the cover is open and does not provide weather protection for the receptacle or plug.

Weatherproof boxes and covers

Made from both plastic and cast metal, weatherproof boxes are used outside the house when the entire receptacle box will be exposed. Because of aesthetics, new construction locates most receptacle boxes for outside outlets inside the exterior walls. This also allows contractors to use less expensive standard boxes with weatherproof covers. When adding an outlet to a finished outside wall, a weatherproof box is normally attached to the outside siding. Even though the box is weatherproof, the receptacle still needs to be protected with a weatherproof cover.

There are two types of weatherproof covers: weatherproof while the cover is closed and weatherproof while the receptacle is being used. These fit onto the front of any type of box in lieu of a common cover plate. Most of us are familiar with the metal spring-loaded caps that snap down over the receptacle or GFCI outlet—these are watertight only when the cap is closed and are no longer code-compliant—a weatherproof-while-in-use cover must be installed. Code-approved covers come in two designs: a bubble type that extends off the receptacle and wall (to house the cord plug while it's inserted into the receptacle) and a type that extends into the wall to allow the cover to be flush with the exterior wall (made by TayMac®). The latter is better (considerably less ugly), but it takes a 2x6 stud wall to house both the box and the recessed cover.

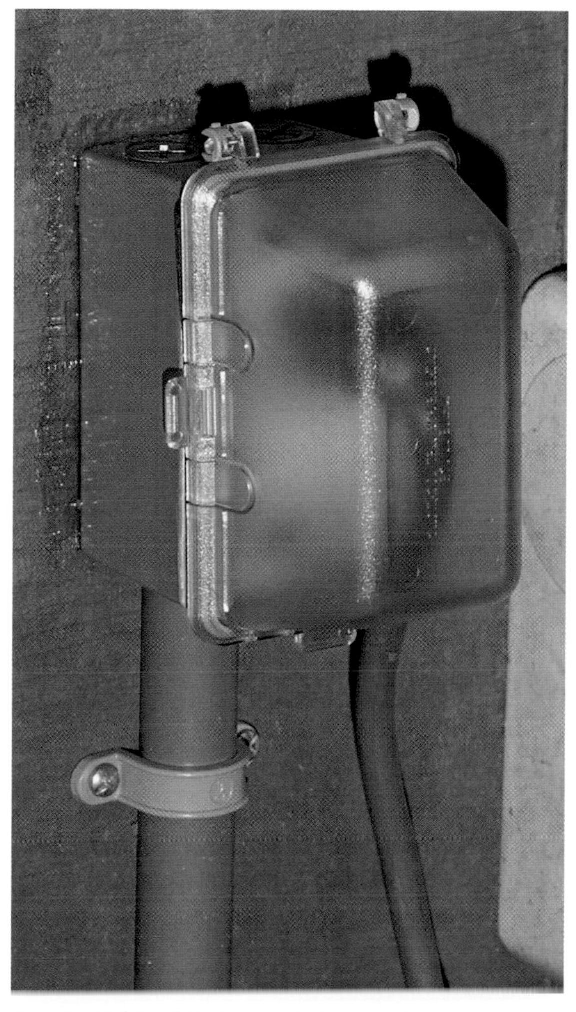

A watertight-while-in-use lid has a bubble to protect the receptacle and plug from rain and snow while a cord is plugged in.

The best performing (and best looking) weatherproof covers are recessed into the wall and have a seal between the flat lid and the box. (See TayMac at www.tay-mac.com.)

The bubble type is best when you want to convert an existing outside outlet by just replacing the lid.

The problem with some watertight-while-in-use designs is that some are not watertight as advertised—although water does not rain in, it can leak in. To make it really watertight during a driving rain, there must be a seal around the sides and top. Another problem I've found is that the bubble does not extend far enough away from the receptacle to plug in a thick extension cord and close the lid. If the lid is not completely closed, the whole design is void. Be aware that most designs only work for light-duty, easily bendable cords.

Receptacles

In residential construction, cheap, low-grade receptacles are typically used for both light- and heavy-duty applications. Because of the poor quality of low-grade receptacles, I don't use them for even light-duty applications. They can break apart, cause open circuits, or

Good ($2 to $3), better ($5 to $6), best ($13 to $15). Higher quality receptacles have more and thicker metal inside them as well as better construction with hard-to-break or unbreakable bodies.

The 2008 Code has mandated that the common 125-volt 15- and 20-amp receptacles be tamper resistant to prevent accidental insertion of foreign objects into the slots. Check with your inspector to verify that this requirement has been adopted in your area before going to the expense of buying them.

Low-cost receptacles such as these are made from easily breakable plastic, are riveted together, and have little metal inside.

even worse, create a high-resistance loose connection that results in a fire. While writing this chapter I was called out on a job where the customer had a burned-out receptacle. He first smelled something burning (the insulation on the wires) and then pulled the appliance away from the wall to see the receptacle glowing cherry red. If he were not home, his house probably would have burned to the ground. His problem was a cheap receptacle and the way it was wired. Instead of worrying about color and style, homeowners and electricians should worry more about quality.

Quality

The overall quality of receptacles has improved over the last few years, especially on the low end. However, stay away from receptacles that have the push-in holes in the back (push in the wire and it locks into the receptacle). These are being phased out and I am happy to say that the last several times I checked the home centers, none were for sale. For quality receptacles, the more you pay the better you get. The lowest-grade receptacle you should let in your house will cost you around $2.50, the best around $5. You can get super-heavy-duty receptacles for around $8 to $10 each. The difference in any receptacle is in the quality of plastic, the amount of copper within, and the yoke.

A single receptacle has only one place to plug in a load. Note the horizontal slot, which indicates that it is a 20-amp receptacle and capable of taking cords from heavy-duty, 20-amp appliances.

Duplex or single

A duplex receptacle has two outlets to plug into. A single receptacle has only one. There were many instances in the past where you might have wanted to use a single over a duplex, for example, to install GFCI protected (or not protected) outlets in basements, garages, and crawl spaces. However, those days are gone along with the exceptions. Thus there is rarely a need for a single receptacle unless you want one and only one outlet on the branch circuit. In this case, that one appliance is loading the circuit to the max and any additional loads might trip the breaker.

A duplex receptacle has tabs on both the hot and neutral sides. When half the receptacle will be switched and the other half will be permanently hot, break off the tab to isolate

Breaking the tab on the "brass screw" side of a duplex receptacle isolates the top and bottom outlets for a "one switched and one not" combo. The neutral tab is normally left intact.

the top plug-in from the bottom one. Insert a screwdriver into the slotted tab and wiggle it back and forth until the tab breaks off. The tab is normally broken only on the hot side, allowing the neutral to be shared.

Current ratings

Receptacles are rated for the amount of current allowed to flow through them. Standard duplex receptacles used throughout a residence are rated for 15 amps—even those that are connected into a standard 20-amp branch circuit. A 20-amp receptacle looks different from a 15-amp receptacle. The 20-amp receptacle has an additional slot going sideways off the neutral slot (the wide one) to allow a 20-amp plug with a sideways prong to fit into it (see the photo on p. 219). Both a 15-amp and a 20-amp plug will fit into a 20-amp receptacle, but a 20-amp plug will not fit into a 15-amp receptacle. The receptacle is designed this way so that you don't put a 20-amp load on a 15-amp circuit, which will overheat the wires and trip the breaker. You are required to use the more expensive 20-amp receptacle when there is only a single receptacle in the entire 12-gauge, 20-amp circuit.

A significant safety problem can occur when a homeowner purchases an appliance and discovers the appliance's 20-amp plug doesn't fit the 15-amp outlet. The appliance salesman may tell you that all you have to do is change the outlet from a 15 amp to a 20 amp so that the plug will fit. Don't believe him! Replacing the receptacle is not enough.

The only way to deal with this situation is to install a new circuit for the appliance using 12-gauge cable and a 20-amp appliance receptacle. (Just hope there is still room in the panel for another circuit!)

Polarity

Plugs and receptacles are polarized when they fit together in only one way. The neutral blade of the plug (the wide one) must fit into the neutral (wide) slot of the receptacle and the hot (narrow) blade of the plug must fit into the hot (narrow) slot of the receptacle. Making the neutral blade and slot wider than the hot blade and slot ensures that the hot and neutral wires cannot be reversed for an appliance whose plug doesn't have a grounding pin.

Plugs and receptacles with grounding pins are automatically polarized because of the grounding pin. When the plug is inserted into the receptacle so that the grounding pin of the plug aligns with the grounding slot of the receptacle, the hot and neutral blades will slide into their corresponding receptacle slots. Some motorized appliances, such as fans and double-insulated drills, normally don't need polariza-

Aluminum Connections

Aluminum wiring (normally 10 gauge and 12 gauge) that connects onto receptacles presents special problems. Because aluminum and copper have different characteristics, no receptacle may accept aluminum wiring unless it is designated as AL/CU or CO/ALR, which means the receptacle is UL approved for both aluminum and copper wiring. Most manufactures make such receptacles and they are readily available at home centers. If you don't have a receptacle that accepts aluminum wire, you'll have to splice a 6-in. 12-gauge copper pigtail of the appropriate insulation color onto the aluminum wire in the receptacle box. Be careful not to put a severe bend in the aluminum wire, which breaks easily, and be sure to use an approved connector. Then terminate the copper wire onto the receptacle screws.

Polarity of Receptacles and Plugs

In an electrical circuit, neutral and hot should be kept separate, which is called polarization. For a plug to be properly polarized, its neutral and hot wires must plug into the neutral and hot slots, respectively, of the receptacle.

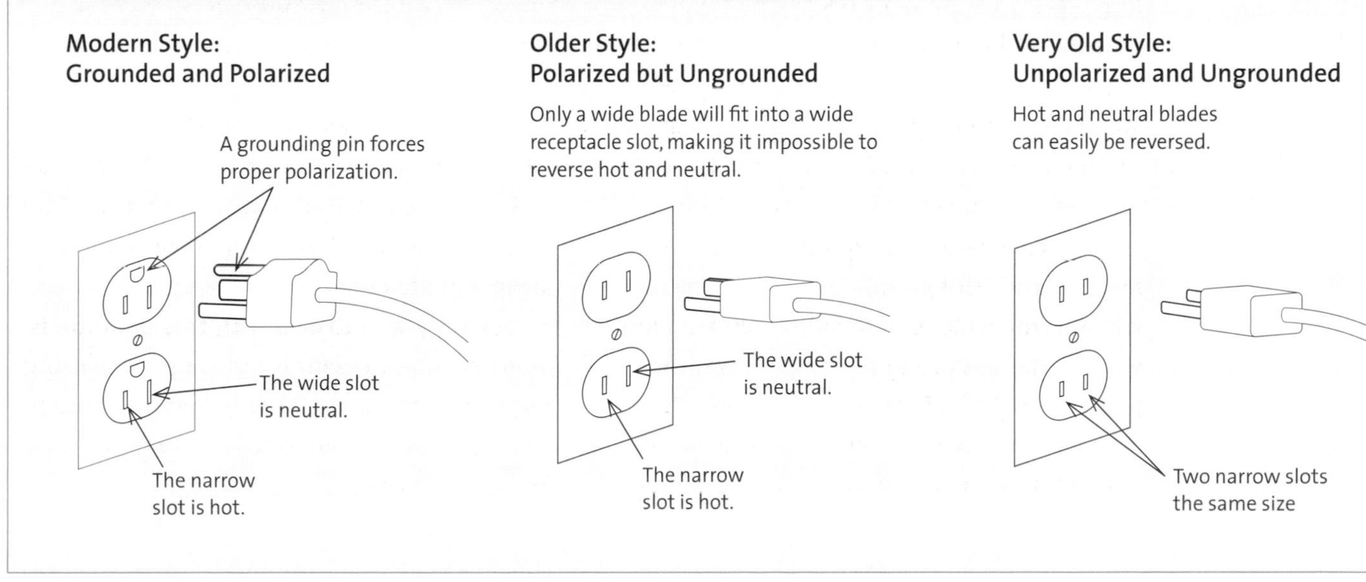

tion because a motor doesn't care which wire is which and the internal wiring is insulated from all metal parts. Such an appliance has an ungrounded, unpolarized plug with both blades of the same width.

Safety precautions

Some disasters are just waiting to happen. Sitting on my desk is a table lamp connected to the receptacle via an old-fashioned two-conductor cord. The plug has two narrow, unpolarized blades; it can be inserted either way into the receptacle and the lamp will work just fine. However, for safety reasons (it's also a code requirement), the center contact of the lamp shell should be connected to the hot wire of the branch circuit and the screw base of the lamp should be connected to the white neutral. But because the plug isn't polarized, you have a 50/50 chance of getting it right—or wrong. If the bulb is out of the lamp and the screw base is hot because of incorrect polarization, someone could easily be electrocuted simply by touching the female threads of the screw base. (I know of at least one child who was killed by such an occurrence.)

To ensure that this doesn't happen to you (or your kids), follow these steps:

1. Always leave a bulb in a lamp if it's plugged in.

Work Safe • In damp areas, try to orient the wire connector's skirt downward to prevent condensation from pooling in the skirt and corroding the splice.

2. Check the continuity between the center contact of the light's screw-in base and the plug blades to determine which blade is connected to the center contact. Once that's determined, put a small dot of paint on that plug's side and a corresponding dot near the receptacle's small slot. You now have correct polarization: Hot to center contact, neutral to screw base. The safest solution is to wire in a grounded cord and plug.

Old-house receptacles

Old houses with unpolarized receptacles present a unique problem. All electricians know there is no simple way to tell hot from neutral when the wiring is so faded that you can't tell black from white. Here's how to tell which is which: Run a wire from the panel neutral to one lead of a VOM and the other VOM lead to one of the receptacle slots or wires. The neutral should read 0 volts and the hot should read 120 volts to 125 volts. If the system is in conduit (you'll know if you have conduit by seeing it enter the panel), the conduit can be used as a ground reference (in lieu of neutral) for the test. (This is also a good way to test if the conduit can be used as a ground conductor.)

Ungrounded old-house receptacles are acceptable and don't need to be changed as long as whatever will be plugged in doesn't need to be grounded. Cheater plugs are not officially accepted because you are taking an appliance that requires grounding and running it on an ungrounded circuit, which is dangerous to the operator. Remember that a cheater plug only allows the appliance to be plugged in; it does not ground the appliance. For safety reasons, you need to run a ground wire or replace the cable with a grounded one.

Cheater plugs allow you to plug in an appliance but don't ground anything unless there is a metal box and metal conduit leading back to the service panel. If the screw attachment isn't connected to the cover plate, it will pull free and expose the hot prongs of the plug, as shown.

Beware old receptacles. This age-old Bryant receptacle was used for both 120-volt (15-amp) and 240-volt (10-amp) loads. Thus, when you plug into it you don't know what you are getting.

Wiring and Installation

The wiring and installation of receptacles is one of the most error-prone areas in residential wiring. The installer assumes that the ground terminals must go down, polarities are reversed, receptacles have the wires attached to them in an unsafe manner, and the wiring is installed to the receptacles in a design akin to cheap Christmas-tree lights.

People always ask me, "Does the grounding terminal go up or down?" and "Is there a bottom or top to a receptacle?" The answer is that there is no "official" right or wrong way to orient the receptacle—the NEC doesn't specify—but you can figure this out for yourself by using logic and common sense. For example, clothes washers, refrigerators, and window air

Positioning Receptacles

Preferred unless molded plug dictates it needs to be reversed

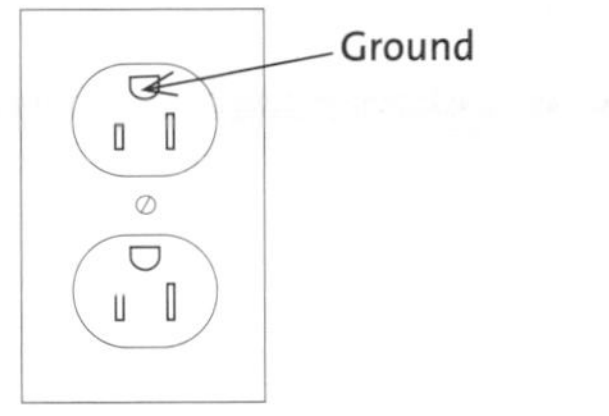

Not preferred unless molded plug dictates this position

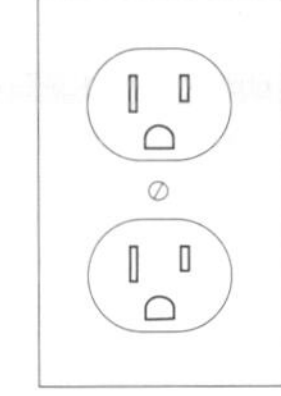

With the grounding slot on top, a falling metal object, such as a picture frame, will bounce off a partially exposed plug with no harm done. With the grounding slot on the bottom, an object could short across the hot and neutral blades of the plug.

Preferred

Not Preferred

Neutral

Hot

A receptacle installed horizontally should have the neutral (wide slot) on top. A falling metal object will make contact with the neutral blade of a partially exposed plug, not the hot.

Which Cable Is Which?

The electrician's nightmare is leaving out the incoming power cable to a particular box such as a string of receptacles. Follow this system to prevent it. If you always bring power into a box via one particular knockout and run outgoing cables through others, you'll always be able to tell at a glance which cables are outgoing power and which one is incoming power. This system also works for switches. If there is no cable in the top left of a switch box then you know there is no power feeding the switch (assuming you followed my Above Code system of always running the feeder to the switch box).

Using Pigtails

When you've got more wires than screw terminals, use the pigtail method to splice wires together, then attach one pigtail to its respective terminal on the receptacle.

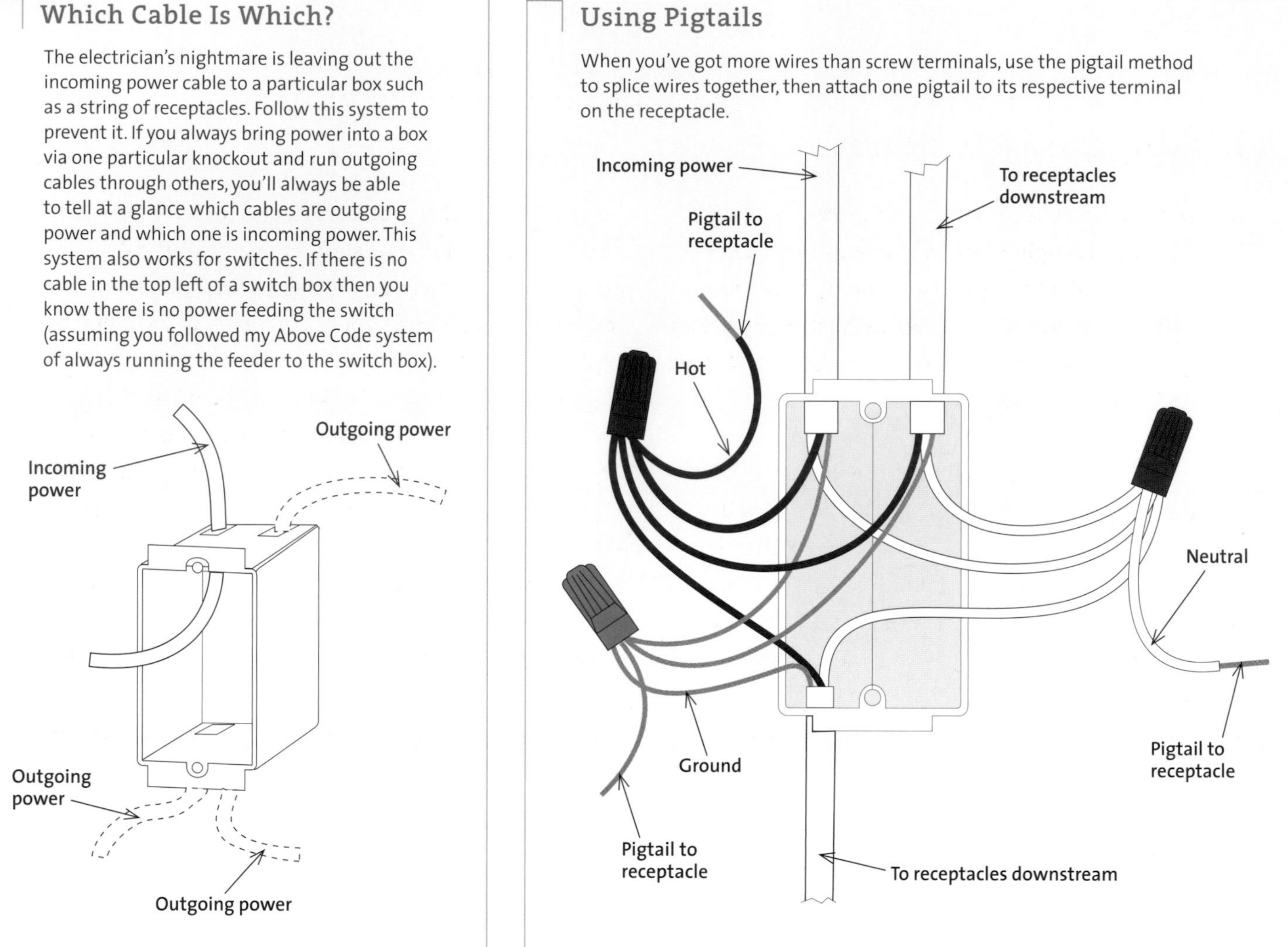

conditioners typically have cords with immediate-turn plugs. For these appliances, orient the receptacle so that the plug can insert into the receptacle without having to loop over itself. Otherwise, the downward pull of the cord will tend to pull the plug out of the receptacle.

Other than for immediate-turn plugs, I prefer to put the grounding slot on top. The reason is that if a plug were partially pulled out of the receptacle, exposing the hot and neutral blades of the plug, and something metal fell on it, a direct short would not occur because the grounding pin would deflect the item away from the hot terminals.

Receptacles can also be positioned horizontally. When this situation occurs, place the grounding slot to the left and the wide neutral slot on top. That way, if a plug has its blades partially exposed and something metal (such as a picture frame) falls onto it, the object will hit the grounded neutral blade, not the hot blade.

Identifying screw terminals

A receptacle has three locations to accept wires:

- A hot side with two gold screws internally connected to the narrow slot of the receptacle.

Making the Perfect Splice

A good electrical splice is first a good mechanical splice. If the splice is not tight, a fire can occur due to arcing. Although a standard shell doesn't support fire, a loose connection destroys the connector to the point where an arcing wire's flash can start a fire someplace else. Making a perfect splice isn't hard, but it does take some practice. Here's how to do it:

First, pull all the wires that are to be spliced together to one side of the box (for example, all black wires to the left, all white wires to the right, and all bare wires to the bottom), then cut them the same length. Make sure the wires come straight out and that they do not intertwine or loop in and out with other wires. Don't make the wires the same length by forcing some of the wire back into the box—the extra wire length takes up too much space in the box.

Once the wires are cut, strip the insulation off the wire ends. The amount of insulation you need to strip depends on the size connector you're using. It's better to strip too much (and cut some wire back) than not enough. Typically, 5/8 in. to 3/4 in. should suffice.

Use electrician's pliers to twist the wires together, turning in a clockwise direction. Each wire should spiral around the others in the group. Make sure that no wire slips out and simply twists around the others. If necessary, cut the twist back 1/8 in. to even out the conductors.

Pick the appropriate connector and twist down onto the group of wires. You will know when to stop twisting when you feel significant resistance. If you don't feel much resistance, use a smaller connector. If bare conductor shows past the wire connector's skirt, use a larger one.

Splicing with a Wire Connector

1. Cut all wires the same length.

2. Strip off enough insulation that no bare wire sticks out of the connector.

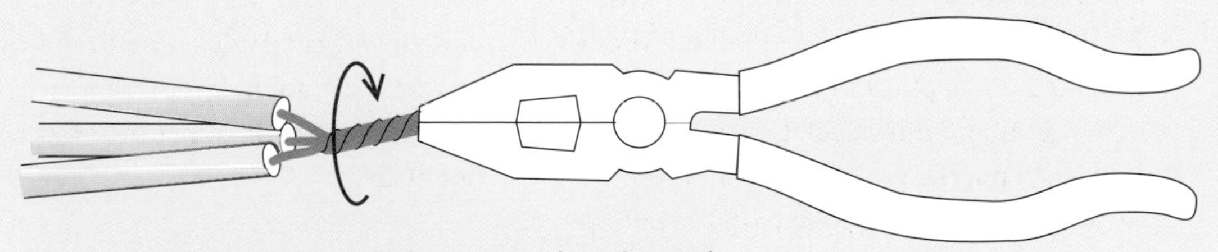

3. Twist the wires together clockwise using broad-nosed electrician's pliers. Then insert the wire connector over the wires and twist clockwise.

- A neutral side with two silver screws internally connected to the wide slot of the receptacle.
- A green grounding screw internally connected to the receptacle's grounding slot.

The electrician connects the black or red (hot) insulated wire of the incoming cable (which can be traced back to the main panel's circuit breakers) to the gold screw. The white insulated (neutral) wire (which can be traced back to the neutral bus on the main panel) is connected to the silver screw. The green screw is wired to the incoming ground wire, which eventually finds its way to earth ground via the main panel. The black and white wires provide a current path to and from the load plugged into the receptacle. The grounding wire (called the equipment-grounding conductor) places the frame of the load at 0 volts, or ground potential, and provides a return path for the current to the panel should a ground fault occur.

What to do with extra wires

Each screw on the receptacle can have only one wire under it. If you've got power coming into the receptacle and the receptacle also feeds additional circuits, do not be tempted to twist two wires together and tighten them both under one screw. This is dangerous because the wires will eventually work loose, resulting in an intermittent circuit—if the receptacle doesn't break apart first.

You may also be tempted to put any extra wires in the push-in terminals—available on some receptacles—or you may find it easier to attach all wires in the push-ins. Don't do it. I once had a service call in which the owners were having so many problems with the receptacles in their home that they thought the house was haunted. When I put my hand close to several of the receptacles that had nothing plugged into them, I could feel heat—enough

Above Code • Always bring power into the box through one particular knockout—I prefer to bring power through the upper-left knockout. Doing so enables you to quickly identify the incoming cables and the outgoing cables.

This Hubble design wires very quickly. Strip off 1 in. of insulation from the wire end, then insert it from the front so that it barely protrudes from the built-in slot.

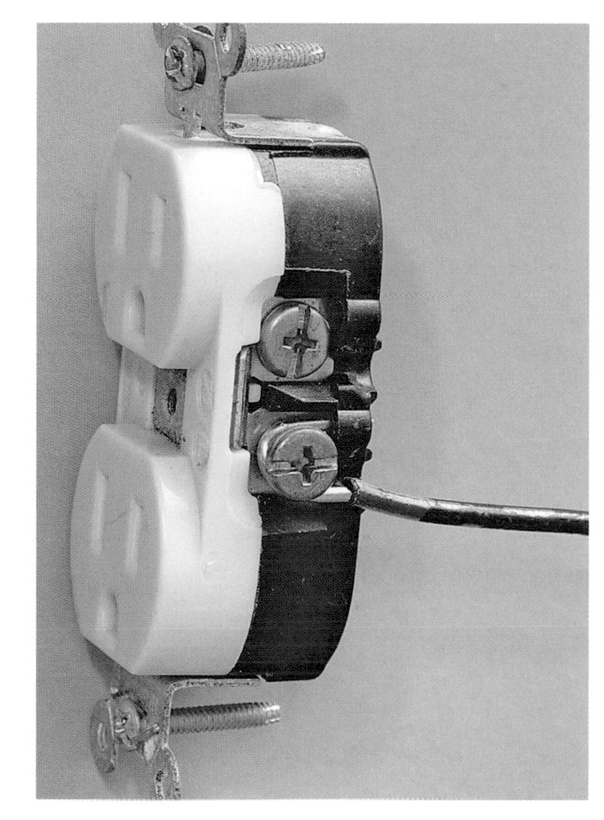
Twist the wire around the screw, then tighten it.

to melt the child-protector inserts. Other outlets were generating enough heat to deform the kitchen-appliance plugs. A few hours earlier, sparks had flown out of an outlet that had nothing plugged into it.

The problem was that the installer used push-in terminals, which seem as though they were designed for lazy electricians and inexperienced do-it-yourselfers. Going by various trade names, such as EZ Wire®, Quickwire, and Speedwire, these systems use the spring tension of a metal "check" (similar to a clamp) inside the receptacle to hold the wire after it is pushed into the hole. They don't always clamp solidly onto the inserted wires (especially when dealing with small-diameter 14-gauge wire), causing receptacles to overheat or a total loss of power to one or all receptacles in a string. In the interest of safety, never use push-in terminals. Since January 1995, push-in terminals have been banned from 20-amp circuits. Receptacles made after that date prevent the insertion of 12-gauge wire into the push-in terminals; however, they are still able to take 12-gauge wire under the screw terminals. In my opinion, this is too little, too late: Push-in terminals should be banned altogether.

So what do you do with the extra wires when there are no extra screws and you don't want to use push-in terminals? One method is to purchase a high-quality receptacle that accepts four hots and four neutrals using screw pressure on a metal bar to hold the wires. Otherwise, the wires must be spliced together mechanically with an insulated cap or a wire nut, with one short wire, called a pigtail, extending from the splice to connect to the receptacle.

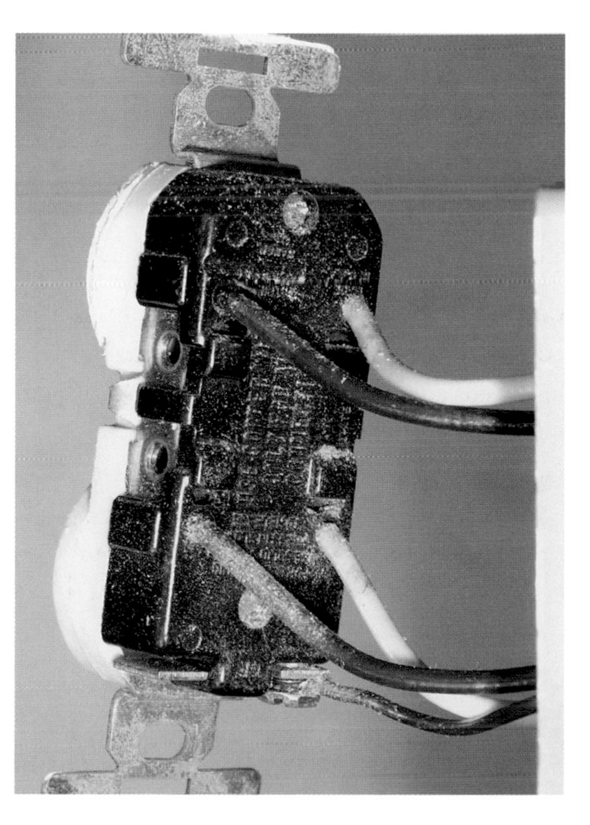
Push-in terminals keep me busy troubleshooting dead receptacle strings. Don't use them—they may become loose and pop out of the receptacle back.

This cutaway view shows the order in which to fold wires back into a box: first the grounds, then the neutrals, then the hot wires. The pigtails to the receptacle are prebent so that they fold neatly into the box.

Pushing wires back in the box

Once wired, the receptacle should not just be shoved back into the box, bending and crumpling the wires in a haphazard manner. If only one receptacle and just a few wires are in the box, there shouldn't be a problem; however, if there are several wires and a couple of receptacles, there's a trick to getting all the wires back in the box without using a hydraulic jack. First, strip all the cable sheaths ¼ in. to ⅜ in. from where they enter the box, rather than having several extra inches taking up box space. Next, cut off all excess lengths of wire; 6 in. from where the cable enters the box is plenty. Bring all the grounds to one side, the neutrals to another, and the hots to still another; don't have wires threading through each other, making the box look like a bird's nest.

Once everything is hooked up, neatly push the wires into the box. Do the ground wires first and use side cutters to push them up against the back of the box. Next, push the neutrals to another part of the box. Push the hot wires in last, routing them along the edge, not the center, of the box. The center of the box will need extra room for the receptacles. If the receptacles don't fit after all this, you'll

This type of wire attachment is extremely secure.

have to increase the size of the box or remove some cables.

Testing

Once the receptacle is wired and powered up, it should be tested with a plug-in tester to make sure it's wired correctly. The basic tester will indicate whether the wires are connected properly. A more sophisticated tester, like the Ideal 61-165 (the old SureTest), can also tell whether there is a bootleg ground (neutral shorting to ground within a few feet) and can electronically create a 15-amp and 20-amp load.

Receptacle strings

There are two ways of wiring a string of receptacles: direct connect or via pigtails. The latter is better--it just takes longer.

Direct connect This type of string is most common. The two wires that the current flows through are physically attached to each receptacle. Thus if the last receptacle in the string is pulling current, that specific current is flowing through the physical body of each receptacle that precedes it even though there may be no load plugged into those receptacles. This could overheat the receptacles if the entire string was loaded up. You want the load current to go through only the body of the receptacle that the load is plugged into.

Pigtail connect In this type of string connection you use pigtails from a splice in the back of the box to connect to each receptacle. Pull the cable into the receptacle box as you would normally do—and strip the wires out as you would normally do. But instead of wiring them direct to the receptacle, use pigtail jumpers instead. This keeps the upstream current out of that particular receptacle if there is no load attached to that particular receptacle.

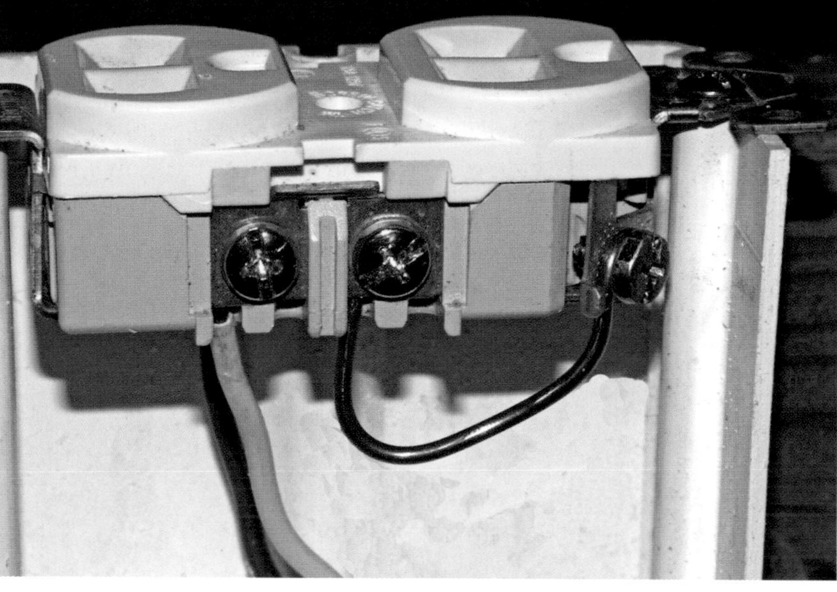

A bootleg ground is a jumper from neutral to ground on a receptacle. It will fool a standard plug-in tester, showing that the receptacle has a ground connection when it really doesn't.

Above Code • Receptacles in a string should be wired via pigtails. In the direct connect method if one receptacle has its hot and/or neutral wires removed (the receptacle breaks or a wire simply pops out) you lose power to the entire string.

Combination system What normally happens when electricians are wiring a house is a compromise between the two systems. As long as there are not more than two black or two white wires attached to the receptacle, the attachment is normally done under the screws (direct connect). If there are more than two wires, they should be pigtailed.

Receptacle String Direct Connect—Not Preferred

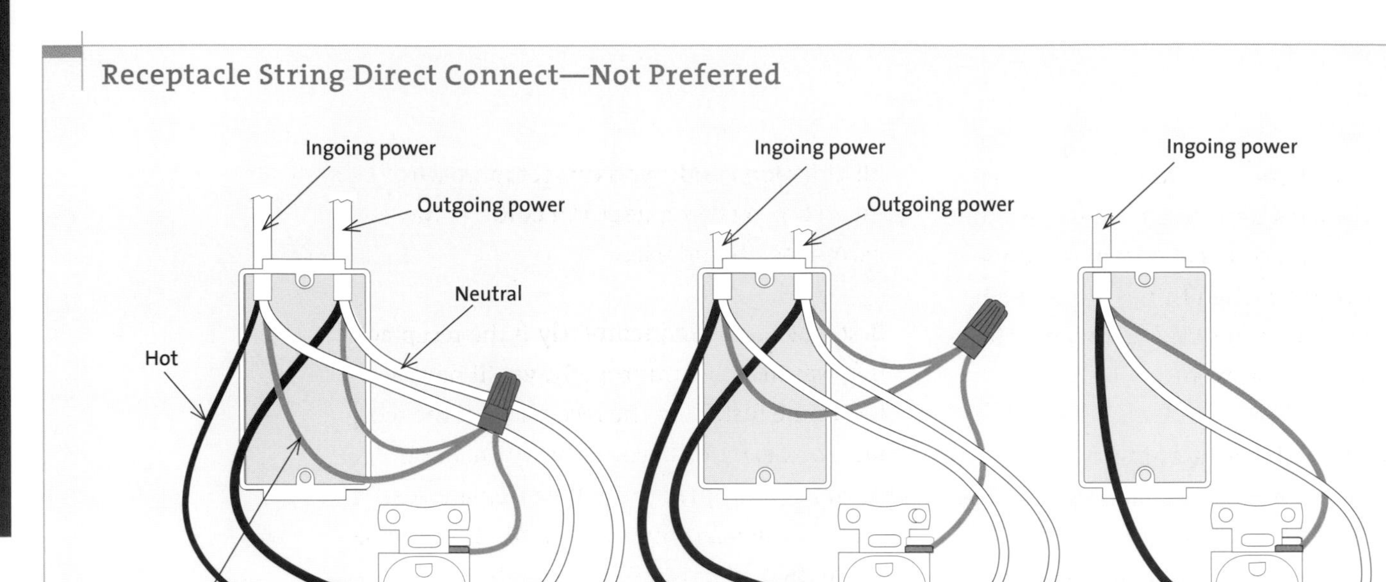

These receptacles feed others downstream. If one is disconnected or removed, any others downstream will lose power, too. Also, all upstream current will be physically flowing through the receptacles downstream, which may overheat the receptacle and perhaps damage the appliance plugs—this has actually happened.

The last receptacle in a string does not power any others.

Receptacle String Pigtail Connect—Preferred

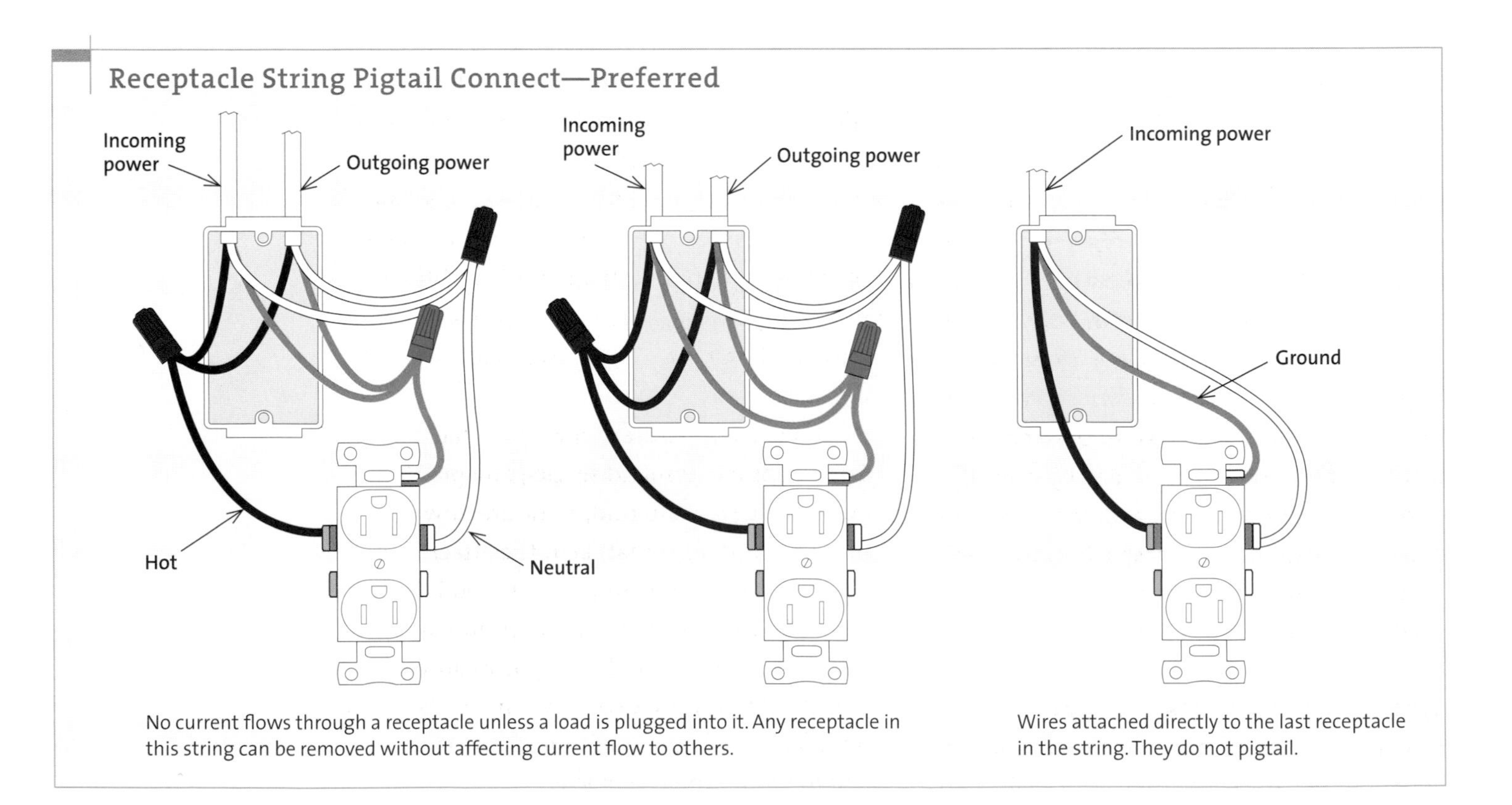

No current flows through a receptacle unless a load is plugged into it. Any receptacle in this string can be removed without affecting current flow to others.

Wires attached directly to the last receptacle in the string. They do not pigtail.

Troubleshooting

Solving problems is the most time-consuming part of the job. A good electrician not only knows what to do when things go right but also what to do when things go wrong. Here's a list of common problems.

Box has stripped-out receptacle-holding threads. One common installation problem is stripping the box's female threads so that the receptacle's screws cannot pull the receptacle up to the box. This normally happens when you do not bend the wires back into the box but rely on the screws to pull the receptacle into the box. (Even when you're careful, this problem eventually happens to everyone.) What do you do about it? Drywall screws work nicely and they're easily obtainable on any job site. A drywall screw has a flat head, so the faceplate will still fit tightly against the receptacle.

Box's receptacle-holding threads have broken away What do you do if the entire female threaded portion that holds the receptacle to the box breaks off? Obviously, you can replace the box, which is not much of a problem if the finished walls aren't up yet. But if they are up, this can be very problematic. Besides, this only happens at the end of a long, hard day when everything else has gone wrong and you have neither the time nor the patience to change a box that is already surrounded by drywall. Rather than pull your hair out, try this: Squirt a dab of construction adhesive into the area where the screw will attach. After it hardens, predrill it and use a drywall screw.

Wires won't fit into box Another common problem is having too many wires in the box and a receptacle that doesn't seem to fit. Cable-fill violations aside, you can get more wires into a box by being neat and cutting away all excess sheath and wire.

Box was installed incorrectly If the receptacle box was installed improperly, you'll have a hard time installing the receptacle. If the receptacle box extends beyond the finished wall, so will the receptacle and the faceplate, leaving an unsightly gap behind both. In this case, move the box in or grind down the front edge so that it's flush with the wall (grinding also removes female screw attachment threads for the receptacle).

If the box is set too deep within the wall cavity, you will be in code violation (the box must be flush with the finished wall) and the receptacle also will be set too deep in the wall cavity. If the drywall is cut perfectly around the box, the plastic ears (I call them Mickey Mouse ears) will affix to the drywall. But it's rare that you get a tight cut; normally, it's just the opposite. Spacers can be placed between the receptacle and the box to extend the receptacle out to the finished wall. I use small washers, nuts, and Mickey Mouse ears with long attachment screws (always break off and save the little ears from receptacles you're throwing away; they make perfect washer-type spacers for just this purpose).

Some manufacturers make adjustable extensions for standard single-gang plastic boxes. Once hard to find, these are now commonplace. Beyond that, I cut the attachment nails with a reciprocating saw and pull the box forward, remounting it with drywall nails and glue. If the box is installed slightly canted to the left or right, the two horizontal slots on the receptacle (one above and one below) allow you to plumb the receptacle.

CHAPTER

Switches

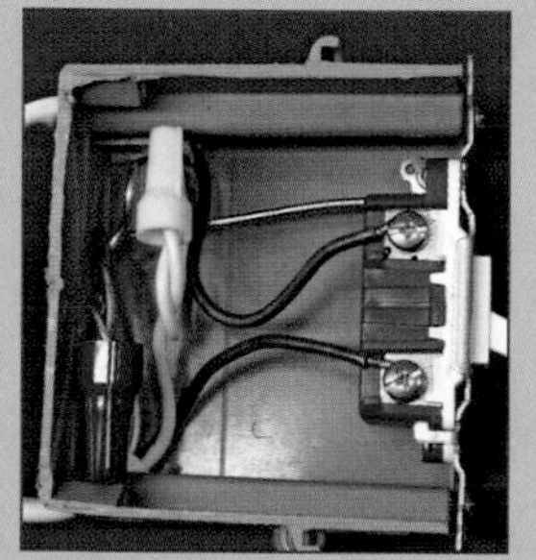

Switches control the power used throughout the house. For proper control, the right switch and box must be installed and the switch must be wired correctly. Installing and wiring switches is not very difficult.

Wiring pictorials usually show only the most common (and correct) wiring methods. In this edition I have added alternate methods of wiring a switch. These aren't methods I recommend but I include them because you might run across these examples in the field. I added them for two reasons: First, so you can appreciate simple, logical, and fast wiring methods; second, so you will have a reference when you come across these methods in the field. That is, so you will know where the power is, where the load is, and how to troubleshoot the system. I also have added wiring pictorials on how to wire multiple lights and specialty switches.

In this chapter, I also discuss types of switches—which grades are right for different situations—and how to wire them correctly. I have added dimmer switches, including how to wire three-way dimmer switches that can dim from two locations. I'll also show

Switch evolution. On left, a screw-in switch and receptacle (very early Hubble configuration). Then came the early snap switches (middle) and then the ever-popular push button (right).

The Old and the New

Switches have evolved for the better. Old switches were called snap switches for a reason. When you turned them on or off they made a distinctive loud clack or snap. For that reason, we old-timers still occasionally refer to them as snap switches. But that is long gone. The new generation of switches are quieter, smaller, and work better.

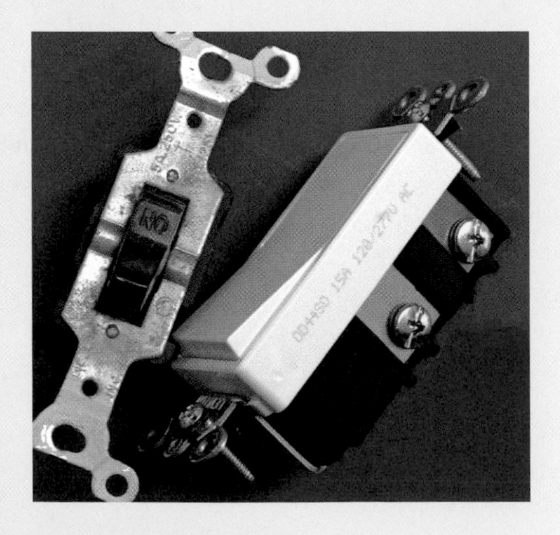

The evolution of the switch, from the old snap switch (left) to the new-style rocker switch (right).

you how to make the job simpler by following a plan and using some common sense. It all begins with choosing the right switch box.

Switch Boxes

Because there are so many boxes on the market, the initial challenge is knowing which type of box is needed for each switching situation. Normally, a single switch requires a single box, known in the trade as a single-gang box. Two switches normally use a double-gang box, three switches require a triple-gang box, and four switches require a four-gang box, or quad. I don't recommend installing more than four switches at a single location, because nonmetallic boxes larger than four gang aren't available. In addition, switch faceplates larger than four gang are hard to find and may be expensive.

There are times when I need to install a switch in a wall in which the studs are too close to allow the correct-size box. For example, suppose I need a triple-gang box, but only a two-gang will fit. In this situation, I use a double switch, which has two switches on the same yoke, or frame. The double switch allows me to use a smaller box—a single instead of a double. I also use double switches when I forget to wire in a switch or when a design is changed at the last minute. Because a double switch looks different from the other switches, I can easily identify what each switch does.

Nonmetallic boxes with integral nails are the most frequently used switch boxes in residences. However, these boxes are easily damaged and so must be kept within the wall cavity. Metal boxes are rarely used within the wall cavity because of their higher cost and extra labor to install (you must ground the box). However, metal boxes are required for

Reaching into a Wall

There are times when you need to replace an old metal outlet box with a new plastic one—without destroying the wall. This normally occurs in older houses where the boxes are too small to accommodate both the wires and a new switch, receptacle, or GFCI. Cut the power and remove the cover plate. Using a mini recip saw, reach the thin blade between box and stud and cut the nails holding it in place, which shouldn't take more than a few seconds. With patience, work the wiring out of the old box and pull the box out of the wall. Once it's removed, increase the hole to the size of the new, larger outlet box. To get the new box back into the wall, I bring the cables into the back of the box, slide the box in, and carefully screw the box to the stud.

With the finished wall removed for clarity, a Makita mini recip saw reaches into a wall cavity and cuts the nails holding the outlet box to the stud.

The mini recip saw, cordless and lightweight, has a blade thinner than a dime, which is invaluable in remodel work.

a surface mount in an unfinished basement or garage wall, where they take a certain amount of abuse, and conduit must be run to protect the wiring at least up to 8 ft. overhead.

One important rule applies to all flush-mounted switch outlet boxes (this also applies to flush-mounted receptacle boxes): They must be installed flush or extremely close (approximately 1⁄16 in.) to the finished wall surface. This prevents a fire within the room from entering the wall cavity—or vice versa. If the box is set too deep and cannot be moved, metal and plastic box extensions are available (the metal ones are extremely hard to install). If the box is set too far out from the finished wall, you must

This switch outlet box was installed flush to the stud and is now in too deep. If you can't move the box forward, install a plastic box extender, which is available from most electrical distributors.

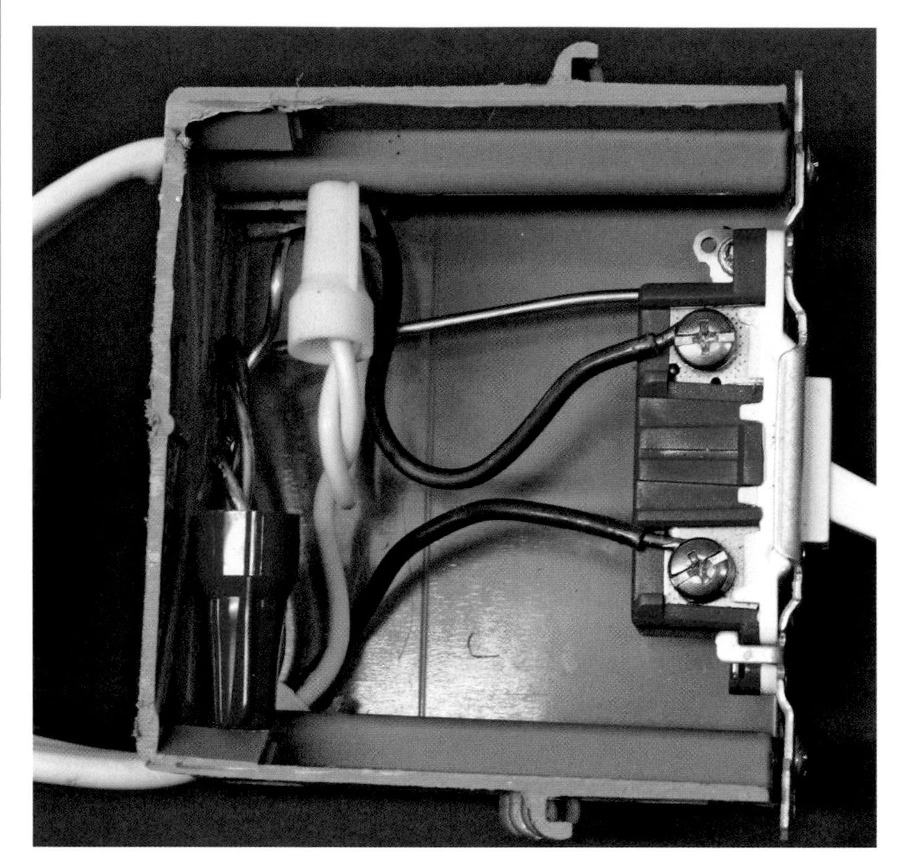

Cutaway view of a switch in the outlet box. Fold the grounds first (pigtail the ground wire over to the switch), the neutrals second, and accordion the hot switch wires back last.

cut it off flush with the wall or somehow set it back within the wall.

Keep in mind that metal outlet covers must be grounded. This is done via the metal attachment screw. The screw grounds the cover to the switch, and the switch is grounded via the equipment-grounding conductor, AC armor, or metal conduit. If the outlet is not grounded, you cannot use a metal receptacle cover plate. In general, I don't recommend metal receptacle cover plates at all. I've seen too many come loose, fall onto a plug's prongs, and create a short.

Single-gang boxes

A single-gang box with one switch is the most common switching arrangement in a residence. Even though switches normally require less volume in the box than receptacles do, you can easily put yourself in cable-fill violation if you purchase a box that's too small. I never use a box smaller than 18 cu. in., and I prefer to use the bigger 20.3-cu.-in. and 22.5-cu.-in. boxes that I use for receptacles, even though they are significantly more expensive than 18-cu.-in. boxes. As with receptacle boxes, nonmetallic switch boxes have their volumes stamped inside; metal boxes do not.

Work Safe • If you must cut the front off a nonmetallic box because it extends beyond the finished wall, be careful not to cut off too much, thus removing the female attachment threads.

Double-gang boxes

Switching via two switches in a double-gang box allows you to control power to two different loads from one location. Typical examples are at the main entrance, where outside lights are switched on at the same location as interior lights, at a junction of a room and a hallway, and at a box that controls both a ceiling fan and a light.

Double-gang nonmetallic switch boxes have a variety of attachment mechanisms (depending on the manufacturer); they are normally available in volumes of 25 cu. in. to 34 cu. in. Smaller boxes should be avoided because they have scarcely enough volume to accept one incoming power cable and two load cables—the bare minimum required in a box with two switches.

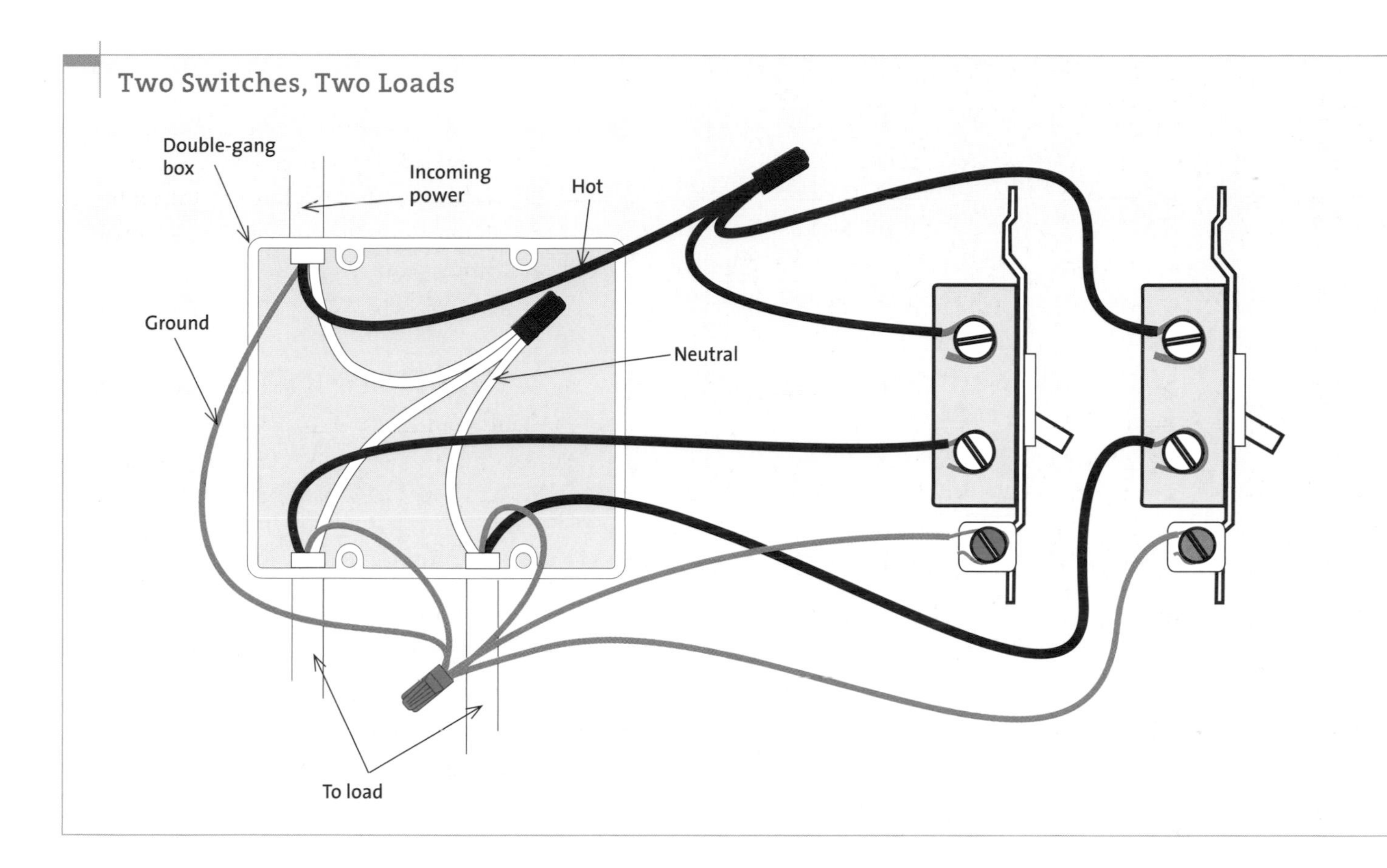

Number of Cables Allowed in a Single-Gang Box

Single-Gang Box Types and Sizes	Number of 12-Gauge Cables with Switch	Number of 14-Gauge Cables with Switch
Metal handy box, 1 1/2 - 1 7/8 in. deep (11.5-13 cu. in.)	0	1
Metal handy box 2 1/8 in. deep	1	2
Metal box with integral nails 2 1/2 in. deep (12.5 cu. in.)	0	1
Metal box with integral nails 2 27/32 in. deep (15.8 cu. in.)	2	2
Metal box with integral nails 2 7/8 in. deep (18.8 cu. in.)	2	3
Nonmetallic box, 16 cu.in.	2	2
Nonmetallic box, 18 cu.in.	2	3
Nonmetallic box, 20.3 cu.in.	3*	3
Nonmetallic box, 23 cu.in.	3*	4

* Cable-fill violation, because 24.75 cu. in. volume required.

Triple-gang boxes

It is very common to have three switches at a house's main entrance to control the outside light, the entry light, and the interior light. Three switches (in a triple-gang box) are common whenever an overhead fan and fan light occupy two spaces in the box and a third switch is needed to control the hallway light or another load. Triple-gang nonmetallic boxes are typically available in volumes of 45 cu. in. and 54 cu. in. I recommend using the larger box—you'll be surprised how quickly it fills with cable. These boxes can be attached via integral nails or metal brackets.

Four-gang boxes

A four-gang, or quad, box with four switches, is the largest switching arrangement I recommend using within a residence. If you must use a box larger than a four gang, make sure the switches are clearly marked: It's hard enough to remember what four switches do, let alone additional ones. A four-gang arrangement can be used at the main entrance to control entry lights, room lights, spotlights, and outside lights from one location. Or if there is a hallway between the dining and living rooms, you may want switches to control a fan and a fan light, as well as lights for the dining room, living room, or hallway. The problem is remembering which switch is which. If this situation arises, you may want to pull the two fan switches away from the others and install them in a double-gang box.

Switch grades

Switches, like receptacles, are available in many grades and vary in terms of safety and quality. The spectrum starts with a thin, low-cost, residential or general-use grade and ends with an expensive, high-quality spec grade.

The most common switch installed in a residence is residential grade, which is normally rated at 15 amps/120 volts AC (switches normally are not available in 20 amps or higher). In theory, a residential grade can have full-rated current through the switch for loads such as fluorescent and incandescent lighting and inductive and resistance loads for heating. For motor loads, a residential-grade switch shouldn't be used for more than 80 percent of its rated current and not beyond the horsepower rating of the switch, which is normally ½ hp at 120 volts. In my opinion, a residential-grade switch shouldn't be used for anything that pulls current close to the switch's maximum rating.

Residential-grade switches are inexpensive, and you get what you pay for. It's not uncommon for this grade of switch to fall apart

Above Code • Most cheap switches are designed to be wired via push-in terminals. Never, under any circumstance, use the push-in terminals—always use the screws. I've lost track of the number of push-in connections that have pulled out of the switch and created an open circuit. Another danger with a push-in terminal is that the wire can loosen over time. A loose connection increases resistance and can lead to arcing and burning. This type of cheap push-in terminal should not be confused with the high-quality switch connections that bring in wires from the back but also have screws that tighten a bar to apply pressure on the conductors.

during installation or to die an early death if a load is pulling close to the switch's rated current. Occasionally, the side screws strip out in the body, even with just minimum pressure, and the cheap plastic body breaks apart as the screws are tightened.

Spec-, commercial-, and industrial-grade switches are also available for those who want something other than minimum grade. These switches have thick bodies—about twice as thick as those of residential-grade switches. They are also available in 20-amp and 30-amp ratings. A typical commercial-grade switch only costs $1 to $2 more than a cheap, residential-grade one. This small extra cost increases the quality tremendously. Those extra couple of dollars buy you the following features: voltage increase from 120 to 277; 20-amp and 30-amp designs that increase allowable current and horsepower rating from ½ hp to 2 hp; the ability to wire to side screws—no push-in terminals; a nylon toggle to resist breakage; silver cadmium contacts for longer switch use; and an internal neoprene rocker for longer switch wear.

For renovation work, you can still buy push-button switches (newly made), which are expensive. Or you can still get a few brand-new antique (right out of the antique box) brass and ceramic push-button switches, which are very, very, expensive.

Types of Switches

There are many types of switches, and your situation will dictate which type to use. (The most common switches in a residence are single pole, double pole, three way, and four way.) The way switches are wired has changed a lot since the early 1900s, when push-button switches were used. Today, there are many more wires going to and from switches, and an electrician must try to keep them straight. As with receptacle boxes, I normally run the incoming power cable into the switch box through the upper-left

Toubleshooting a Possible Bad Snap Switch

If the light isn't working, first check the bulb. If the bulb is good, check the switch—throw it back and forth to see how it feels. A bad switch won't feel "normal." Next check voltage at the center pin of the light socket to the metal female thread. If you have voltage (switch on and power to the circuit), it has to be the bulb or the center pin in the socket is corroded or bent down. Use a pencil eraser to clean the center pin (power off). Also, you may have to pull it up if it is squished down so far the bulb does not make contact with it. If there is no power to the light, look to the switch. Check power across the switch—120 volts if the switch is off, zero volts if it is on. If you do not get this, replace the switch.

Controlling Multiple Lights via One Switch

Although there many ways to splice multiple lights together so they turn on as one, there are only three good ways of doing it. Either bring all the cables from the individual lights to the first light, bring them all to a single splice outlet box (make it big), or daisy chain them. The bottom line in splicing together multiple lights is to put them all in parallel, which you do by matching all the same-color wires.

Controlling More Than One Light

1. Bring all conductors to the first light's outlet box and wire all there.

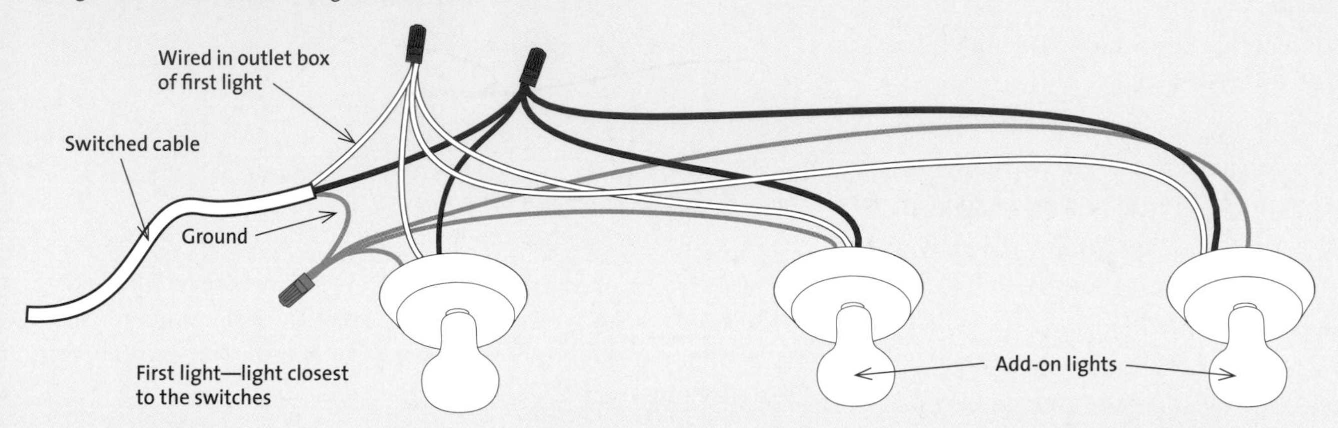

Controlling more than one light is quite simple. Ignore all lights except the closest one and wire it into the switching system. To get the rest of the lights working off the same switching system, simply wire all others to the first light—matching the conductor colors.

2. Or, have one large independent outlet box as a master splice box and bring all light conductors there.

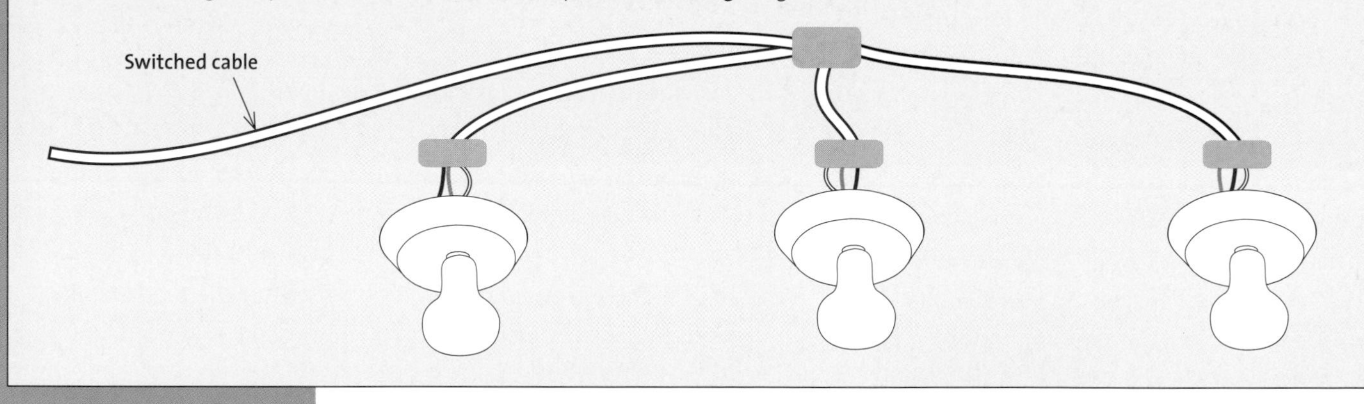

knockout and run loads out through the bottom. This way, you can tell whether the power and load conductors are wired to the switch just by glancing at it. Even if you prefer to run the incoming cable through another knockout, be consistent.

Single-pole switches

The basic on/off snap switch is referred to as a single-pole (SP) switch, because only one contact is opened or closed at any one time. It is the most common switch in a residence and is used exclusively to turn lights on and off from one location. An SP switch allows electri-

3. Connecting lights in a daisy chain.

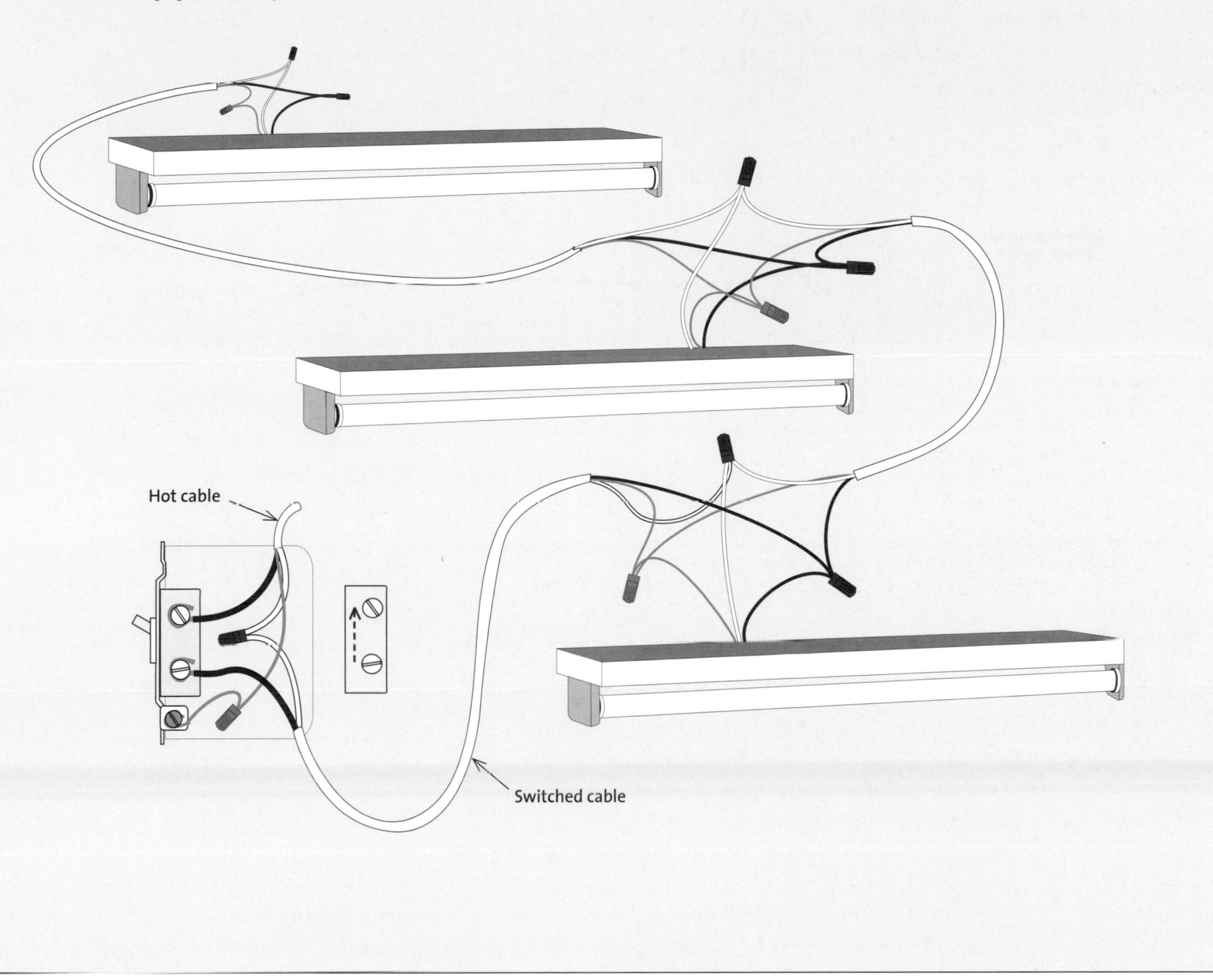

cal current to flow through it in the "on," or "make," position and opens the connection in the "off," or "break," position. The most common mistake in installing a simple switch is installing it upside down, making up "off" and down "on."

An SP switch has two terminals or screws: one terminal for the incoming hot wire and one for the wire called the "return," which takes the power to the load when the switch is on. As far as making the circuit work, it makes no difference which wire connects to which terminal. But I prefer to follow a system: I

Which Switch Is Which?

- Controlling one or more lights from a single location? Use a common single-pole switch.
- Controlling one or more lights from two locations? Use a three-way switch in both locations.
- Controlling one or more lights from three or more locations? Use a three-way switch on both ends. All switches in between are four-way switches.

Above Code • I prefer bringing both the power cable and the load cable (from the light) into the switch outlet box. I splice the neutral from the power cable straight through to the load, take the hot wire to the switch, and then run a wire from the second switch screw to the light. It's simple, logical, and easy to troubleshoot. If you want to tap into the power to feed a receptacle, simply splice onto the hot cable as it enters the box.

always run the incoming hot wire to the top terminal and run the wire to the load off the bottom terminal. This way, I can tell at a glance which wire runs to the switch. It also is an extremely effective troubleshooting aid.

How you wire an SP switch depends on whether the power cable runs to the switch box or the load box. If a load, such as an overhead light, runs to the load box, a two-

Work Safe • Wiring multigang SP switches is identical to wiring single poles—you just have more wires and switches to install. Simply pigtail the main black hot wire to power each switch and connect all the neutrals together. (Do not use the push-in terminals behind the switch as jumpers from switch to switch.)

conductor cable must connect the light to the switch. One wire, the white (taped black), takes the power from the light outlet box to the switch; the black wire brings the switched power back to feed the light. I don't like this system because it places the power cable in a location that's hard to access for troubleshooting.

Double-pole switches

A double-pole (DP) switch can turn two independent circuits on and off at once because two contacts, or poles, are opened and closed at the same time. A DP switch is wired as two independent SP switches—one on the left and one on the right. In this situation, the white is hot, not neutral, so it must be taped black to indicate that it's hot. A DP switch is normally used to control a 240-volt appliance, such as a submersible pump or other piece of equipment without a plug that can be disconnected for maintenance. The appliance therefore must have a cutoff switch adjacent to it or its controls to protect the person performing the maintenance. I wire a DP switch the same way I wire an SP switch: hot to the top, load to the bottom.

Two Methods of Wiring a Single-Pole Switch

Preferred: Power Cable to Switch Box

Power is brought through the black (hot) wire through the switch to the load. White neutral wire from incoming cable is spliced directly to white from load. Ground wires from both cables are spliced together with ground from switch.

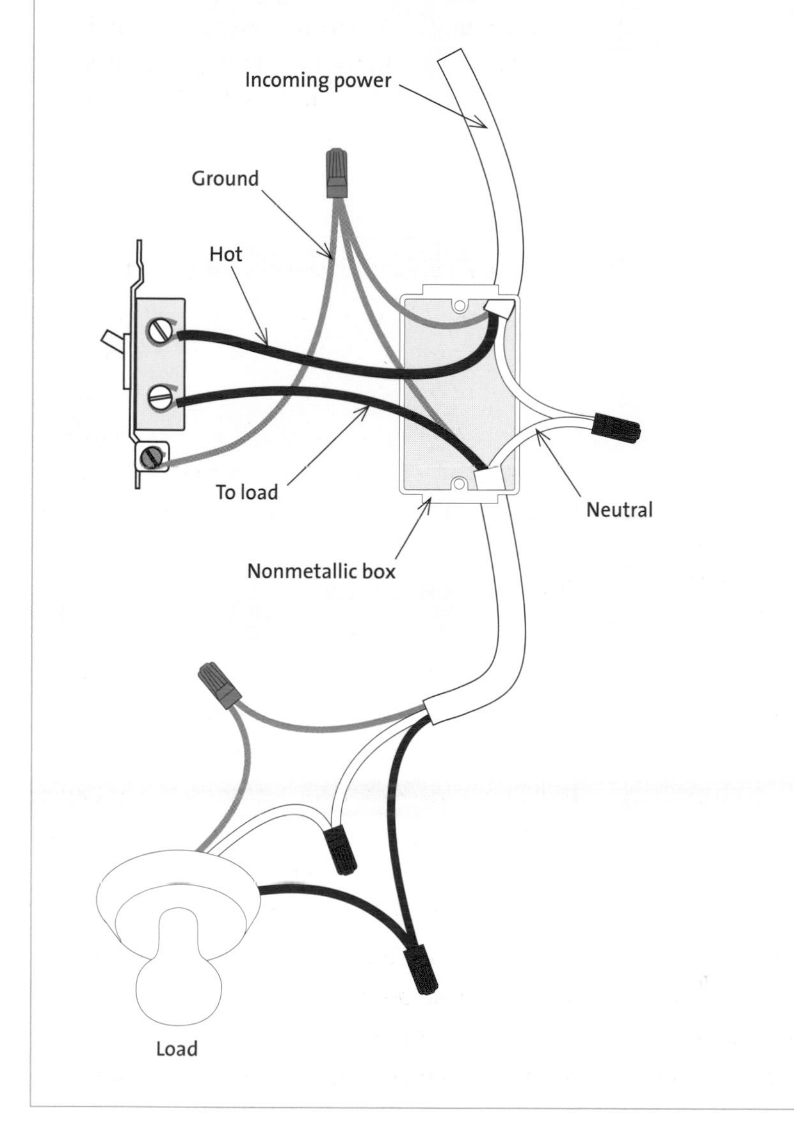

White wire (taped black) becomes hot feed to the switch because white is not allowed to provide power directly to the load.

Not Preferred: Power Cable to Light Fixture

Incoming power cable (feeder) is brought directly to load, or light. Splice is made at load, not in switch box. From splice, one cable goes to switch. White wire to switch should be hot, not black wire, because black wire of same cable must be hot (switched) feed to load. In this case, adding black tape to hot white wires to indicate they're hot is required.

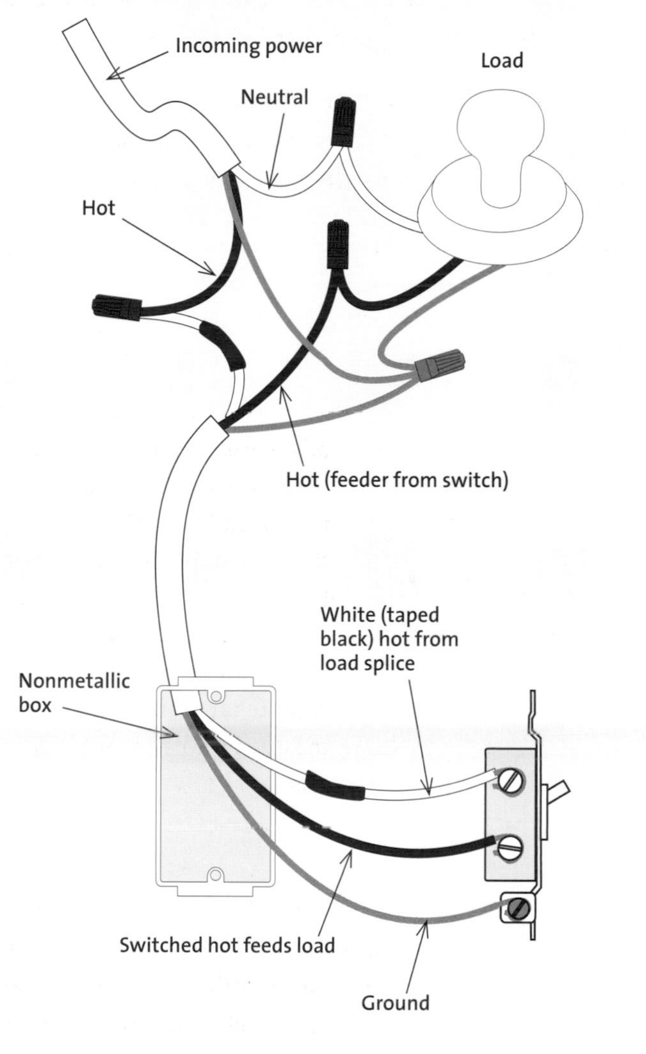

Three-way switches

A three-way switch controls a load from two locations, such as at the top and bottom of a staircase or at each end of a hallway. A three-way switch has three terminals: a COM (common) terminal, noted by a dark-colored screw, and two terminals for the hot traveler wires that have no polarity (the switch has no on or off). The COM terminal is also called the "tongue," which explains how the switch

Troubleshooting Three-Way Switches

Problems in three-way switches normally come down to improper wiring. Forget trying to find out what you did wrong. All you need to know is how to wire it correctly. Let's assume we have two three-way switches and a light and it's obviously wired wrong. Follow these steps:

1. Remove power and pull the switches from their boxes.
2. Remove all the wires from the switches and set the switches aside. Bend the wires off to the side, grouping them to their cable. Do not let bare wire tips touch anything. Remove the load and pull its wires off the feed cable (you can now see all the wires that are in the circuit—the load's wires and the wires in both switch outlet boxes). The job now is to see which cable is which.
3. Locate the power cable, the load cable, and the cable to the opposite switch. To identify them, throw the power back on. Measure across each cable's white to black conductors. The cable reading around 125 volts indicates the hot or power cable. Remove the power and mark the power cable.
4. To find the load cable, take a continuity reading. A low reading across two of the conductors would be like the filament of a bulb. Typically the neutral of the load should be one of the white wires. If that fails, simply try to physically match the type of cable to what is going into the load and do a continuity check for verification. For example, twist together the two wires of a two-conductor cable and take a reading at both outlet boxes. Whichever cable gives you a "short" reading is it. You now know where that cable goes. Do this to any cable you are not sure of. Once all cables are known, untwist wires and draw a picture of what you have.
5. Follow my three rules of wiring a three-way switch and you will have it wired properly. Take the hot wire to the COM terminal of switch one. Take the neutral to the load neutral and take the black wire from the load to the opposite switch's COM terminal. Connect the two extra screws of the first switch to the two extra screws of the second switch—no polarity. Connect ground wires. Place wire connectors over any left over wire ends. You're done.

works: It laps up to connect to one traveler terminal and down to connect to the other.

Although there are myriad ways to wire a three-way switching system, I only pick from two. Either run the three-conductor cable from switch box to switch box (and wire according to the drawing) or run the three-conductor cable from switch box to light to switch box (and wire according to the drawing). Always use three-way switch cable, not standard two-conductor cable, and run the cable from the first switch to the load and then to the second switch.

There are alternate ways of three-way switching that are worth mentioning only because you might run across them (see drawings on pp. 245–246) and need to know how they are wired. Others I ignore due to the sheer illogic of their wiring. The alternates involve bringing the power cable into the light as

Wiring a Three-Way Switch

Three-way switches control a load from two locations. Here, the hall light could be controlled from either end of the hallway. Taping the traveler white wire black is required.

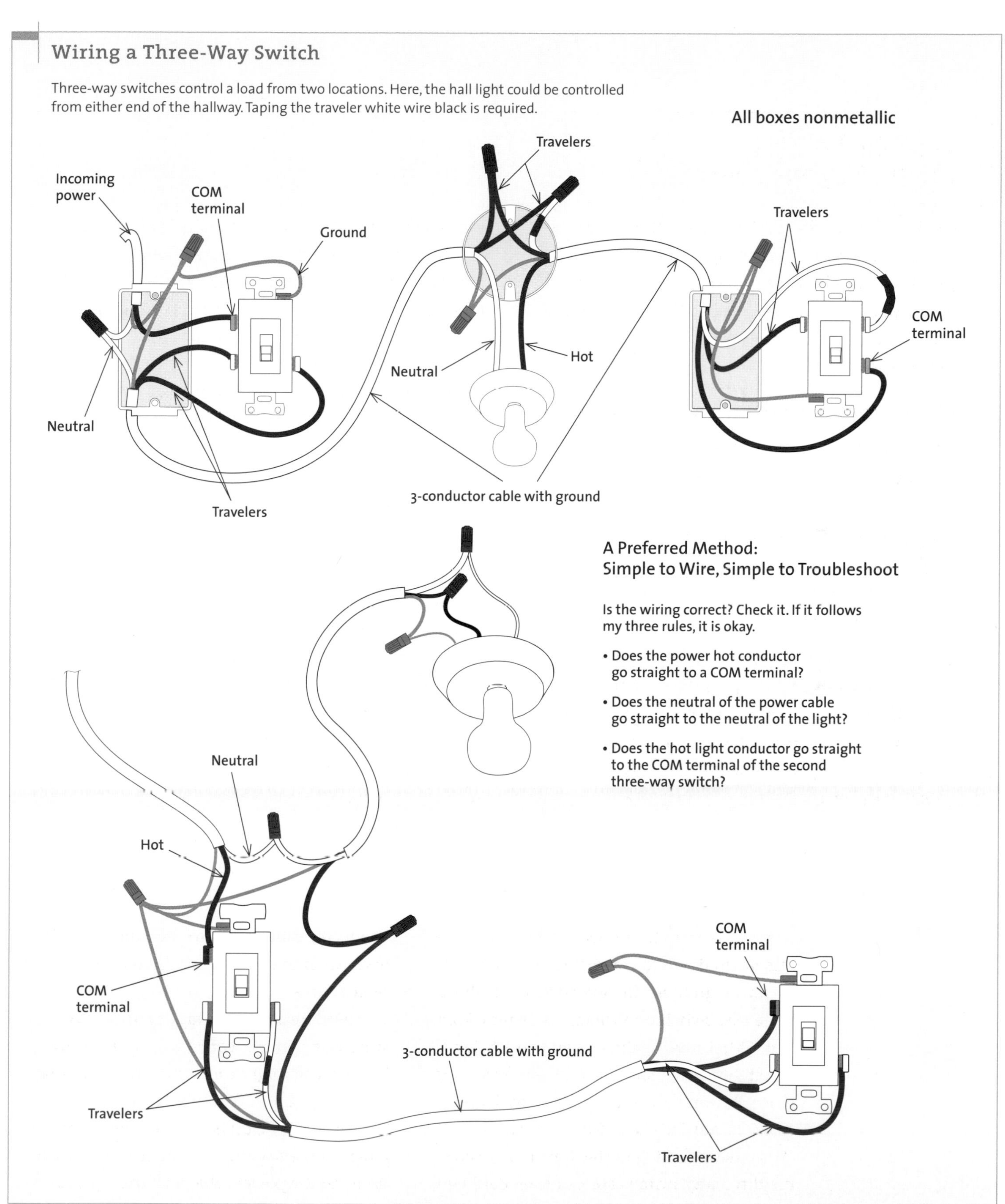

Wiring a Three-Way Switch Alternate Method

A preferred method only if light is immediately adjacent to distant end switch. Easy to wire, easy to troubleshoot.

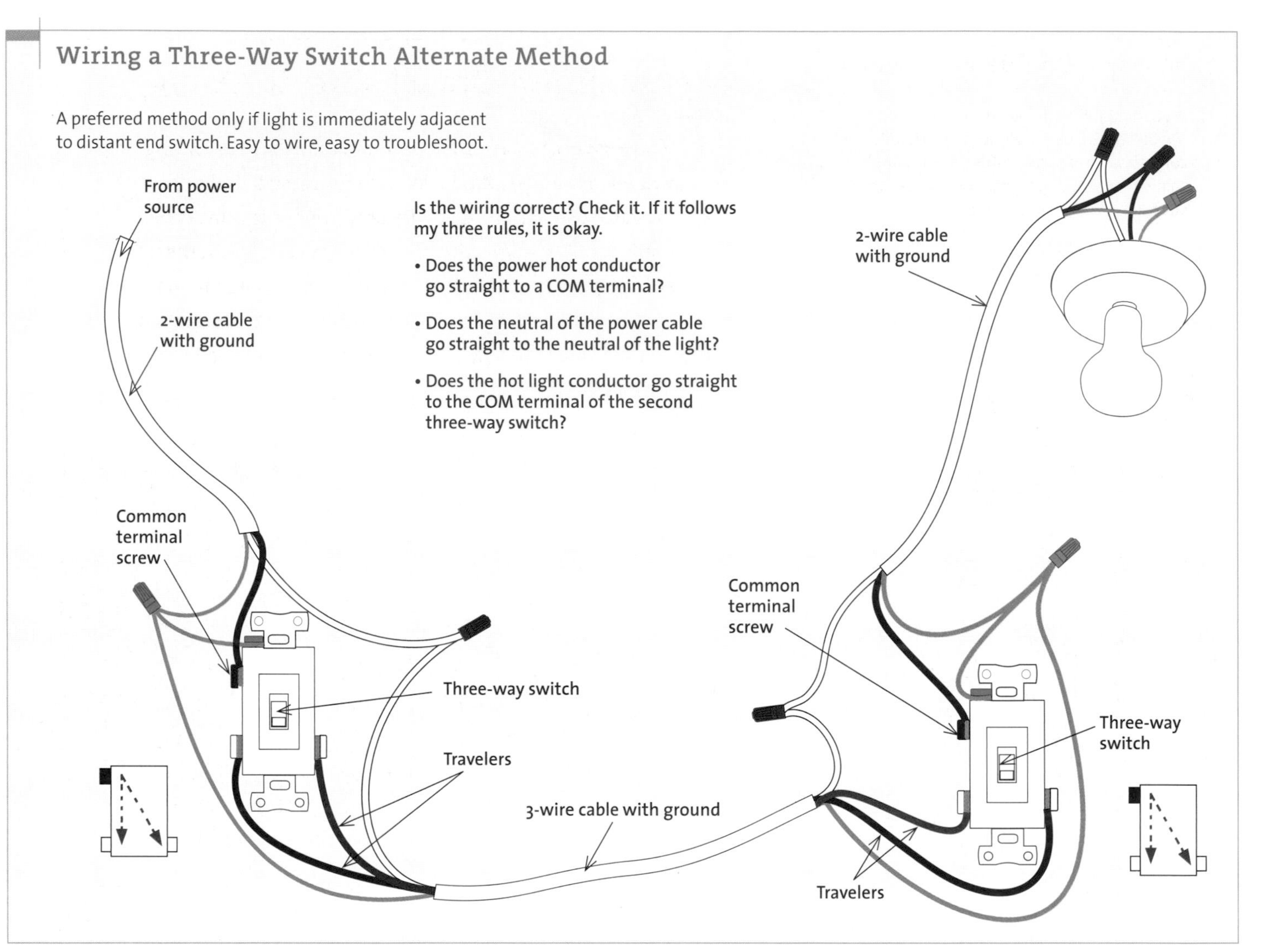

opposed to the smarter method of bringing the power cable into a light switch so you don't have to tear a light apart to verify incoming power.

Four-way switches

For additional point-of-load control between three-way switches, install a four-way switch. There is no limit to the number of four-way switches that can be installed, as long as they are installed between the three-way switches (a three-way switch must always be on each end of a multiple-point switching circuit). Four-way switching is used for rooms with three or more entries with switched lighting at each one.

The switch's box may have a schematic on how to wire the switch. It's easy to wire four-way switches incorrectly, so use a continuity tester to verify the switching arrangement (see the drawing on p. 249). I connect the incoming travelers to the bottom switch terminals and the outgoing travelers to the top terminals

Wiring a Double-Pole Switch

A double-pole switch controls a 240-volt appliance and uses both legs of a circuit. Its most common use is as a cutoff switch for appliances that cannot be unplugged for maintenance, such as a submersible pump.

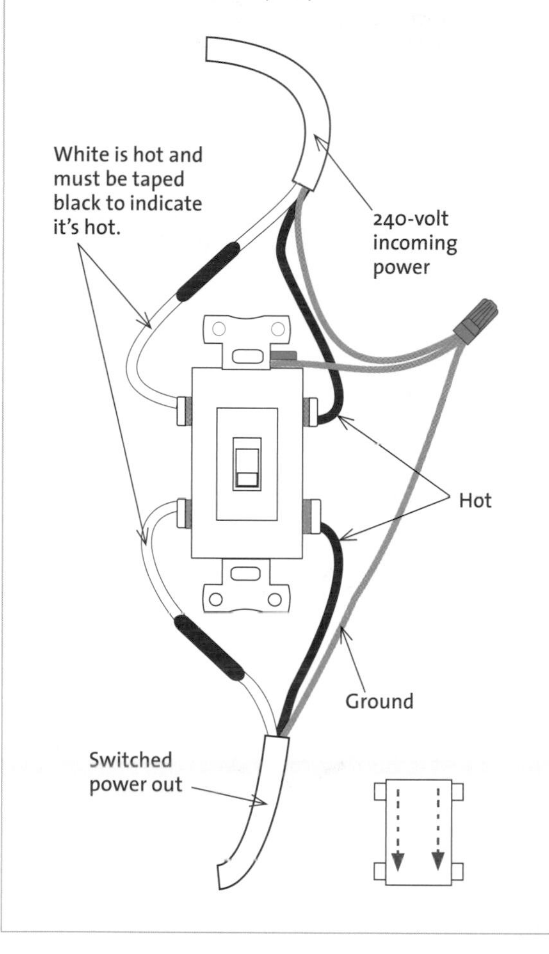

(verify the terminal connections). Don't attach each set of travelers to one side of the switch. Doing so will throw a short on the circuit.

Specialty switches

Switches are available for just about any switching condition that exists. Lighted toggle switches that glow when off are perfect for dark locations or even a child's room

Aluminum Branch Circuit Wiring

If you have a house with aluminum branch circuit wiring, you are not allowed to install a common switch that's rated for use with copper wire only. Your switch must be made for aluminum wire as well as copper. Luckily, these switches are available off-the-shelf at most large home centers and electrical suppliers. Look for CO/ALR stamped on the front of the switch (on the metal yoke) and in raised lettering on the back.

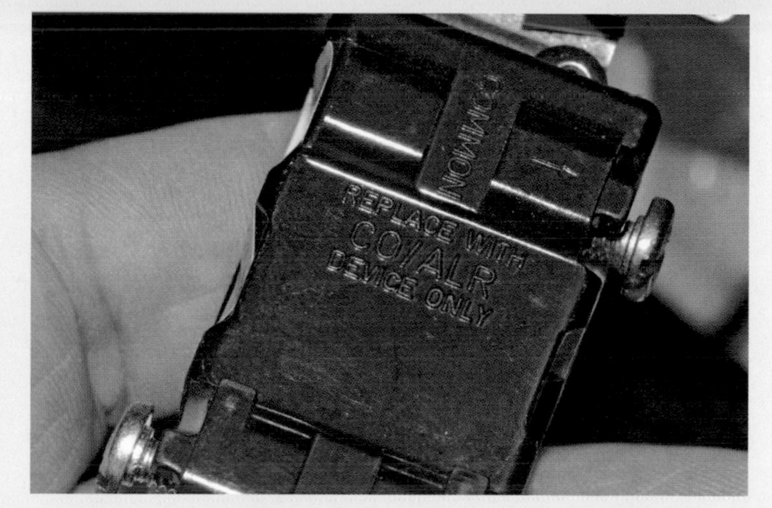

Switches for aluminum branch circuit wiring. Such switches will be marked front and back with CO/ALR.

Three-Way Switch Rules

Worried about whether you wired the three-way switch correctly? All three-way splicing situations, no matter how complicated, follow three rules. Get them right and the wiring is right.

1. **Take the power cable hot wire (normally black) straight to one COM switch terminal.**
2. **Take the power cable neutral straight to the light neutral.**
3. **Take the black light wire to the COM of the second switch. Then connect the travelers—no polarity.**

Although more expensive, the larger 1,500-watt dimmer (left) has a large heat sink, enabling it to handle more lights and last longer than the standard 600-watt dimmer (right).

so the child always knows where the light switch is. Pilot-light switches that glow when the load is turned on are very useful when a switch is at a distant location from the load. Keyed switches are available for situations that require extra security. Timer switches are available for temporary loads, such as bathroom fans. Programmable timer switches are available to turn on lights, a radio, or a television at preset times so that people will think someone is home when the house is empty.

No matter what the use, however, most specialty switches are wired the same way and have the same two terminals that an ordinary switch has. The one thing to look out for is whether the switch is polarized, which means each wire must go to a specific screw or location on the switch. With a common snap switch, it doesn't matter which wire goes where.

Light Dimmers

Dimmer switches were developed to allow more control over lighting than just on and off. As a side benefit, they also save energy and extend bulb life by reducing the amount of electricity used. Dimmer switches are available in single-pole and three-way models. High-tech electronic multilocation switches allow you to dim the lights at a number of locations from one central dimmer switch.

Single-pole dimmers only can be used for incandescent lights. Never use a dimmer to control a ceiling fan or any other motorized appliance—you'll burn up the dimmer and the appliance. Although a table lamp will work fine, you will damage all other appliances plugged into that receptacle.

Four-Way Switching

A three-way switch must always be on each end of a multiple-point switching circuit. Taping the traveler white wire black to indicate it is hot is required.

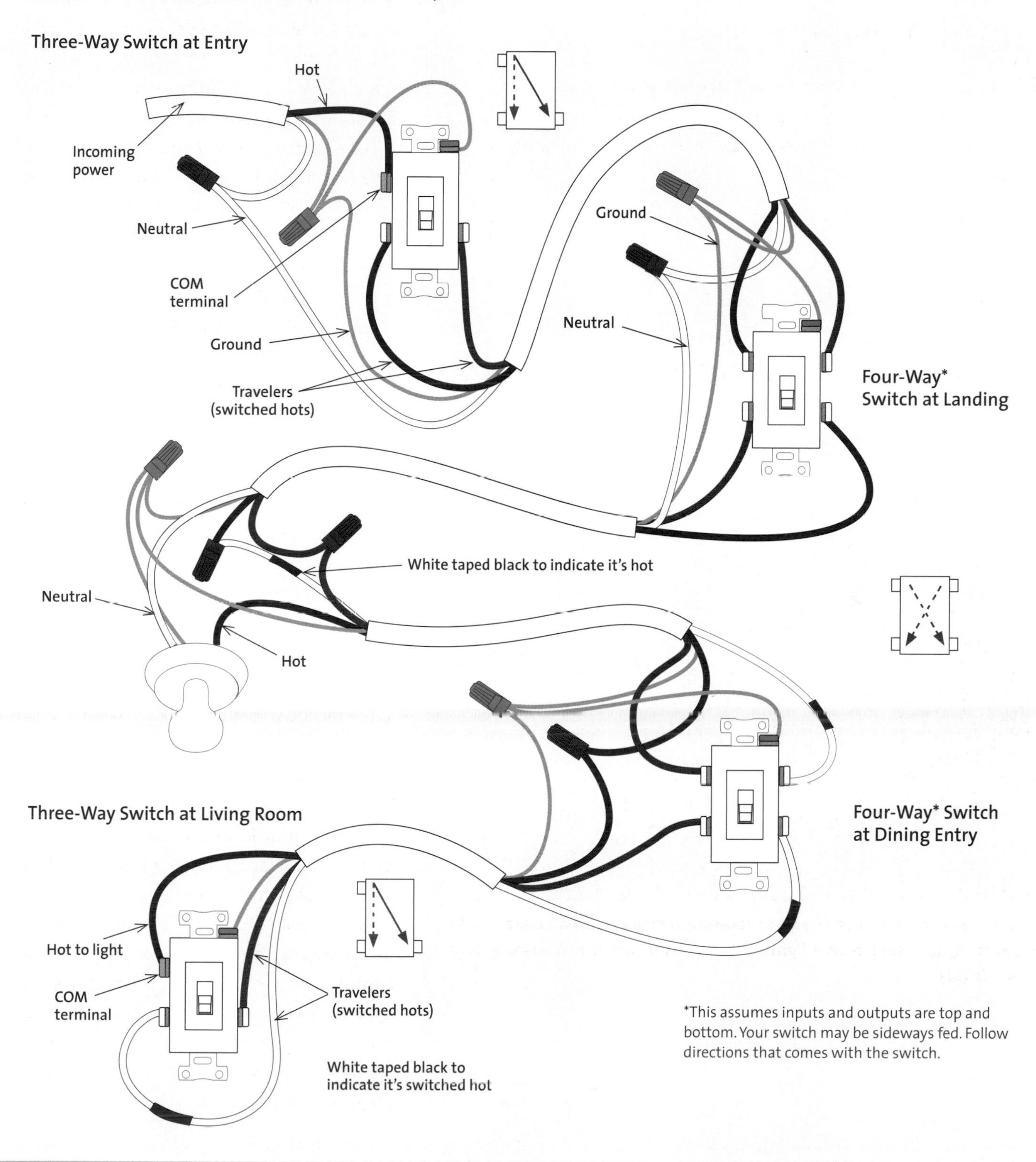

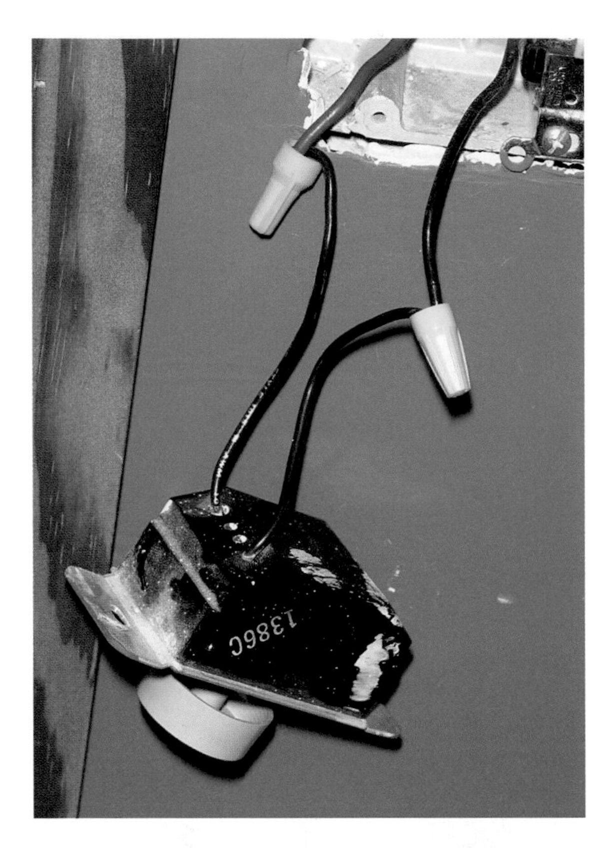

Dimmers are quite easy to wire. Take the two wires that went to the switch to the two dimmer wires. Newer dimmers may have a ground wire. Normally, no polarity is observed.

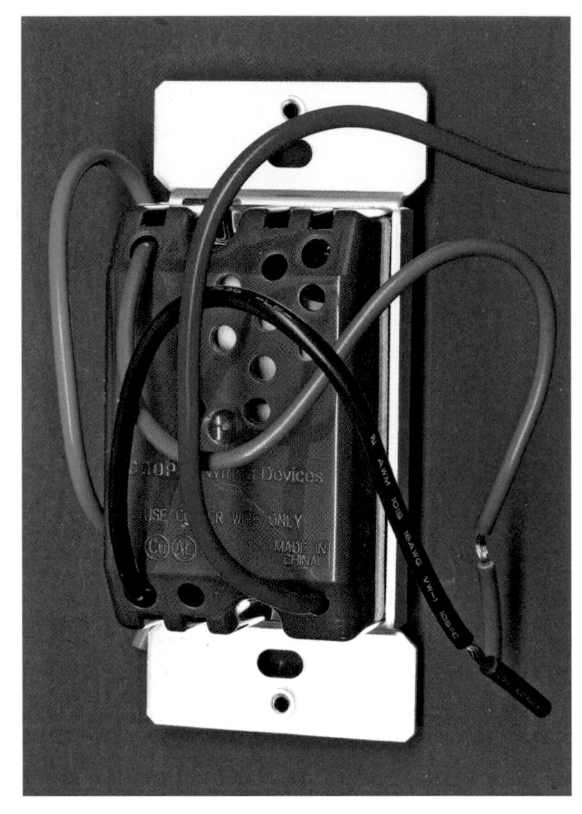

In electronic dimming, the master switch may have three wires plus ground. However, the COM terminal no longer exists.

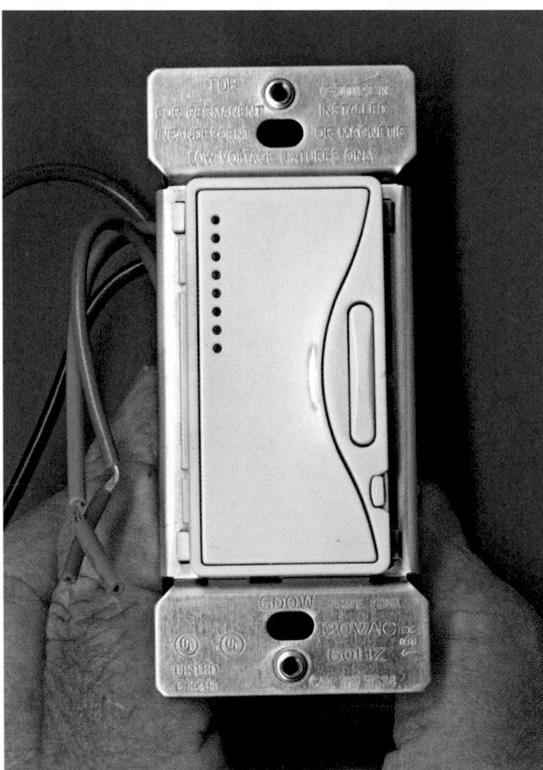

The master switch of a three-way switch dimming system (Cooper Wiring Devices) where both end switches can be dimmers. Install this one in the box where the power cable comes in.

Wiring dimmers

Single-pole dimmers can be installed by simply replacing a single-pole switch. However, three-way dimmers cannot simply replace every three-way switch. Only one of the two three-way switches can be replaced with a dimmer; the other must remain a standard three-way switch.

For standard dimmers, the rough-in wiring is done the same way as for standard switches. If there will be only one switch, the two wires that would have gone to the standard switch simply connect to the dimmer. If a three-way system will be dimmed, choose the switch side on which to install the dimmer. If a four-way

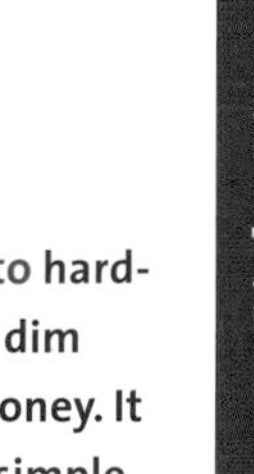

You do not have to hard-wire a dimmer to dim lights and save money. It can be done by a simple plug-in dimmer. This table-top dimmer plugs into a receptacle outlet and the lamp plugs into the piggy-back. A hand-held device at the lamp allows touch-slide control at any light level.

or larger system will be installed, the logic is the same—only one dimmer in the system. Dimmers with two black wires have no polarity, so it doesn't make any difference which wire gets hot or neutral. If the dimmer has a black and a red wire, the black normally goes to the incoming power source; the red to the load.

Fluorescent dimmers

You cannot dim standard fluorescent lights because their ballasts are not designed for dimming. Instead, special electronic ballasts or magnetic dimming ballasts are required. Magnetic dimming ballasts work with T-12 fluorescent bulbs. This type of ballast uses large magnetic components that may emit a buzzing noise, and the lamps may visibly flicker at low light levels.

Work Safe • Never wire a dimmer hot. That is, never hook up a dimmer to an electrically hot wire—turn off all power first. The surge will burn up the dimmer's electronics.

Common problems with dimmers

No product is problem-free, and dimmers are no exception. Specific problems include noise, buzzing, overheating, and a short life due to surges coming from light bulbs that blow.

Noise Dimmers are noisy—sometimes it's noticeable, sometimes it's not. They put radio-frequency interference (RFI) on the power line on both sides of the dimmer circuitry—the load and the entire branch circuit—through conduction and radiation. Any load on the

Take Note • Dimming fluorescent fixtures is not logical or cost effective on the residential level. It is extremely expensive and has to be customized to the specific lamp. Even in commercial and industrial settings these typically are only cut back a fixed 1 percent, 5 percent, or 10 percent.

Work Safe • When remodeling, look for overhead fans with remote-control units. The remote is convenient and keeps you from having to rip out a finished wall to install a switch. Fans come with a remote or a remote can be purchased separately and installed later.

Take Note • Dimmer switches, especially electronic dimmer switches, have a top and a bottom and cannot be set sideways in the wall, but must be vertical.

dimmer branch circuit receives this noise. For example, you may hear it through stereo equipment, broadcasting equipment, intercom systems, public-address systems, and cordless telephones.

There are several ways to eliminate or reduce this noise. The first option is to buy a dimmer with a built-in noise filter. Next, isolate the lighting branch circuits from the receptacle branch circuits. It's also important to make sure all ground connections are firm. A poorly grounded or ungrounded light on a dimmer circuit generates noise like a radio antenna. Last, look for dimmers that are advertised as being quiet.

Lamp buzz As the current to the lamp is being adjusted, the lamp's filament will buzz (it's more noticeable with bulbs above 100 watts). All lamps buzz to some extent, but the problem can be minimized by installing high-quality light bulbs. The difference is in the filaments—for instance, rough service bulbs have heavier filaments that reduce or eliminate vibration.

Heat and overloading This problem is typically created by designers or homeowners who create lighting systems that pull too much current through the dimmer. The solution is simple: Don't exceed the dimmer's wattage rating. This problem is extremely common with track lighting and with several light banks ganged together. If extra lights are added to the track lighting or if the bank's bulbs are replaced with higher-wattage bulbs, the dimmer can easily become overloaded.

A standard low-wattage dimmer uses its aluminum mounting plate as a heat sink. Because heat is the bane of dimmers, the thicker the heat sink, the better. Higher-wattage units, such as 500, 1,000, and 2,000, dissipate heat through heat sinks mounted at the wall plate. If dimmers are ganged together in the same box, some of the fins on the mounting plate may have to be broken off so that they can overlap, reducing the size of the heat sink. If this is the case, the wattage of each unit should be reduced by 20 percent to 50 percent (for example, a 1,000-watt dimmer should be reduced to 800 watts, or eight 100-watt lights).

Is The Switch Killing the Bulbs?

Turn on the switch and bang!—the bulb blows. Is there a problem with the switch? No. Is the problem with the way the switch releases the surge current in the wires? No.

Manufacturers will give you a whole list of reasons, including these two, but none of them points the finger back to the bulb makers and to their thin, tiny, overheated, fragile filaments. I tested one of the new daylight-type incandescent bulbs when they first came out—the ones that appear blue when you buy them. Most didn't last a week. A few lasted only a few hours. I contacted the manufacturer and sent them a box full of bad bulbs (at their request so they could find the problem). And find the problem they did: Microscopic holes in the glass around the base. They knew they had this problem but thought they had it fixed. They replaced the whole bunch. I must admit I was surprised at their honesty.

It's true that longer, thicker filaments will make a bulb last longer, but still, the bulb has to be made without holes! Also, 130-volt bulbs will last significantly longer than common off-the-shelf bulbs. Bottom line: Keep the receipt when you buy bulbs and take them back if they blow within the free take-back period. If beyond that time, and you have several, contact the manufacturer.

Fan Control Speeds

There are two types of fan speed controls: fully variable, also called full range, and step control. Both types of controls may be used on a paddle fan, regardless of the number of speeds the fan has—simply set the fan's speed on the highest setting as it's installed. The wall-mounted speed controller can then vary the fan through all its speeds. Be sure to read the directions that come with the fan. A manufacturer may indicate a specific control that should be used with its fans.

Noise is a problem with fans as well. I remember the first variable-speed control I installed and how loud the humming noise was when the fan was on. I've been trying to find quiet controls ever since. Step controls normally do not hum, but they can be used with only one fan and have a maximum operating current of 1.5 amps. Fully variable controls are noisier but are able to control the full-range speed of a single fan or multiple fans wired in parallel. The maximum current depends on the model, but 2-amp to 12-amp models are pretty common. As a fan's speed is reduced, the fan itself can emit an annoying hum. Make sure the controller addresses that issue as well.

Wiring Fixtures

Light fixtures help make a house a home. They add a personal touch to any house and should only be picked out by the homeowner—never by the contractor.

Planning is important. Even though fixtures are installed after the finished walls are up, you should decide beforehand which fixtures you want. Each fixture has advantages and disadvantages that you should know about before installation—even specific height and width requirements. All this knowledge is needed so that you use the correct rough-in ceiling box and put it in the proper location. Bottom line: Order the lights ahead of time and keep them on site so that you can verify that they will physically fit in their locations. On one job the owner didn't get the outside lights until the house was built. It was only when I started installing them that I discovered they were too tall to fit the locations without cutting into the soffit. All the lights had to be returned.

Choose the Right Box

In general, ceiling boxes are round, not rectangular like receptacle boxes. Old-style boxes are 3½ in. in diameter, whereas the

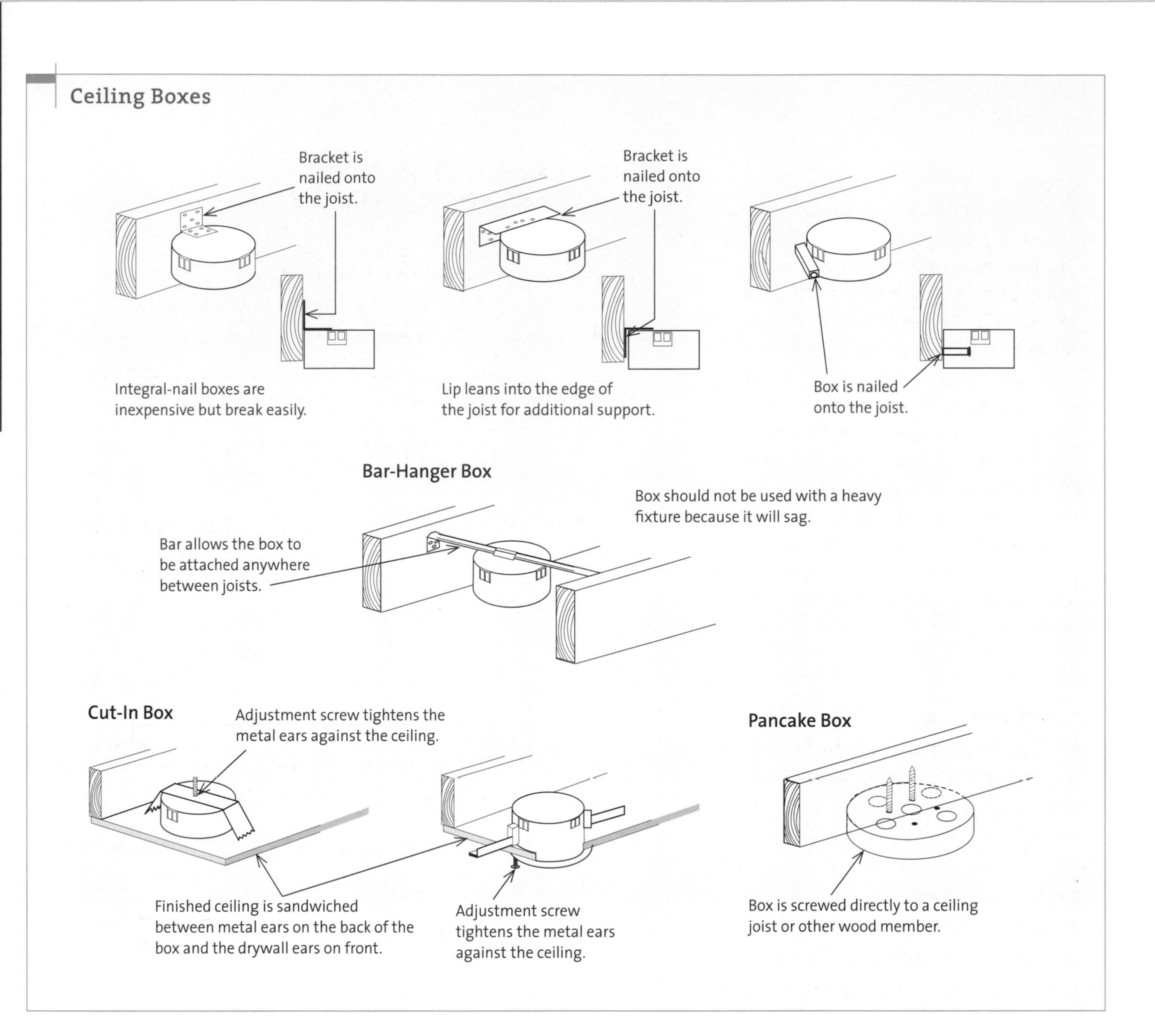

new-style boxes are 4 in. in diameter. The basic difference between a round box and a rectangular box is the size of the screw that holds the light fixture to the box. Round boxes use a thicker screw (a #8 rather than a #6 for common rectangular receptacles or switch boxes) and thus can support a heavier weight (fixture) before ripping out the box's female threads or stripping the threads on the screw. However, for lightweight fixtures that weigh no more than 6 lbs., such as smoke alarms and small outside light fixtures, rectangular boxes may be used.

Many designs, many problems

There are many types of round ceiling boxes from which to choose. Standard cut-in boxes are made for renovation work but are limited to lightweight fixtures. Thus, if you want to

Above Code • Now that you know to use round ceiling boxes to support heavy lights, the next logical question is how much weight can they hold? The answer: 50 lb. Attach lights heavier than 50 lb. directly to the joist and use the outlet box only to house the wire splice. The only exception is when the manufacturer specifies that the box can support more than 50 lb. However, it's been my policy not to hang a light on an outlet box if it weighs more than 25 lb. If the screws pull out of the box and the irreplaceable light fixture is destroyed—or falls and hits someone on the head—you don't want to find yourself arguing about why it wasn't your fault.

hang something heavy you'll have to cut into the ceiling and attach a support. Bar hanger boxes are commonly used in renovation and new construction. These use 16-in. to 24-in. bars to span joists, allowing the box to slide anywhere along the bar. However, they tend to sag with even moderate weight, and I don't use them to hang anything. The more common boxes use captive nails or brackets for a fast, easy installation. But those have problems, too: One too many hits on a captive nail and the plastic arm snaps. And the nails bend easily. If you are nailing into oak or locust, forget it—you'd better predrill before nailing. That leaves boxes with brackets. Such boxes come with both narrow and wide support brackets. Always use the wide ones, which are much more stable and provide a more secure attachment. Typical ceiling boxes, regardless of design, have volumes ranging from around 13 cu. in. to 23 cu. in. Always try to buy the deepest ones.

Three Ways to Use Bridging to Support a Ceiling Outlet Box

If you can't attach the box to a joist, install 2x6 bridging or a fan-mounting bracket between joists.

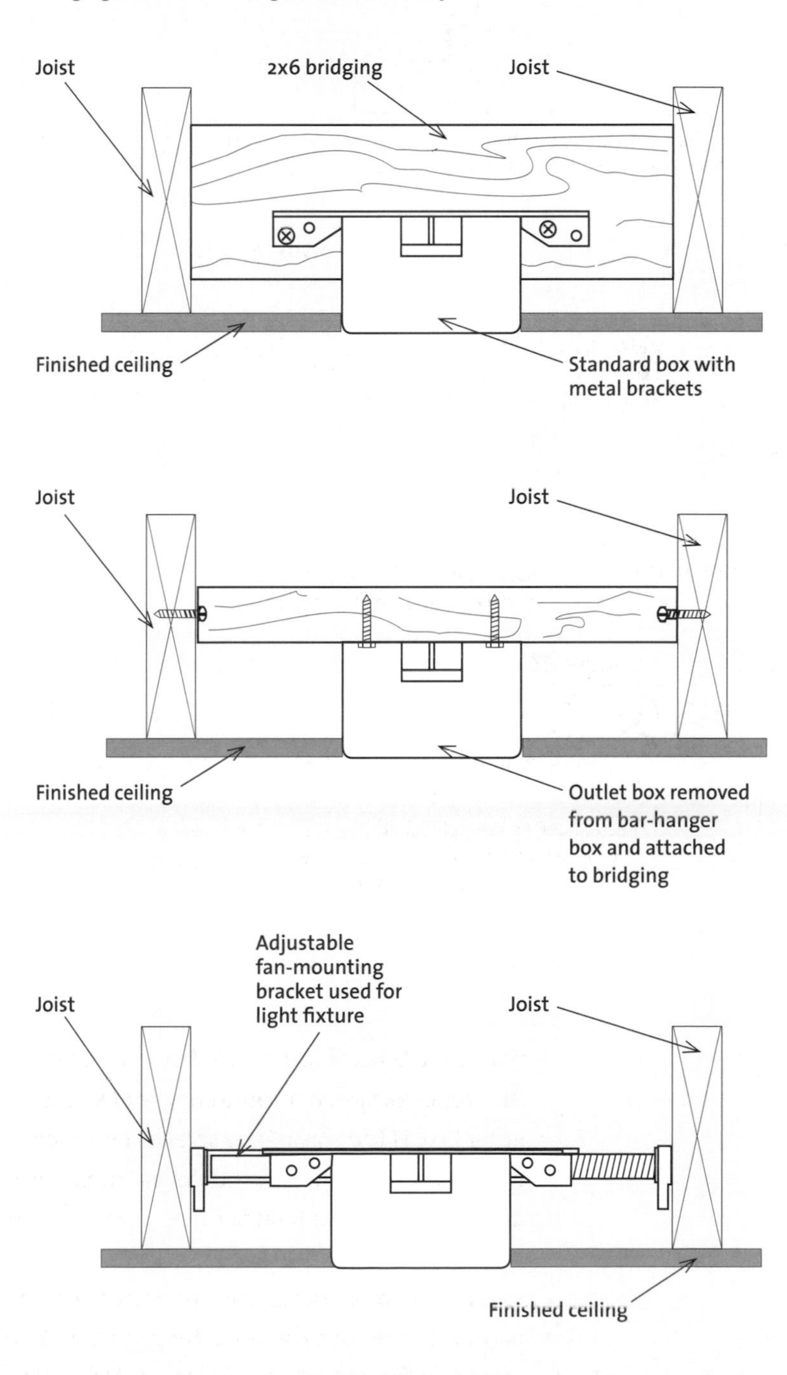

Instead of forcing an overhead light to be only where a joist is located, you can install the light fixture wherever you want. If you are using a bracket box, nail a crosspiece on edge from joist to joist and attach the box to it. Even better, nail a scrap 2×6 or 2×8 flat across the joist span, remove the box from the bar-hanger, and screw it directly into the wood (use two screws). Before nailing the board into place, remember to recess it into the joists at a depth that allows the outlet box to be flush with the finished ceiling. Last, an adjustable fan-mounting bracket can be used to hold a light fixture (as well as a fan) between joists.

Wiring an Incandescent Light Fixture

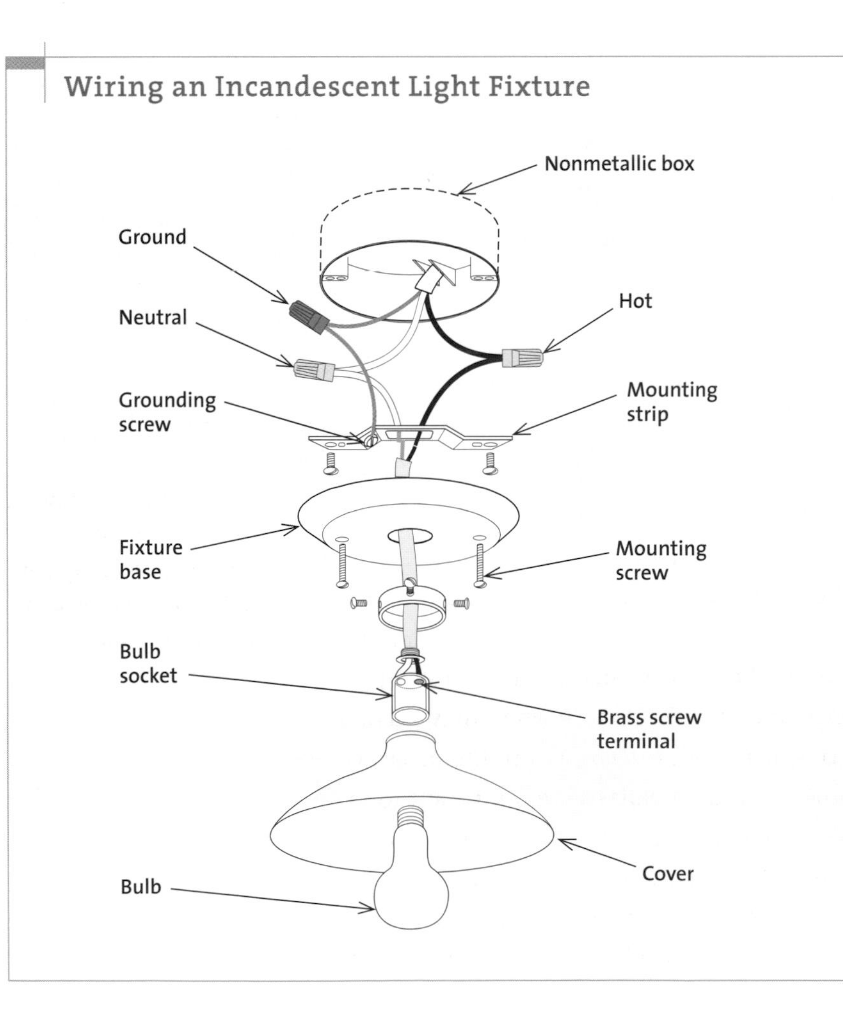

Metal pancake boxes are used to support light fixtures when you simply don't have the physical space to install a deep box, such as when the box must be surface mounted. Pancake boxes are normally ½ in. thick and 3¼ in. to 4 in. in diameter, with volumes ranging from 4 cu. in. to 6 cu. in. Light fixtures attach directly to them. When using this box, make sure that you do not create a cable-fill violation, which is easy to do because of the box's shallow thickness and low volume—it can only take a single 14-gauge cable. If volume is a problem, use a box with knockouts for the cable, not a box with cable clamps built in, to gain a bit of volume.

Pancake boxes are also used to support light fixtures with domed canopies, which are usually seen on the outside of a house adjacent to entries and exits. They're also commonly used to mount on solid wooden supports, such as exposed beams. Drill straight through the beam so the cable enters the center of the pancake box. The problem here is that sometimes the light fixture's base is not deep enough to cover the box. If this is the case, you will have to rout out or chisel a hole for the box to sit in.

Incandescent Lights

I do not care for incandescent light fixtures that totally encase the bulb. The heat generated by the bulb transfers to the light fixture and sometimes fries the wiring—especially if you install high-wattage light bulbs. Most standard incandescent overhead light units are 60 watts maximum per bulb. If you need more illumination, look for a light fixture that is UL rated for higher wattage. In addition, look for a light that is easily disassembled for bulb replacement. Standard bulb bases are

Work Safe • Stay away from any fixture that uses plastic female threads to screw the bulb into the fixture. To make connection to the bulb, they have a tiny metal strip in one of the female threads. These tend to make the bulbs flicker or not burn at all because of poor contact with the metal strip.

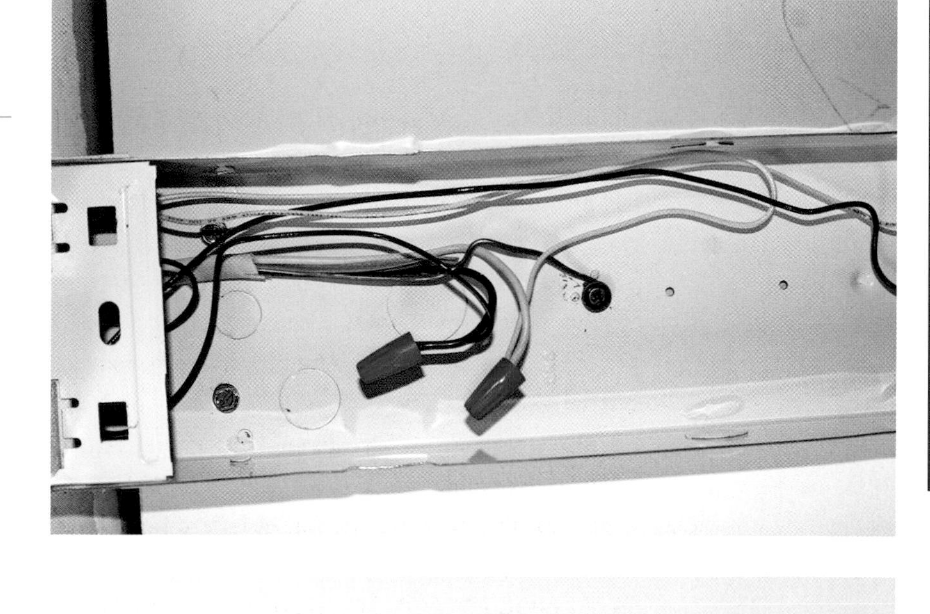

sturdier than small, fancy, specialized bases. Avoid large, round bulbs with extremely small, threaded bases because they snap easily at the base/bulb junction. When installing the fixture, do not throw away the insulation above the base (as is commonly done because it gets in the way of the attachment screws), because it keeps the heat of the fixture from coming through and frying the wires and wire connectors.

Rough-in installation consists of mounting the box and inserting the cable to power the light. After the walls are painted and the fixture is added, attach the incoming neutral conductor to the silver screw or white wire, and then attach the incoming hot conductor to the brass screw. A ground wire or screw on the fixture connects the metal frame to the equipment-grounding conductor.

Fluorescent Lights

Fluorescent lights provide a lot of illumination at low cost and are therefore very popular. Standard fluorescents normally provide sufficient illumination, but if you require more, you can buy high-output and very-high-output fluorescents. Some typical problems with fluorescent fixtures include flickering and buzzing. In both cases, the only solutions are to change the ballast, which sometimes doesn't work, or to install an expensive electronic ballast. Standard ballasts sometimes have a short life span and may short cycle when they become hot, cutting out the lamp until the ballast cools down. Standard ballasts are not meant for cold weather, either. If you plan to install fluorescent fixtures in an unheated shop, garage, or basement, you may have to buy cold-weather ballasts. Corrosion is a problem in damp areas, such as a garage or basement. Both the male

A trick when assembling a fluorescent fixture is to tape the wires up out of the way when you put the cover back on.

Fluorescent fixtures use different bulb designs depending on the bulb's length. These include (from the left): high output (this is what I have above my computer desk), dual pin (used in 4-ft. and smaller bulbs—I hate these), and single pin (very dependable).

pins on the bulb and the socket will corrode over time and produce intermittent problems. Last, some units seem to be so cheap that they have loose connections from day one.

Fluorescent lights come in either enclosed or bare-bulb fixtures. Your choice will be based on aesthetics: Enclosed fixtures are more attractive than bare-bulb fixtures, but the plastic enclosures may yellow over time. The most common fixture lengths are 2 ft. and 4 ft., and these fixtures take bulbs with two pins on each end. For really large rooms, you might consider installing 8-ft. fixtures. The bulbs for 8-ft. fixtures have a single pin on each end, which makes them much easier to change than the smaller two-pin bulbs. Personally, I prefer 8-ft. units no matter what the room size, because there are too many problems associated with 4-ft. fixtures.

Installing fluorescent fixtures is easy. For a hardwired unit, simply take the NM cable through a knockout in the box (use an NM connector) and screw the fixture into the ceiling joists or a flat piece of wood installed between the joists (drill a large hole through the wood to run the cable). Wiring the fixture is simple: black to black, white to white, and ground to frame. Some fixtures are equipped with a plug and cord. These units require a standard 120-volt receptacle.

Never attach a fluorescent fixture directly to the grid frame of a drop ceiling. If the ceiling frame bends even a little, the fixture will crash to the floor (I've seen it happen). Fluorescent fixtures are heavy and should be independently and securely mounted to ceiling joists or 2×6 bridging.

Undercabinet kitchen lighting

Kitchen countertop lighting often features fluorescent lights mounted under the cabinets. Such illumination removes the shadows that are invariably present when using ceiling lights. Try to get a fixture that is narrow—about 1½ in.—so that it can be hidden. Such fixtures normally have built-in switches. Some can be hardwired and others come with a plug and cord. A plug-and-cord fixture requires a 120-volt receptacle nearby. For a hardwired unit, pull the rough-in wiring to just under the kitchen cabinets and mount the lights against the wall so the wiring isn't exposed. Hookup is the typical black to black, white to white, and ground to frame or green wire. The units also can be butted up against each other for a continuous lighting arrangement.

Both open-bulb and covered-bulb designs are available. Open bulbs provide more light, but if the bulb breaks, the glass can go into your food. At my house, I installed covered lights but wound up taking the covers off to get more light—I decided the bulb-breaking situation was too remote to worry

Buying Fluorescent Lamps

If you're tired of the harsh, gray light given off by standard fluorescent bulbs, look for a lamp with a higher Color Rendering Index (CRI). The CRI rates light sources on a relative scale from 0 to 100 (sunlight = 100). Lamps with a higher CRI make people and objects look more realistic.

I am not a CFL (compact fluorescent light) fan—at least in this phase of their evolution. They sometimes die early, their ballasts may not work well in cold environments, they give off a depressing color, and are too dim. They also pollute when they are broken, and require a costly, sophisticated energy-hog recycling system when discarded.

about. Low-voltage halogen lights are also available for undercabinet lighting, but they are better suited for task lighting.

Ceiling Fans

Ceiling fans have been very popular since Hunter® started making them in the late 1800s. Hunter took advantage of the fact that blowing air down into a room makes people feel cooler in the summer. (Moving hot air also helps keep away mosquitoes—another big plus.) In addition, reversing the motor to pull air up to the ceiling circulates warm air down exterior walls, making people feel warmer in the winter.

Before you think about installing a fan, you need to consider a few things. First, you must make sure that you have enough overhead room. Don't put a ceiling fan in a bathroom, where people will raise their arms to undress or dry off. You need at least 7 ft. of clearance from floor to blades; any less and you'll be giving people crew cuts as they walk by. For a low-ceiling room, you can buy a special kit that allows the fan to be installed flush with the ceiling. And you'll also need at least 18 in. from any wall to the ends of the blades for efficient air movement. Another important consideration is the ceiling pitch. The fan must

Summer Cooling, Winter Warming

Summer Cooling

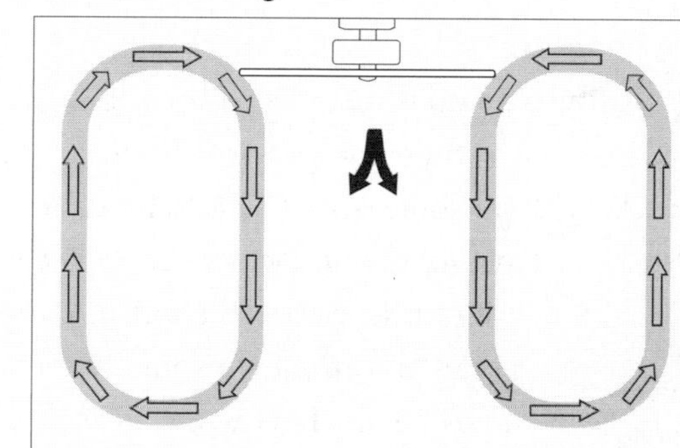

The fan cools by pulling air up along the exterior walls, bringing it to the center of the room, and then forcing it down.

Winter Warming

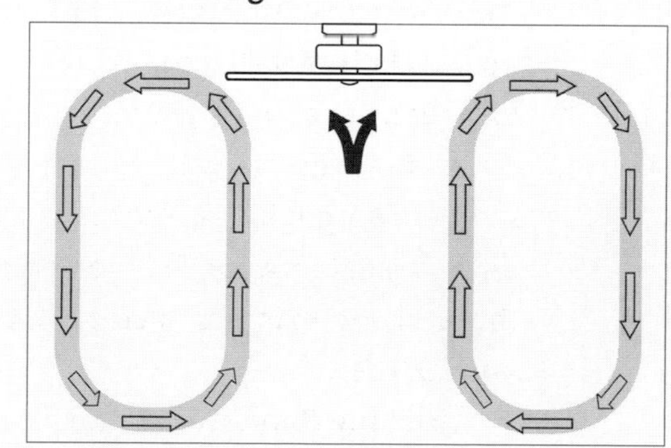

With the fan in reverse mode and set on low speed, warm air trapped at the ceiling is circulated downward along the exterior walls.

hang down far enough so that the blades don't hit the ceiling.

Fans range from about 32 in. to around 52 in. in diameter. To get the most air movement and comfort, buy the appropriate-size fan for the room in which it will be installed. Here are some guidelines for rooms with 8-ft. ceilings:

- For rooms up to 64 sq. ft., buy a 32-in. fan.
- For rooms up to 144 sq. ft., buy a 42-in. fan.
- For rooms up to 225 sq. ft., buy a 44-in. fan.
- For rooms up to 400 sq. ft., buy a 52-in. fan.
- Larger rooms may require more than one fan.

I don't recommend buying a ceiling fan with light fixtures unless you first know the problem you may have with fan wobble. Most fans wobble a little; adding a light fixture accentuates the problem; however, many times the wobble doesn't develop until a year or two later. Why does this happen? Fans from the factory are almost perfectly balanced, but if you put a light kit on it and the light is imbalanced—or you accidentally bend one of the blades—the fan may wobble.

The bottom line is that a good-quality fan/light combination should not wobble, provided that it is installed correctly. Make sure that the fan outlet box is securely attached via screws to a secure mounting, such as a joist,

An old metal splice box in the middle of the kitchen ceiling will not support a ceiling fan.

Rather than attach the fan to the old splice box, screw an approved-for-fan pancake box into the stud over the old splice box and run the wires through the center hole.

and that the fan is tightly mounted to the outlet box. Last, be sure to assemble the light per the instructions, tighten all screws, and do not bend the blades.

Buying guidelines

Try to buy a fan that's designed intelligently. Make sure the fan allows you to hang it at an angle while you wire it, and then, once it's wired, swing it up into place. Nothing is worse than trying to wire a fan that has no swing-up bracket. In the old days, I hung a fan from a homemade rig while I wired it and then removed the rig to install the fan once I had the splices made.

Good fans need not be the most expensive on the showroom floor—just don't get the cheapest ones. For example, better-quality fans have sealed bearings so dust can't get in—

Fan-mounting boxes must be UL approved. The box on the left is a pancake box. The one on the right is a standard ceiling box.

cheaper ones don't (some are shielded on one side only). In my house, I installed three $50 paddle fans and the bearings wore out on all of them within five years (I live on a gravel road and dust is a problem). Here are some other things to consider:

- Low-cost fans have cheap, low-wattage motors (some of which are advertised as energy conserving). To compensate for this, the pitch of the blades is reduced (made more parallel to the floor) so that the motor doesn't burn out. As a result, not much air is moved. Look for pitches between 13 and 25 degrees. Blade length and the motor's rpm also affect air movement. All this adds up to cubic feet per minute, or cfm, of air moved. Look for the highest cfm listed on the box.
- Check the box for bragging points, such as "double the size of most fan motors," "sealed bearings," "sturdy steel motor," and "quiet operation." Also, listen to the display models. If it sounds like the motor is humming, keep looking.
- Look for units that have sealed, permanently lubricated, oilless bearings. Cheaper exposed bearings collect dust and wear out quickly.
- Look for a fan from a high-quality, brand-name manufacturer that comes with a long warranty—so that you never have to use it. The problem with most warranties is that they're so restrictive. Not having the original receipt and/or the box the fan came in can sometimes void the agreement. And don't forget the warranty doesn't cover labor. Still, the warranty provides some indication of the manufacturer's faith in its product.
- Fan design can add to the theme of a room. Don't be satisfied with a standard design if you want something different. I changed one of my typical ceiling fans to a replica of an 1886 cast-iron oval design. Manufacturers make a wide variety of fan styles, including replicas of antiques and even cartoon themes. They also make fans for wet environments, such as an outside porch.

Hanging a fan

When installing a ceiling fan, follow the manufacturer's instructions. For maximum comfort, position the fan as close as possible to the center of the room. The fan must be mounted securely via an approved plastic or metal box (I prefer metal because it is more stable). The inspector may look for an approval sticker as well.

No matter which box you use, make sure it is securely attached to a stable ceiling joist or

Above Code • I install metal UL-approved boxes for paddle fans for all overhead light fixtures in bedrooms and living rooms—whether or not the homeowner is installing a fan. That way, if future homeowners decide to install a paddle fan, they'll already have the right kind of box in place. Fan-approved boxes are easier to install during rough-in than as a retrofit.

Supporting a Heavy Fan

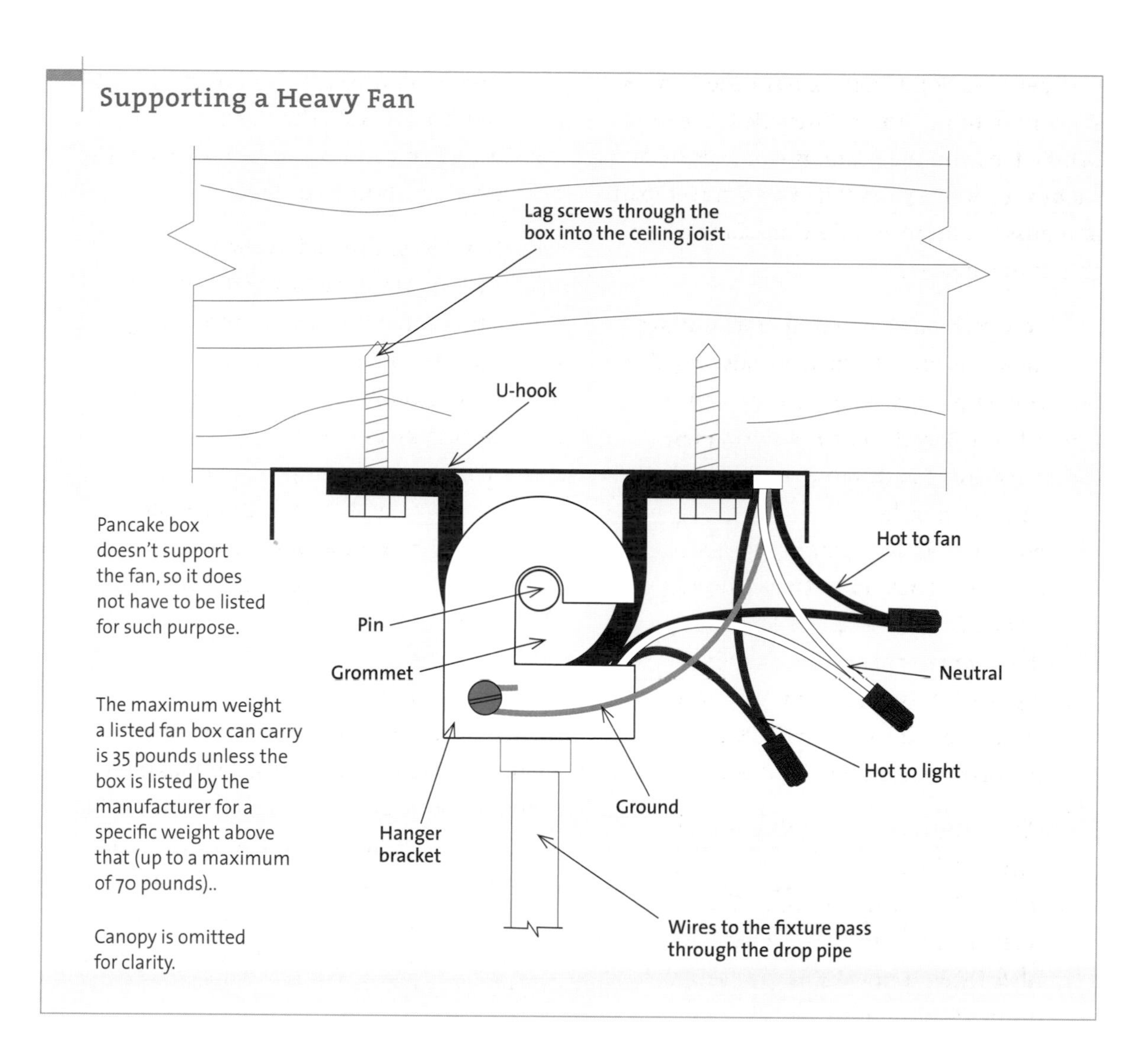

cross support. If the joist wobbles, so will the fan. If a joist is not close to where you need to install the fan, nail a 2×6 bridge from ceiling joist to ceiling joist and mount the fan to it. Approved fan-mount kits are also available so that you can mount a fan from joist to joist without having to go into the attic.

The weight of a ceiling fan is an important consideration. First, the box must be designed and listed especially for the purpose of fan support. Second, only fans that weigh 35 lb. or less, with or without accessories, can be supported directly from a ceiling box. Fans that weigh more than 35 lb. must be supported independently of a ceiling box unless the box is listed by an agency for increased weight. (Some boxes can support up to 70 lb.) For such an installation, the fan manufacturer supplies a U-hook that screws into the wood beam in the ceiling. A rubber grommet fits through the center of the U-hook, and a hanger bracket, attached to the down rod, fits over the grommet pins on the outer edges of the grommet.

The top fan support will hold a medium-weight fan. The bottom support should only be used for light duty.

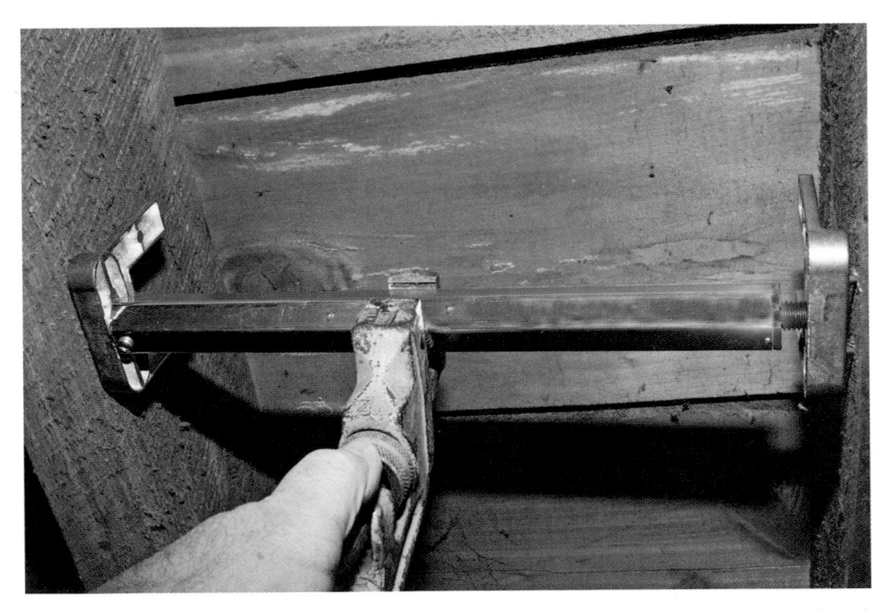

For a heavy-weight fan use a heavy-weight support. Always assume it is your head under the fan if it comes crashing down.

Plastic boxes are now listed for ceiling fans but I am wary of them.

Wiring a fan

Wiring a ceiling fan is quite simple. Just follow the manufacturer's instructions. There is a white neutral, a green or bare ground, and two colored wires—one for the light (if there is one) and the other for the fan (normally black). If you want to control both independently, the rough-in wire must have three conductors plus a ground—a three-way switch cable. The power cable can be brought to the fan, but I prefer to take it to the switch box for easier wiring and troubleshooting. Also, do not use a light control or dimmer as a variable-speed control for the fan motor. Instead, use a speed control that's designed for ceiling fans.

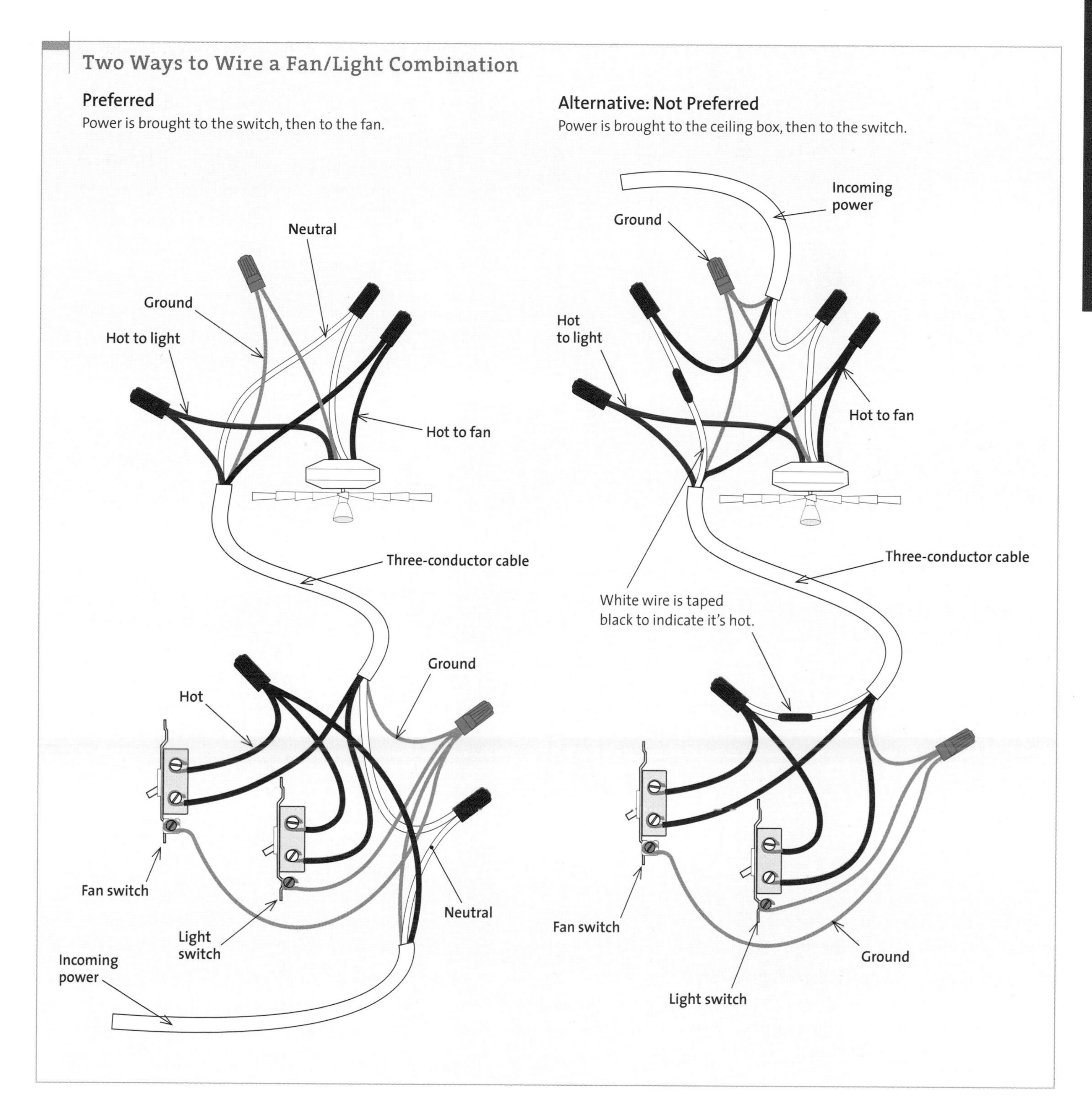
Two Ways to Wire a Fan/Light Combination
Preferred
Power is brought to the switch, then to the fan.
Neutral
Ground
Hot to light
Hot to fan
Three-conductor cable
Ground
Hot
Fan switch
Light switch
Neutral
Incoming power
Alternative: Not Preferred
Power is brought to the ceiling box, then to the switch.
Incoming power
Ground
Hot to light
Hot to fan
Three-conductor cable
White wire is taped black to indicate it's hot.
Fan switch
Ground
Light switch

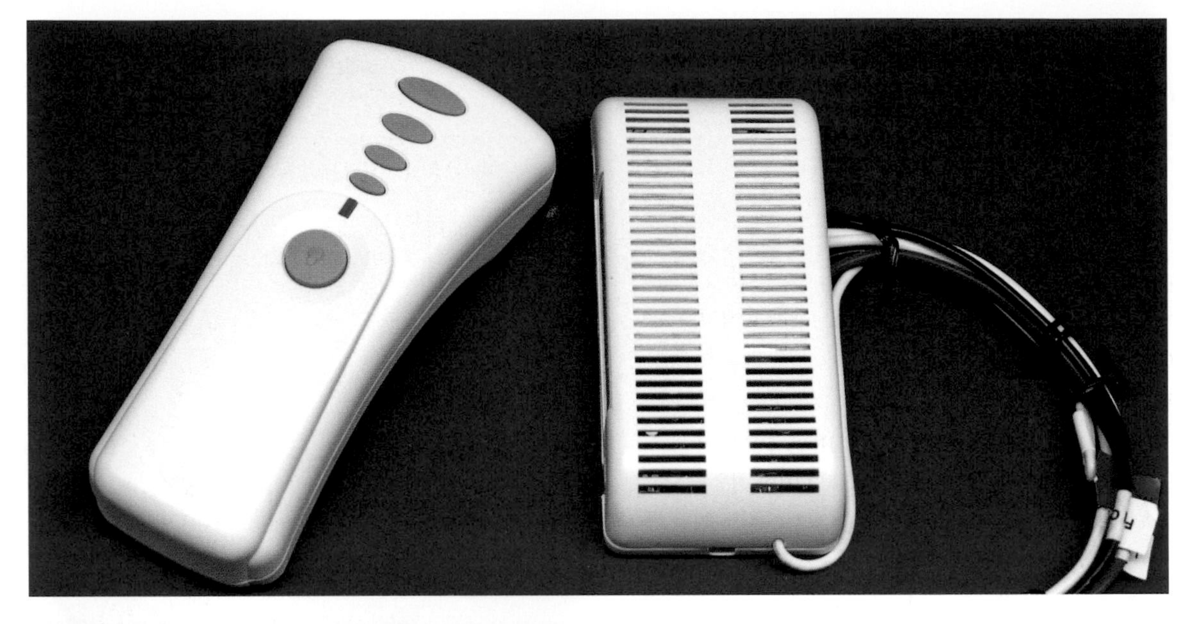

New fans can have remote controls, which means you do not have to have a wall switch.

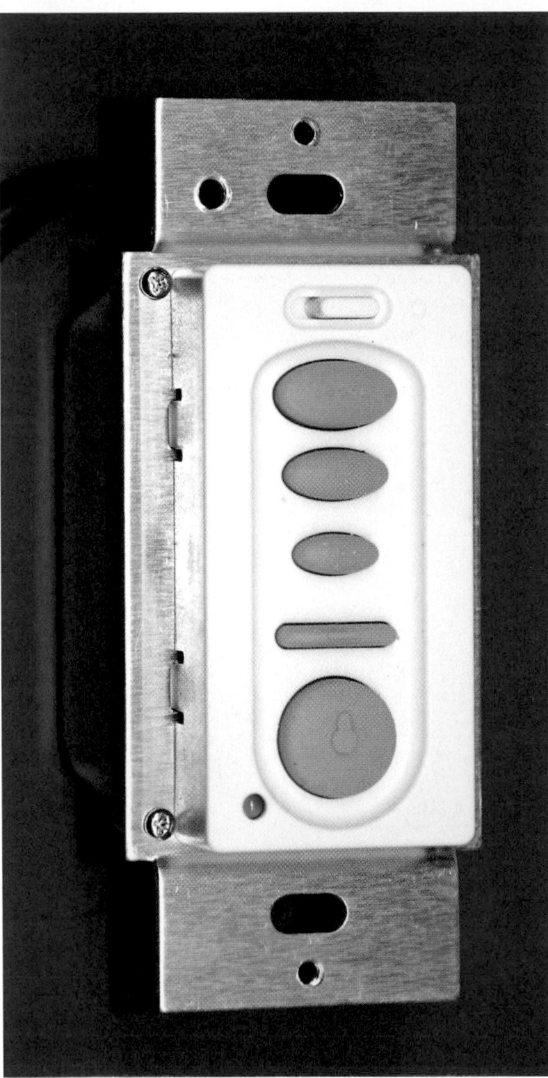

All controls including light and all speeds can be on a wall switch.

Some fans come with a remote control; others you wire the remote later. Either way, a remote is a nice thing to have, because you can control lights and fan speeds without getting out of bed.

Smoke Detectors

When installed properly, smoke detectors save lives. Smoke detectors are required by code to be installed in the hall outside bedrooms. This does not save you if the fire is in the bedroom. Ideally, smoke detectors need to be both inside and outside bedroom doors (this is sometimes required by building codes). If installed in the bedroom, the circuit must have AFCI protection.

Detectors must be hardwired for new construction, and many municipalities also require a battery backup system within the unit. Most areas allow battery units for houses already built when the homeowner just wants to have some protection. Code also requires that all hardwired units alarm at the same time. To do this, all units must be wired together with three-conductor wiring—a three-way switch cable, normally 14-3 with ground.

Track Lighting

Track lighting can provide accent lighting in various areas with only one cable bringing power. The problem with track lighting is that the lamps for one track may have a number of different wattages, which can quickly add up. It's very easy to go from a simple 60-watt circuit to over 1,000 watts by simply plugging a few more lights into the track. Thus, it's easy to overload a circuit. There is no set per-foot wattage in housing other than that the connected load cannot exceed the rating of the track and branch circuit.

Track lighting needs a flat mounting surface. This is not a significant problem with one short piece of track, but it is a problem with a long track or several short tracks if the ceiling has both low and high spots or a sprayed-on compound to make it look wavy. In these cases, you may see gaps between the track and the ceiling or wall.

Wiring a Smoke Detector

If a smoke detector weighs less than 6 lbs., you can use a common rectangular box on the ceiling or wall if desired.

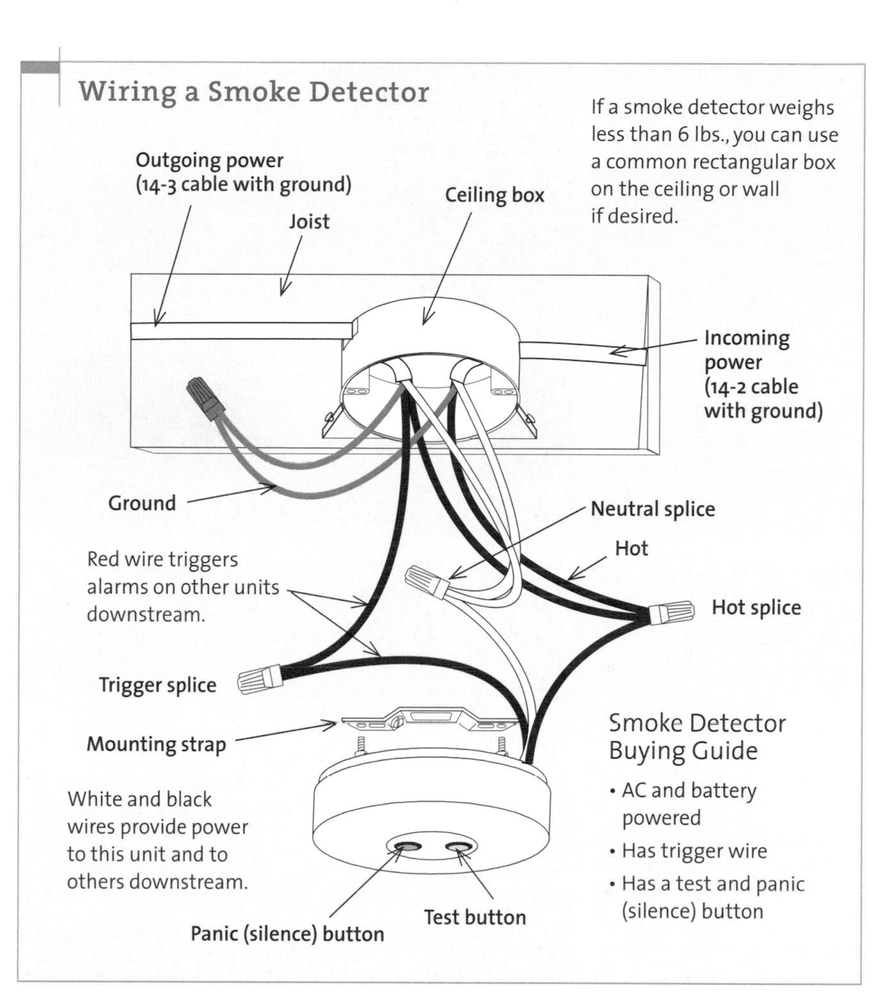

Smoke Detector Buying Guide

- AC and battery powered
- Has trigger wire
- Has a test and panic (silence) button

There is not much room inside the narrow track for a lot of wires, but the wiring logic is still the same: black to the brass screw, white to the silver screw, and bare wire to the frame.

Track lighting can also have end attachments allowing them to be cord and plug. These can be installed separately or bought already installed.

Rough-in wiring for track lighting needs to be as precise as possible to make sure the cable enters the exact spot where the wires insert on the narrow track. Sometimes the rough-in wire can terminate in a standard outlet box. Most times, the splice is done in the end of the track or in the connector, so the cable must be brought through the finished wall at the exact spot where the track begins or ends. Make a hole no larger than the diameter of the cable through the finished wall. The power wires connect to screw terminals within the track.

Outdoor Light Fixtures

The most common outdoor light fixtures on a home are flood lights, entry lights, dusk-to-dawn lights, motion detectors, and low-voltage landscape and security lighting. The wiring for these fixtures is pretty basic, but installation can be complicated. In some cases, planning the locations can save a world of trouble later.

Floodlights

Floodlights come in wattages ranging from 75 to 1,500 and are normally mounted outside at a high point on the house. Standard flood-lamp bulbs can be mounted up to three in one place, all facing different directions for maximum coverage. Don't mount these lights in an inaccessible location because you'll need to change the bulbs at some point. If possible, mount the light above or adjacent to an upstairs window so that you can easily change the bulbs. Always point the bulbs down to keep water from running into the base. Try to find bases with copper threads, rather than aluminum ones, to keep the bulb threads from seizing to the base threads. If you can't find bases with copper threads, lightly wipe an anticorrosion compound onto the bulb's threads. Quartz lights can also be used for outside lighting, but I don't recommend them. These fixtures get hot enough to burn the skin right off your fingers and a bulb's life can sometimes be measured in minutes (another one of my mistakes that will save you time and money).

For rough-in, simply bring the cable out of the building. Install a watertight box and the fixture after the finished exterior is up. If you try to install the box ahead of time, you may not be able to match the exterior surface, and

Work Safe • All outside fixtures have a gasket between the base and the cover. The trouble with gaskets is that they never stay put as you try to position the cover, resulting in a less-than-perfect seal and a corroded splice, which can lead to intermittent power or burned-out wiring. To prevent this, use a few dabs of hot glue to hold the gasket in place.

you will spend a lot of time trying to place a cover on the box that doesn't conform to the finished wall.

Entry lights

Entry lights can be troublesome if the building has beveled or irregular siding. If the fixture is placed flat against the siding, the light will slant along with the siding. If placed plumb to the wall, the light base will have a large gap underneath it because of the bevel of the siding. In this case, it's best to have the carpenter make a fancy base and install it for you. If that's not possible, grab a piece of siding and invert it on the installed siding to make a flat area. If the house has vinyl siding, buy an adapter that allows the fixture to seat on the exterior with a minimum of problems.

Once you've got a flat mounting surface, install a pancake box to mount the light. Double-check the light base to see if it is deep enough to hide the box—if not, recess the box. Roughing in simply involves sticking a cable through the exterior wall where the light will be mounted.

Dusk-to-dawn lights

Normally installed as security lights, dusk-to-dawn lights can be bought as an entire unit or as individual lights. Dusk-to-dawn lights normally have a plug-in photoelectric cell, or eye, which senses when it's dark and then turns on the light. These units are normally mounted in areas for parking or security. Packaged units normally use mercury vapor lights, whereas individual sensors can turn on a light for any purpose, such as a sign to be lit after dark. These cells don't last forever (neither do the bulbs), so locate the fixtures where you can easily reach them when it's time to replace the bulb or cell.

Typical wattages range from 70 to 175. Lower wattages may be cheaper, but they may not throw enough light. These lights are normally hardwired on a 120-volt circuit. The disadvantage of such a fixture is that it could take several minutes for the unit to turn on. Separate photoelectric cells can be wired in series with any light to turn on after dark, but make sure the cell matches the wattage of the bulb, or it could burn out.

Motion detectors

Motion-sensing devices are very popular nowadays. Some units detect just motion, whereas more sophisticated models use infrared rays to detect body heat. The units turn on house lights or flood lights as someone comes near the house. You can also adjust the sensitivity of the unit to compensate for dogs, raccoons, and other small animals. Loads for such devices are typically limited to around 500 watts. You can buy them as individual sensors, which turn on lights that are already installed, or as complete sets with a light-attachment assembly.

Locating a Motion Detector

Mount the motion detector no more than 12 ft. above the ground and locate it so that motion cuts across the lobes of detection. Locating the unit straight into the target area may allow someone to walk up to the house undetected through a dead zone.

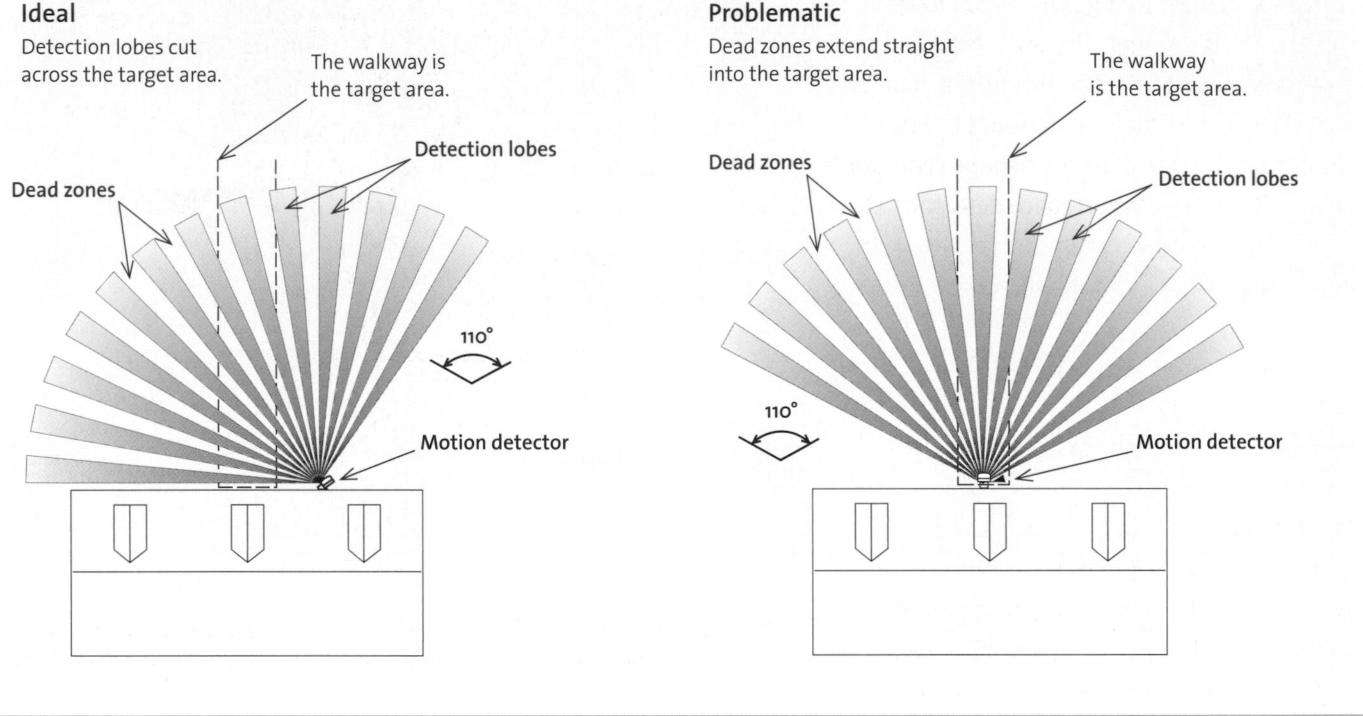

Installing such a device can be problematic. First, make sure the unit you buy can detect at the distance you want—30 ft. to 60 ft. is common (most units have trouble detecting motion at farther distances). Mount the detector no higher than 12 ft. above the ground.

Work Safe • Avoid motion detector units that don't come with instructions (there is at least one) or units that don't have sensitivity adjustments. Also, avoid units with a quartz light. The glass front becomes too hot and the light tends to die an early death.

The higher the unit is installed, the less ground surface will be covered and the less responsive the unit will be to motion on the ground. Position the unit away from heat-producing sources, such as a heat pump, and away from reflective surfaces, such as windows and pools. Trees and ferns that wave with the wind will also create problems.

It may seem logical to face the unit in the direction of the object that will turn the unit on, but this is not what you want to do. These sensors—both infrared and motion—do not saturate the area. Instead, they project finger-like beams (called lobes). If you're facing the unit, you can sometimes slowly walk toward it (between the detection lobes) and not set it off.

The directions on the unit tell you to install the unit at right angles to the area you want to track so that people will pass across the lobes of detection, not between them.

The rough-in wiring only needs to be a cable sticking through the wall. Attach a watertight box to the wall after the siding is up and pull the cable through. Make sure the box shape matches the detector's shape (you don't want a round box for a square detector fixture). Once the box is up, simply follow the wiring instructions included with the unit.

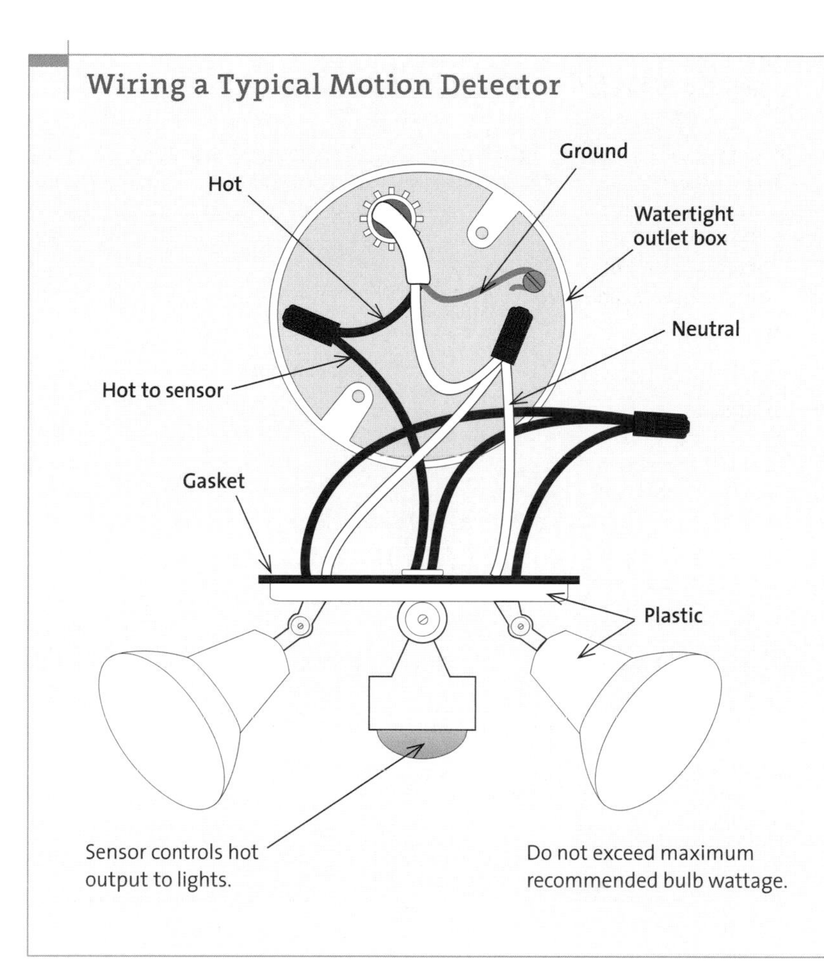

Low-voltage landscape and security lighting

Low-voltage landscape and security lighting has become popular in the last few years. Low voltage means it runs off a transformer (plugged into a wall outlet) that provides around 12 volts to power a string of lights. Compared to their 120-volt cousins, low-voltage systems are safer to use and easier to install. In some cases, installation requires little more than stabbing the lights into the ground. The cable can be laid just under some mulch or buried under a few inches of lawn. Such lighting is generally suited for short distances (less than 100 ft. from the transformer).

You cannot plug a transformer into an outside outlet and leave the outlet's door open. You must use a watertight-while-in-use outlet or outlet cover. If the area to be illuminated is a considerable distance from the house, bring the AC power to the location via a UF cable and install an outlet onto a post.

Although generally economical, landscape and security lighting can get expensive if you buy fancy lamps. Commonly available path lights come in the shapes of mushrooms, tulips, shells, pagodas, mini bulbs, and micro bollards (shaped like a pencil with 360 degrees of illumination); they can also provide indirect light (reflected off the bottom of the fixture). These fixtures run from $50 to $100 per light. If you want landscape and security lighting, follow these procedures:

- Determine what will be illuminated (trees, walks, etc.).
- Determine the type of illumination (spot, shadow, silhouette, overall, etc.).
- Determine the number and location of lights (hidden in trees, spaced along walk, etc.).

- Determine the types of mounts (stakes, outlet boxes, building into structure, etc.).
- Determine bulb wattages (27 watts to 50 watts is typical).
- Add the bulb wattages together and add 10 percent to select the transformer. Go to the store to see if they have a prepackaged kit that meets your needs. Kits come with properly gauged cable. If you buy everything separately, opt for the largest-gauge cable so you can add more lights later.

Recessed Lights

Recessed lighting provides a variety of illuminating effects throughout the house, adding light to an area without affecting other design criteria. It is commonly used in kitchens and closets and for accent and task lighting.

The housing is the main factor in a recessed-light installation. You can buy housings for new construction or remodeling. There are two basic types from which to choose: T-housings for uninsulated ceilings and ICT-housings for insulated ceilings (commonly called non-IC and IC). Non-IC housings must be kept at least 3 in. from insulation in the ceiling or walls. IC housings are preferred, because you can insulate around them and they normally have a lower wattage. All housings are equipped with a thermal protector that shuts the light off if insulation material is installed too close to the housing or if the lamp wattage is exceeded. The type of housing is written on a sticker where the bulb screws into the socket.

IC fixtures have an ICT-housing (standard round), a UICT-housing (double-wall rectangular), or an ICAT-housing (airtight). Housings are also available for special applications,

Insulation and Recessed Fixtures

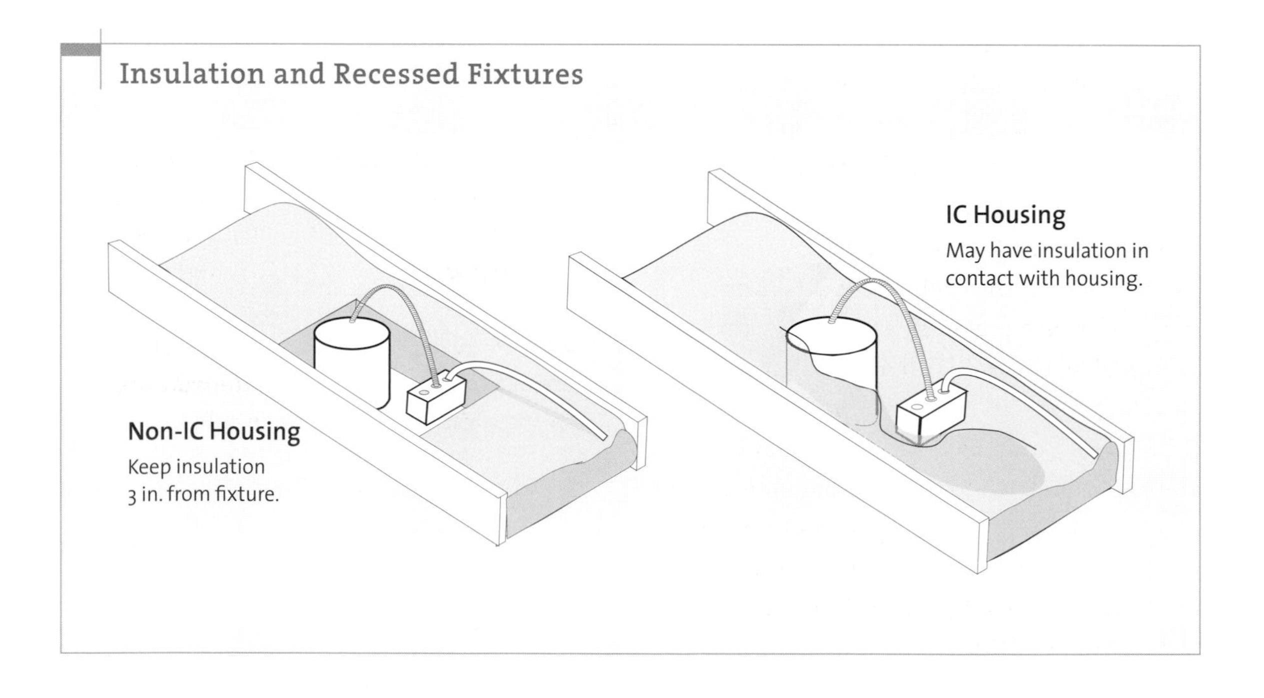

Wiring a Recessed Light

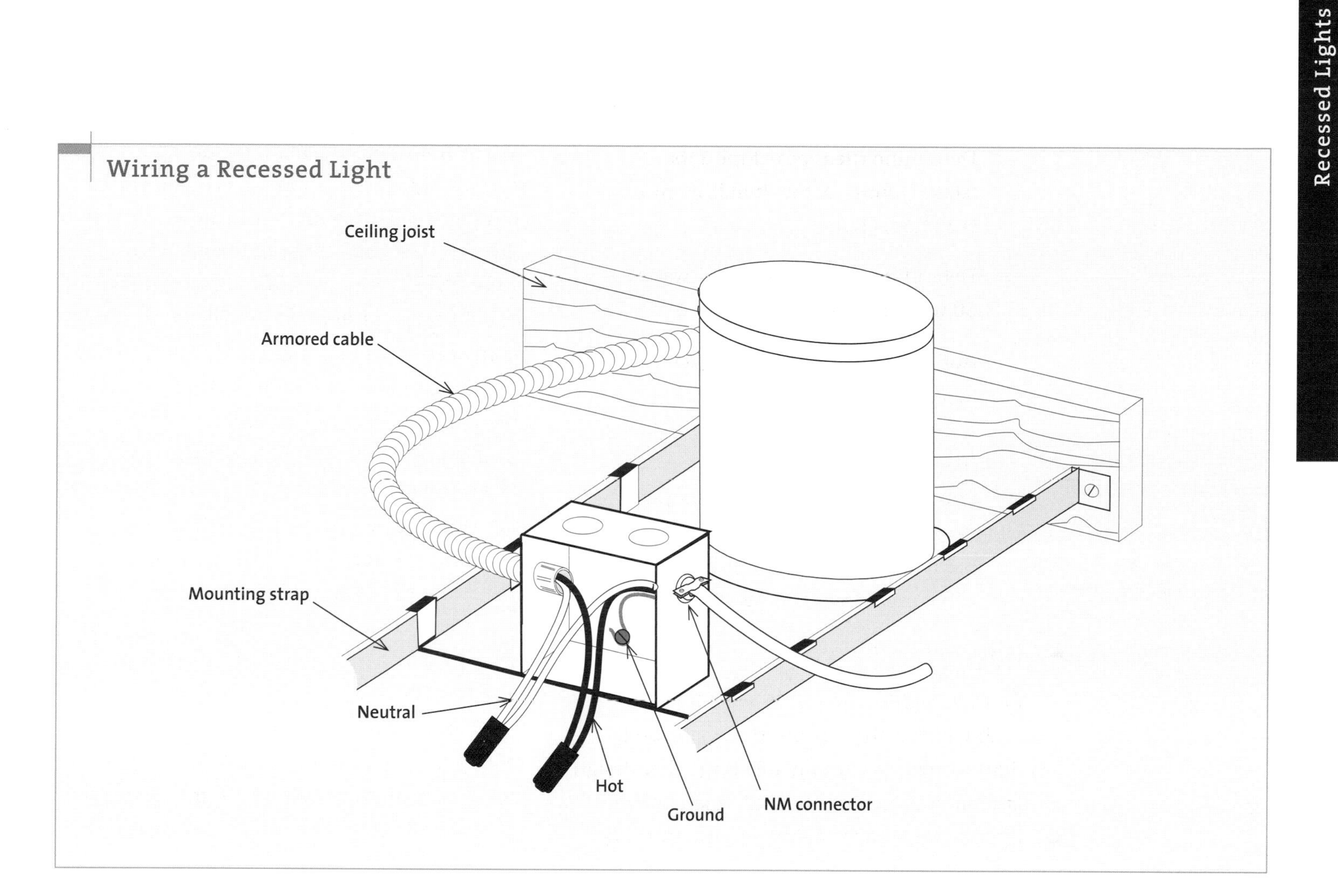

including sloped ceilings and closet lights. There are also housings with deep baffles designed for superior visual cutoff and others for high-wattage lamps (up to 200 watts). For remodeling jobs, make sure that there is enough room in the ceiling height for the fixture. If height is a problem, shallow models are available. Use only trim that is compatible with the particular fixtures you're installing.

Wiring is normally done via a metal splice box that is attached to the housing frame. Rough-in includes mounting the housing but not the trim plate. When mounting the housing, be sure to allow for the finished ceiling (in new construction).

Work Safe • If a recessed light occasionally turns itself on and off, it's probably due to the heat sensor. This frequently happens when the wattage of the bulb is too high. Look at the sticker inside the can to learn which bulb designs and wattages can be used. If you haven't exceeded the wattage, try installing a lower-wattage bulb.

CHAPTER 13

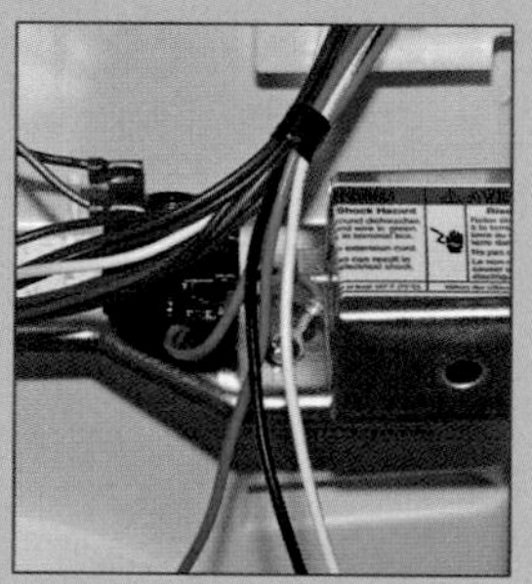

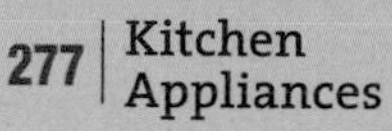

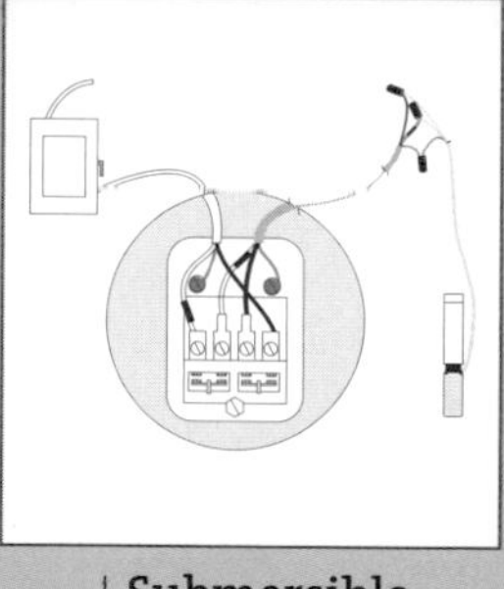

Wiring Appliances

Our homes are becoming more appliance-oriented every day, and the trend doesn't seem to be slowing. With the increasing number of appliances in a house, appliance wiring is taking more of an electrician's time on the job site.

Gone are the days when you simply plugged in a new appliance and that was that. Today, larger, wattage-hungry appliances might require their own circuits. Others must be hardwired, meaning that the branch-circuit wire is connected directly to the appliance. Also, voltage varies: Some appliances are 120 volts, others are 240 volts, and still others are both 120 and 240. And they all wire differently.

Kitchen Appliances

The kitchen typically houses the greatest number of electric appliances, including a dishwasher, refrigerator, garbage disposal, microwave, stove, and a slew of others. Thus, it has the greatest amount of wiring and wiring problems. Here are some approaches to wiring appliances.

Wiring a Dishwasher

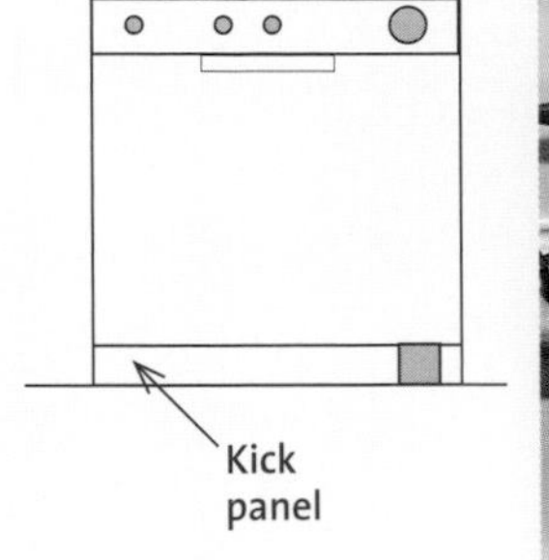

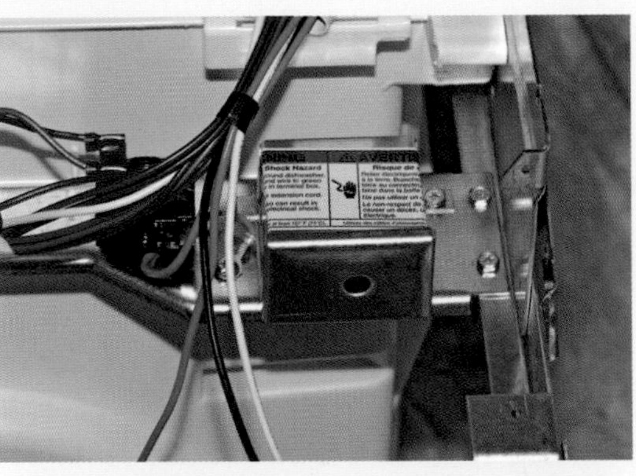

In a dishwasher, look for the metal splice box behind the bottom right of the kick panel. To wire in a dishwasher, connect the incoming black conductor to hot, white to white, and green or bare wire to ground—typically under a ground screw.

For add-ons, InSinkErator® makes a disposal on/off switch that works on air pressure. You do not need to rip into the walls to install a switch.

Dishwasher

Residential dishwashers are normally 120 volts and require a dedicated 20-amp branch circuit using 12-gauge wire. From behind the dishwasher, the cable is brought through the lower part of the wall or up from the floor adjacent to the wall, and then (under the dishwasher) to a metal splice box located in the front just behind the kick panel.

The first step is to install an NM connector, or strain relief, on the splice box, and then bring in the cable. The dishwasher's wires come into the splice box via a prepunched hole. Wrap a few turns of electrical tape around the wires at that spot. The wire connections are simple: hot to hot (black to black), neutral to neutral (white to white), and ground to box or the appliance's green wire. If the house wiring is old and has no ground, you'll need to run an equipment-grounding wire to the main panel or install a new cable. Never install a grounded appliance on ungrounded wiring.

Work Safe • Some residential dishwashers allow 14-gauge cable and 15-amp protection, but don't use them. Someday the dishwasher's replacement may be more heavy-duty, and once installed may overload the wiring and kick the breaker.

Work Safe • When running the dishwasher cable, bring in enough slack that the dishwasher can be pulled out and serviced. Make a large loop and circle it under the appliance to the splice box.

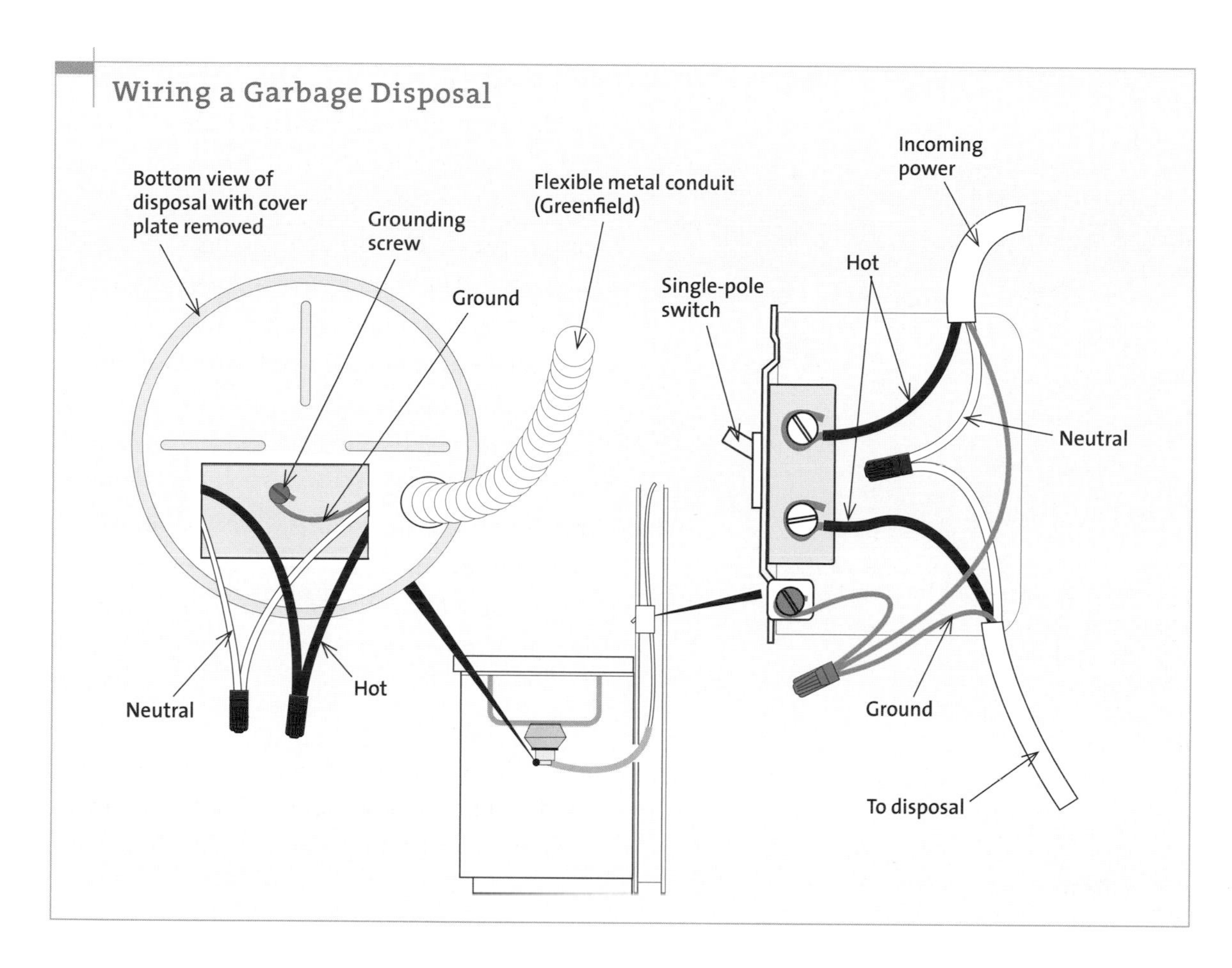

Garbage disposal

Garbage disposals are wired from the bottom and are either hardwired or wired through a plug and cord. If the unit is hardwired, the cable comes right out of the drywall under the kitchen sink or straight up from the crawl space through the bottom of the cabinet. This is the preferred way because it takes less material and less time. I have never seen any reason to attach a garbage disposal via cord and plug. Be sure to keep the wire out of the way of the plumbing. The voltage is usually 120 and most disposals do not need to have their own dedicated branch circuit. The electrical connection is made under the electrical cover, or plate, on the bottom of the disposal. The splices are made with wire nuts: black to black, white to white, and ground to the frame screw on the unit. Be sure to install a switch on the wall near the kitchen sink to control the power to the disposal.

Electric stove

Electric stoves are 120 volts and 240 volts. The burners and bake unit require 240 volts, but the timer, clock, and buzzer use 120 volts. The NEC provides specific guidelines for wiring stoves, including choosing the right size conductors.

For a wall-mount oven and a drop-in range or cooktop, the NEC requires you to use the nameplate rating to determine the correct

Stove outlet wired via 6 AWG NMB copper. The cable contains an insulated white neutral, a bare ground, and two insulated hot conductors.

Above Code • Don't use 8-gauge cable and a 40-amp breaker for a free-standing range/oven even though you can. Having the relatives over for Thanksgiving and Christmas or just having family/friends get-togethers means running the stove and oven on all burners. And lets not forget that a replacement stove may pull significantly more current than the old one. I've had to replace many 40-amp free-standing stove circuits with 50-amp circuits because excessive current kept kicking out the 40-amp breaker on those occasions. To keep this from happening to you and ruining your holiday festivities, design the stove circuit for 50 amps (use 6-gauge copper) instead of using the NEC's derating standards.

size conductor. However, the NEC allows the typical free-standing stove to be derated. That is, if the nameplate rating is not more than 12 kilowatts, the rough-in wiring can be designed for 8 kilowatts. The NEC allows this derating because the organization feels it is unlikely that all elements will be on high, or full power, for an extended period of time.

A standard free-standing stove plugs into a 50-amp female receptacle. The outlet box can be surface-mounted or recessed. I normally attach a nailer (a scrap piece of wood) between the studs, bring the cable through the nailer, and screw the receptacle to the nailer after the finished wall is up. A more complicated installation is to use a recessed box and install a flush-mount receptacle. Because of the large cable required, you'll have to use a deep square box (42 cu. in.), put a large cover plate over it, and mount the receptacle to that.

Because a stove pulls both 120 volts and 240 volts, the neutral must be insulated (this applies to hardwired units as well). As of the 1996 code, three-prong receptacles, both surface and recessed, are no longer allowed. The ground and the neutral cannot be connected, meaning that the neutral must be an insulated conductor separate from the ground. Now you need a male plug and a female receptacle with four conductors: two insulated hots, one insulated neutral, and one ground.

If the electric range or oven is a drop-in type, there should be a pigtail of wires, called a whip, with flexible conduit hanging from the unit. These connect to the incoming power cable in a large splice box that you supply. Simply bring the four-conductor cable into the box to splice to the range's wires. If the whip is not already installed, do not be tempted to use a plug and cord on a drop-in cooktop (some models even come with labels warning against

Wiring an In-Wall Electric Oven and Drop-In Cooktop

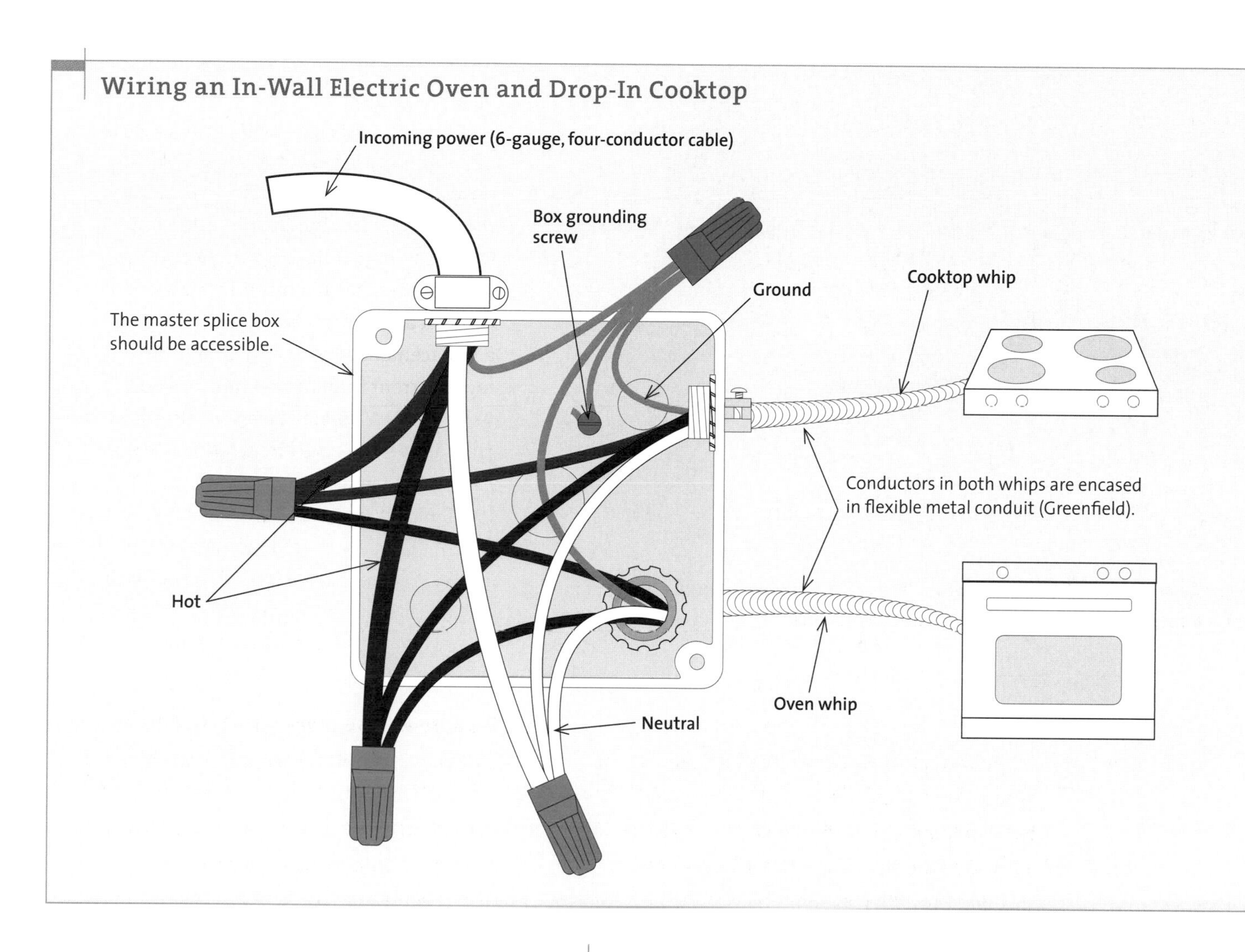

its use). Instead, make your own whip with THHN wires (12 or 10 gauge, depending on the load) and run it through a Greenfield conduit. You can use a leftover piece of wire.

Range hood and fan Range hoods can be almost any size, from 30 in. up to 72 in. wide. You can get everything from a standard economy hood (I'd avoid this type because it doesn't pull enough air) to a restaurant-grade canopy hood. Also available are high-tech models, with features such as blower speed control, heat-sensitive components that automatically turn the blower to high speed when

Work Safe • If you are mounting the stove outlet in the floor, keep it dead center. This keeps the feet from hitting the outlet and allows you to disconnect the plug by pulling the stove's bottom drawer.

Take Note • The more electronics in an appliance, the greater possibility of problems. Use some sort of surge protection device, as lightning and power surges lower the life expectancy of electronics.

Work Safe • Don't do what I did when I once attached a fan hood to a cabinet bottom. I used screws that were too long, and they went all the way through the cabinet bottom and into a fancy menu holder and some Tupperware. This is one of those errors you don't make twice—use short enough screws so that they don't poke through the wood.

excess heat is sensed, warming lamps to keep food hot, high-heat alarms, and dual lighting.

Most fans are 120 volts, with the rough wiring connecting to a box within the unit. Because it pulls little current, the fan can be wired into the kitchen light circuit. The problem with wiring the stove fan is that during the rough-in stage the wire must be brought out of a bare stud wall before the fan and everything else in the kitchen is installed. Once the rough wiring has been run to the correct spot, staple it in place on the stud, leaving about 18 in. of cable sticking out beyond the stud wall. Then install the drywall and the kitchen cabinets. After that, mount the hood and fan unit to the bottom of the cabinets.

The rough-in wiring must be brought out exactly under the proposed bottom of the overhead kitchen cabinets, but not so far down as to be exposed once the fan unit goes in. If you run the cable too high, cut the drywall a bit to get the cable down to its correct location (you can do this only if the fan housing and overhead cabinets will cover the hole). If this isn't the case, you'll have some patching to do, which could get sloppy. The wiring is pretty simple: black to black, white to white, and ground to frame.

The standard installation height for fans is 54 in. to 60 in. from the floor to the bottom of the unit. Add another 5½ in. to 9 in. for the height of the unit. Try to bring the cable out of the wall about 7½ in. to the right of the duct's centerline. Make sure the drywallers do not move the cable location or cover it. If you're responsible for running the ductwork, make sure there is no stud dead center on the proposed location—you normally need at least 5 in. of clearance on each side of dead center.

Baseboard Heaters

Electric baseboard heaters are a fast and economical method of supplying supplemental heat. They are normally installed below windows and on adjacent walls. I can't tell you the number of units or the total wattage you'll need, because that depends on too many variables, such as how cold it gets and how well the house is insulated. A general rule is to mount units under windows and have them run at least the length of the windows.

Electric baseboard heaters can be installed quickly and easily, as long as you know what to do and which problems could occur. The first decision to affect the wiring is whether the thermostat will be inside the heater or on the wall.

Work Safe • When selecting the location for a baseboard heater, remember that furniture cannot be placed in front of it, because the furniture will block the flow of air.

Outlet Placement and Electric Heaters

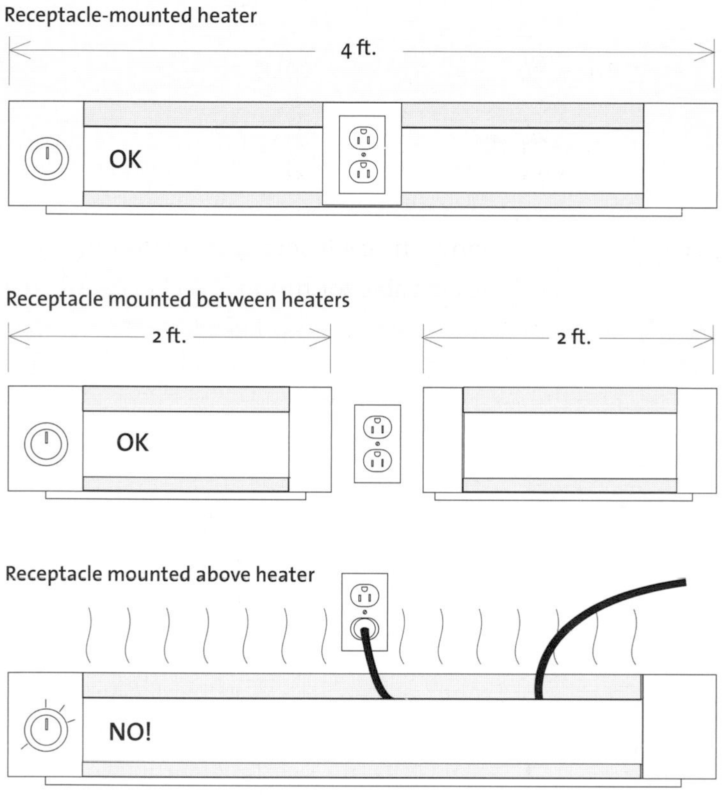

Receptacle-mounted heater

4 ft.

OK

Receptacle mounted between heaters

2 ft.

2 ft.

OK

Receptacle mounted above heater

NO!

Allows appliance cord to fall onto hot elements.

Both in-heater and wall thermostats are acceptable, according to the NEC, but I only recommend in-heater thermostats, with one thermostat per heater. An in-heater thermostat is more economical and faster to wire (wall-mounted thermostats literally double or triple the labor time of baseboard-heater installation). A wall thermostat turns all units on at once, which wastes a tremendous amount of energy. Heaters with integral thermostats turn themselves on and off independently to heat only the area needed, resulting in less overall power consumption. Having the thermostat along the floor also allows a faster response time to incoming cold air. I normally mount heaters 1 in. or so off the floor to allow room for the finished floor and to maintain an air gap between the two. Because they're really the only smart choice, I will limit this discussion to in-heater thermostats.

Although 99.9 percent of all baseboard heaters are rated to use 240 volts, 120-volt models are available. However, I don't recommend them, unless only 120 volts is available. Chances are high that the heater could accidentally be installed on a 240-volt circuit,

Smoking Heaters

When you first turn on a new electric heater, it may smoke a little—or a lot. Don't worry—it's only burning off the factory coating on the elements. This is normal and should stop after a few minutes. It's best to open the doors and windows and let the room air out when a heater is turned on for the first time. (Note that there will be more than enough smoke to trigger the smoke alarms.)

Older heaters also smoke if they have not been used for an extended period of time. Dust and cobwebs frequently collect on the elements. Remove the front cover of the baseboard (it should snap off) and vacuum the elements before turning on the unit.

resulting in a fire (this has happened). In addition, a 240-volt heater, which draws half the current of a 120-volt heater, is more efficient and requires less expensive wiring. Assume that all heaters in this discussion are 240 volts.

The wattage of the baseboard heater determines the heat you desire. Most panels are 250 watts per ft., or about 1 amp per ft. To figure the exact amperage, divide the wattage by the voltage. For example, say you're putting a 10-ft. unit (2,500 watts) under a picture window in a dining room. By dividing 2,500 by 240, you'll find that the amperage is just above 10.

Several small baseboard heaters can be fed from the same cable coming from the main panel as long as the current doesn't exceed 80 percent of the cable's rating: 12-gauge cable cannot take more than 16 amps of continuous current. This translates to a maximum of 16 ft. of baseboard per 20-amp breaker, 12-gauge circuit. Do not use 10 gauge (it is expensive and hard to run) or 14 gauge (it is too small).

In your desire to install the required number of receptacles, it's possible to have a receptacle located immediately above a proposed baseboard heater. This is not allowed. A cord that plugs into the outlet might fall into the hot burners and short out. Even if the cord doesn't fall directly into the elements, the heat will eventually damage the cord. To solve this problem, some baseboard-heater manufacturers have end, corner, and center sections that allow you to insert the outlet within the baseboard. However, the receptacle must not tap into the 240-volt circuit powering the baseboard. It must have its own 120-volt cable.

Wiring an in-heater thermostat

Installing the thermostat in a baseboard heater is quite simple. Simply pull off the cover plate on one end of the unit (it makes no difference which end) and mount the thermostat. Some mount to the heater, others to the removable panel. The rough-in wiring can be run directly to the heater's location and through the wall and the knockout in the heater (be sure to install an NM connector). You'll have a choice of a single- or double-pole thermostat (two thermostats ganged together). Go for the double. A double-pole thermostat cuts power to both legs, not just to one, and offers the convenience of an "off" position on the unit (single poles only turn down to low).

Once the cable is in the heater, splice one hot wire to each half of a double-pole thermostat or one hot wire for a single-pole thermostat; the other end of the thermostat connects to the heating element. This places the thermostat in series with one or both legs of the incoming power. If you splice the incoming power across the thermostat, a direct short will be created when the thermostat engages, and

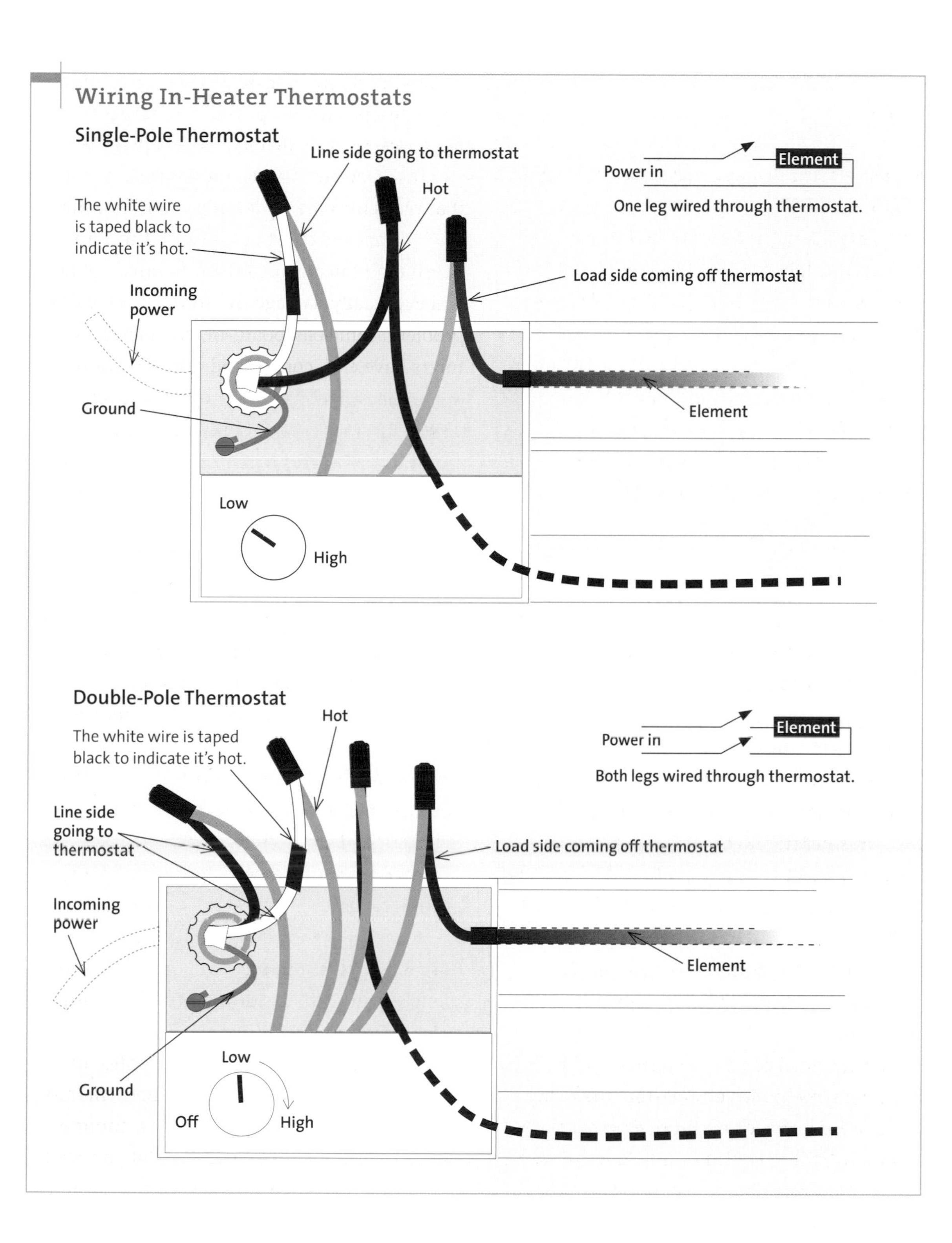
Wiring In-Heater Thermostats
Single-Pole Thermostat
Line side going to thermostat
Hot
The white wire is taped black to indicate it's hot.
Incoming power
Load side coming off thermostat
Ground
Element
Low
High
Power in
Element
One leg wired through thermostat.
Double-Pole Thermostat
Hot
The white wire is taped black to indicate it's hot.
Line side going to thermostat
Load side coming off thermostat
Incoming power
Element
Ground
Low
Off
High
Power in
Element
Both legs wired through thermostat.

Typical in-wall bath heater right out of the box. The chrome front pulls out of the frame so that it can be mounted to the studs.

the thermostat will be ruined. (If you switch on the thermostat and hear a snap, crackle, and click of the breaker kicking off, and then see smoke, this is what you did. I won't tell you how I know this.)

Electric Wall Heaters

Electric wall heaters come with and without fans and in both 120-volt and 240-volt versions. Always buy the fan-operated unit because it better distributes heat around a room. This is especially important if there is plumbing in the room; because heat rises, you'll warm only the top of the room and allow the pipes at foot level to freeze. Another rule: Always buy a 240-volt model, which draws half the amount of current that a

Secure the frame on at least two sides. Extend the frame out from the stud the depth of the finished wall.

The power cable enters the frame's integral triangular splice box via an NM connector (bottom right). Tape the white wire black. The ground wire (not seen) attaches to a grounding screw mounted on the frame.

120-volt unit draws. You are also allowed to wire more than one heater in a branch circuit, as long as the wiring isn't overloaded. Do not exceed 16 amps on a 20-amp circuit. A typical 240-volt, 1,000-watt heater pulls around 4 amps, whereas a 1,500-watt unit pulls a little over 6 amps.

Always get a heater with an integral thermostat for the same reasons discussed earlier: one cable, one hookup, and you're done. A small, single-wattage heater is ideal for a small room, such as a bathroom. Large, variable-wattage units with fans are better for heating large rooms or even an entire house, but heating a house with this type of heater only makes sense if the areas are open. A large unit is about 6 ft. tall and fits between studs or can be mounted on a wall. It can also heat an adjacent room if you install a length of duct from the heater to that room.

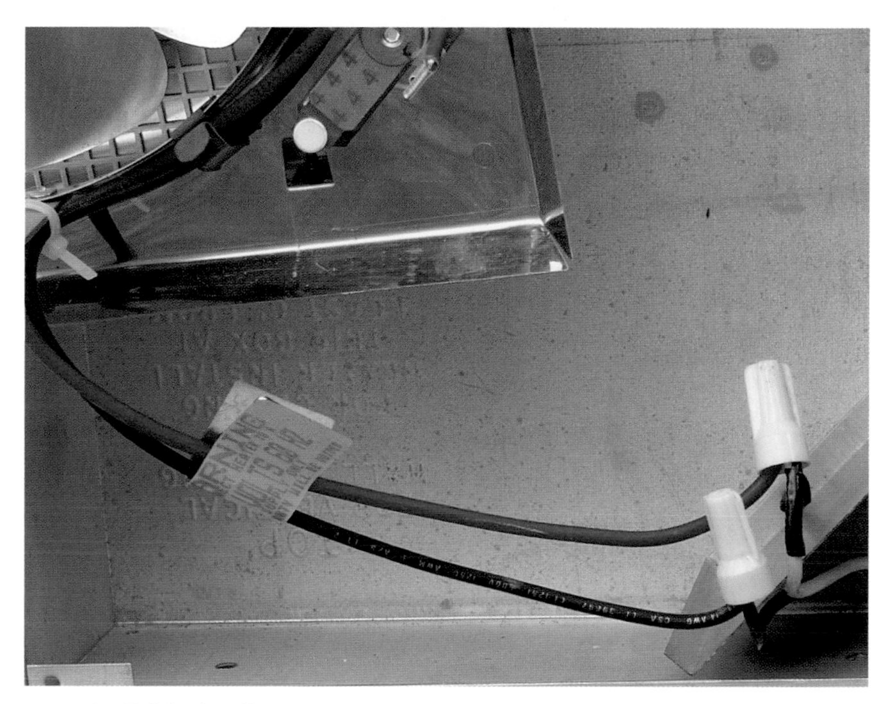

Once the finished wall is up, splice in the 240-volt heater. Attach one incoming wire to one heater wire—no polarity is required. Once spliced, stuff the wire connectors into the splice box and reattach the chrome front.

Utility-Room Appliances

Wiring the appliances in a utility room can be tricky. Many of those appliances are specifically covered by the NEC, and each situation is different. In general, the appliances in a utility room are large, such as a water heater, heat pump, and dryer. Some may use 240 volts only or both 120 volts and 240 volts.

Clothes washer

Although a clothes washer just needs a receptacle, it's still possible to screw up the installation. Place the clothes washer on a separate 20-amp circuit and terminate the wiring on a high-quality, unbreakable ($5 to $6) duplex receptacle (which normally are made from nylon or another durable material). If you install a cheap receptacle you run the risk

Instead of mounting a separate washer and dryer receptacle, opt for the ones that are integrated with the hot/cold/drain plumbing assembly. This saves time and money and places the outlets at a convenient location.

that the receptacle will crack and break. In addition, the receptacle orientation must match that of the cord's plug so the cord can hang straight down.

Though it is not commonly done, I suggest running the washer off a GFCI, because especially in basements the user will be standing on a damp concrete floor when washing clothes.

Clothes washers are required by their manufacturer to be grounded. However, many are not—especially in rental houses where the owner will not upgrade the wiring. This places the tenant at risk, especially if the washer is in the basement on a concrete floor. I suggest using a product like Bryant's Circuit Watch (see the GFCI chapter), which is a plug-in adapter that gives GFCI protection.

Water heater

Water heaters come in both 120 volts and 240 volts. Buy a 240-volt heater because it pulls half the amount of current and costs no more than the 120-volt models. For these reasons, I will focus my attention here on 240-volt heaters only.

Of all the appliances, a water heater is usually the simplest to install. The standard electric water heater is a 50-gal., 240-volt, 4,500-watt, storage-type appliance fused at 30 amps. It is hardwired with 10-2 w/g NM cable run from the main panel to the heater or to a disconnect adjacent to the heater. A disconnect is required if the heater is out of sight of the main panel, and the heater should be as close as possible to the switch. If the heater

Wiring a Water Heater

The cutoff switch can be a fused or nonfused disconnect rated for the load of the heater—normally 30 amps. It must be close to the heater.

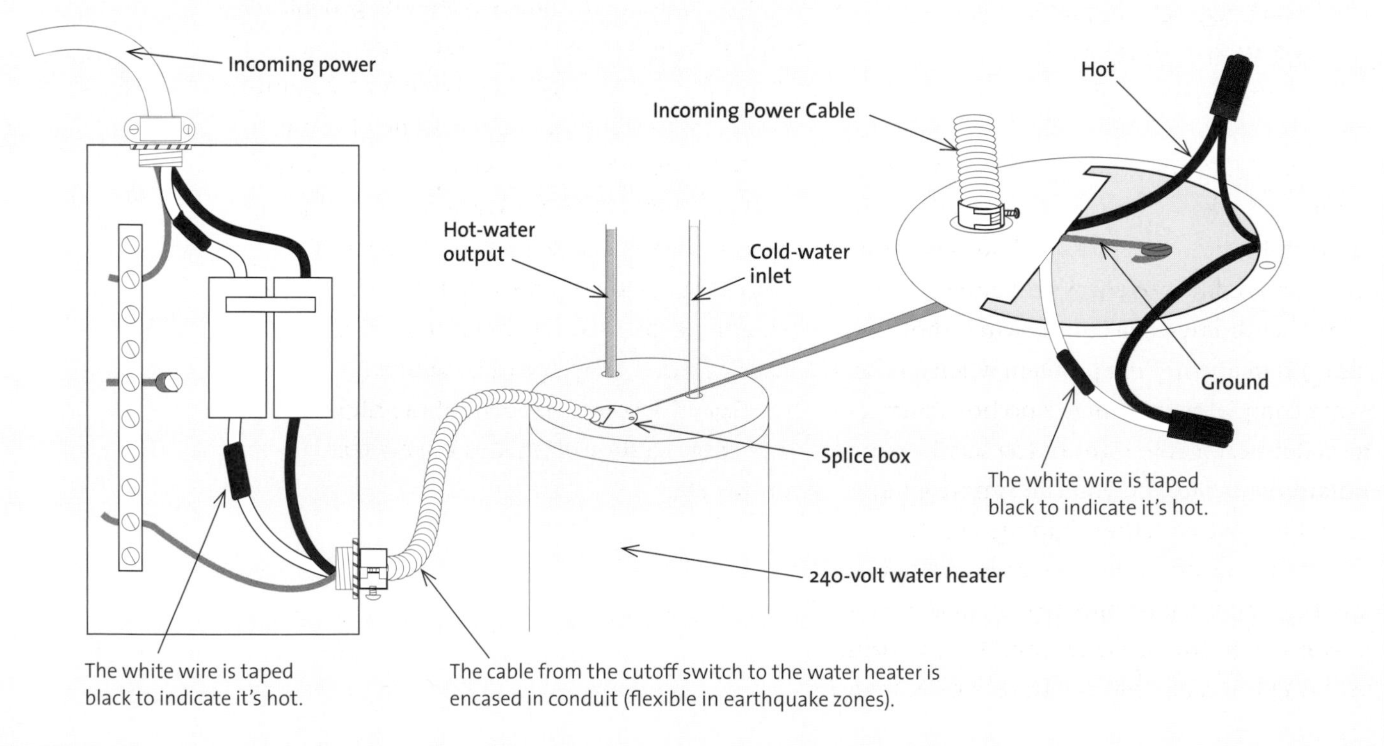

Above Code • If you want to wire a 240-volt water heater with 12-gauge cable, the heater needs an element that is 3,500 watts or less. There is nothing wrong with using a smaller-wattage element; it simply takes longer to heat the water. However, because the water heater has a label on its jacket that lists the specs of the unit, there is a possible safety risk. When the plumber or do-it-yourselfer replaces a blown element, he or she may accidentally replace the 3,500-watt element with a standard 4,500-watt element. The large element will overstress the 12-gauge cable with excessive current and could start a fire. The same thing will happen if the entire water heater is replaced with a larger 4,500-watt unit. Therefore, be sure to write on the water heater—in large letters—that the 4,500-watt element has been replaced with a 3,500-watt one. Personally, I don't think it's worth the risk. For that reason, I recommend using 10-gauge wire.

is in a habitable area, the NM cable from the switch to the heater must be enclosed in conduit (flexible conduit or Greenfield is best for this situation but not always required).

In older homes, the cutoff switch (not the heater) is frequently the problem when a homeowner complains that there's no hot water. If the water heater is in sight of the service panel containing its breaker, you can simply wire around the switch rather than replace it.

Wiring is quite simple, because there are only two connections plus ground and there's no polarity on the hot wires for 240-volt loads. Just bring the cable into the water heater via an NM connector and splice it into the two water heater wires via wire nuts. Tape the white wire black and attach the ground to the frame.

Heat pump

Heat pumps are normally wired by their installers, but sometimes an electrician needs to run the feed wires to the unit outside. The wire gauge will vary according to load, but the size I use most often is 10-2 w/g NMB. The size of the cable for the inside unit depends mostly on the size of the electric backup heating. Disconnects for new installations are required at both the outside and the inside unit if one is not already in the unit. If the inside unit is in an isolated area, it needs switched lighting and a receptacle outlet nearby for maintenance and troubleshooting.

Do not be confused when the amperage on the breaker does not appear to match the gauge of the cable going to the heat pump. There are many instances where the general rule of thumb of using 10-gauge wire for a 30-amp breaker, or 12-gauge wire for a 20-amp breaker, is not followed. This is one of those cases because the compressor has an overload that basically allows it to take care of itself. Typically what you will see is 10-gauge wire coupled to a 40-amp breaker as well as to a 30-amp breaker. And perhaps more often you will see a 12-gauge cable with a 30-amp breaker. The compressors need the higher amperage so they don't kick the breaker during start.

Though the cable may enter the dryer outlet from the top, the dryer's molded four-prong plug cord will point down. Don't reverse it.

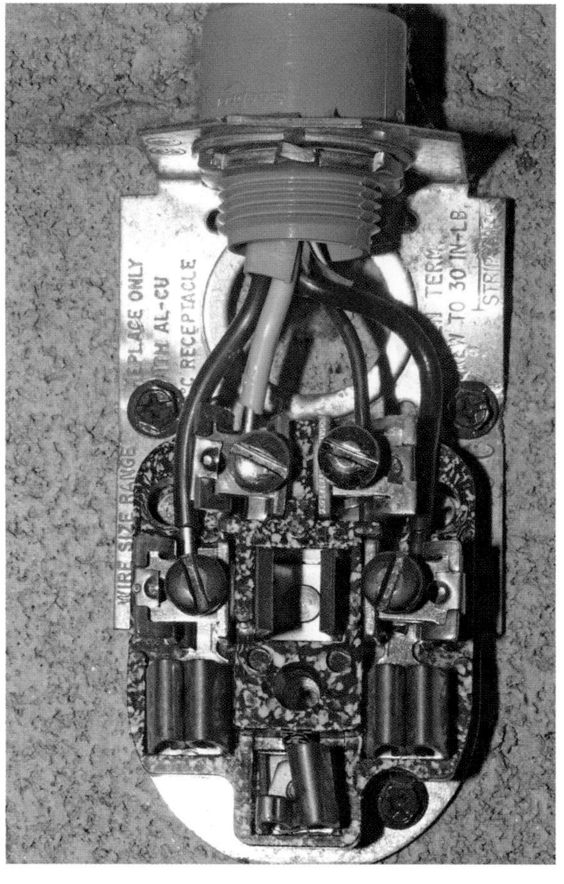

Electric dryer

Dryers are the most miswired of all the appliances because they are both 120 volts and 240 volts. And it's all because of the neutral and ground. Until 1996, the NEC allowed the neutral and ground to be the same conductor. This put the frame in the conduction path. Although this configuration still had a good safety record, it is now a code violation. Therefore, a dryer should always have a three-conductor cable—three insulated hot wires and a ground—such as 10-3 w/g NMB.

The design load must be at least 5,000 watts or the nameplate rating, whichever is higher. For a standard 240-volt electric dryer, that's a minimum of 21 amps of current, so the rough-in wiring must be 10 gauge and the circuit must have a 30-amp breaker. The wiring terminates in a four-slot (ground, neutral, hot, hot) receptacle.

Work Safe • To minimize the risk of fire, always try to position a dryer along an outside wall so that the vent run is as short as possible. Also, use smooth-walled (rigid) metal duct instead of lint-catching corrugated (flexible) duct work.

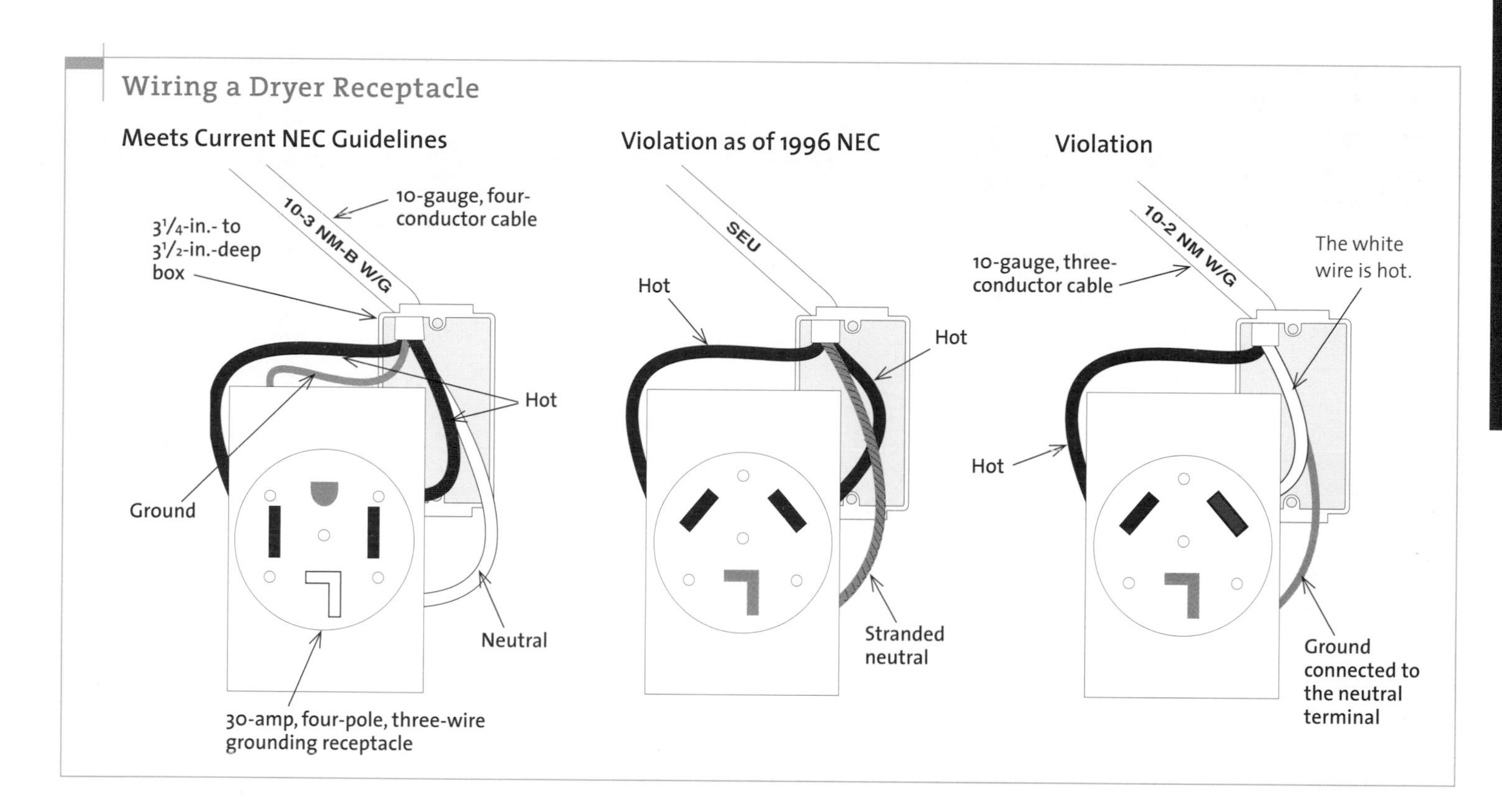

Submersible Pumps

Submersible pumps come in many horsepower ratings and in two wiring configurations: a two-wire system that needs only 240 volts and a three-wire system that needs a control box to provide a start and run wiring to the pump. Whether you use a two-wire or three-wire system, it is code to ground the pump and the well casing, if it is metal.

The wiring to a well must be approved for direct burial. It cannot be the standard yellow submersible-pump wire or even the twisted wire so commonly used within a well. In other words, use common UF cable from the pressure switch or control box to the well. Conversely, the wiring within the well should be listed as pump wire, not UF cable. The gauge for both cables depends on the depth of the well, the

Garage-Door Openers

Although most garage outlets must be GFCI protected, the outlet for the garage-door opener may not. I recommend using a GFCI in this instance. People have been known to use the outlet for general purposes. In addition, a GFCI may help keep lightning off the opener's circuit and protect its electronics.

The receptacle and opener must be installed directly behind the end of the proposed door location in the upright and raised position. That means you can't install the rough-in wiring for the opener until you know the height of the garage door.

Wiring a Two-Wire Submersible Pump

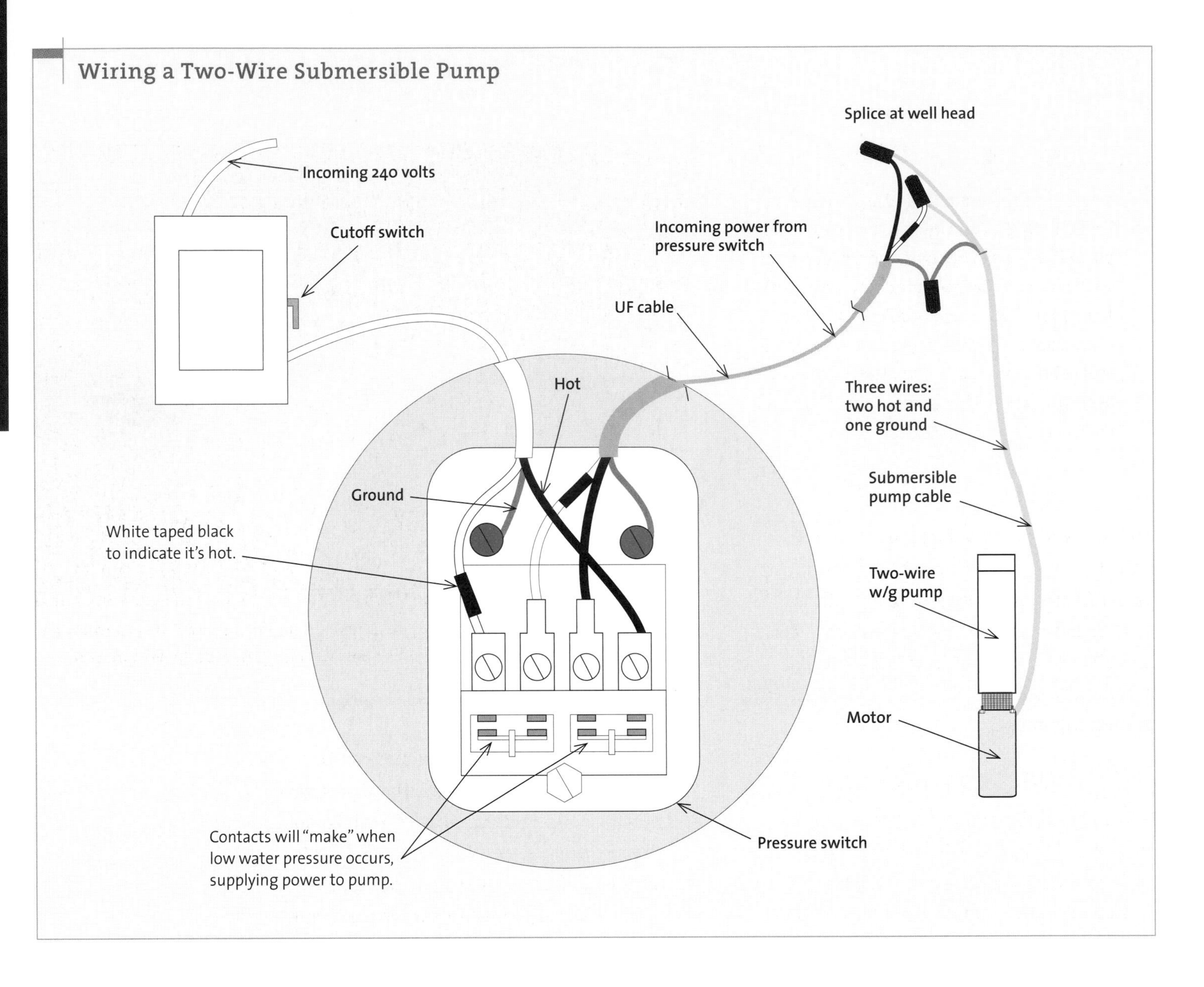

distance to the well, and the horsepower of the pump. A 140-ft.-deep well located 100 ft. from the house and powered by a standard ½-hp pump can get by with 12-gauge wire. A wire chart in the pump's instruction manual lists the gauges required.

It does not matter whether you decide to use a two-wire or a three-wire pump—they both work just fine. I recommend two-wire pumps because they're easier and cheaper to install (no control box installation and one less conductor in the cable). For a two-wire pump, the branch-circuit wiring terminates first at the pump's cutoff switch and then jumps to the pressure switch. From there, run UF cable to the well casing, where the splice is made under the well cap to the pump cable. This splice can be either insulated crimp (heat shrink) or wire connector. If using wire connector, point the skirts downward so that condensation does not pool in the connectors.

Do not automatically match the breaker size to the cable gauge run to the submersible pump. Many times a large-gauge wire is used

Work Safe • When installing the water line and UF cable from the house to a well, place it in black corrugated pipe (use pipe with slits or holes so water can drain out). The flexible pipe comes in both 3-in. and 4-in. diameters. The pipe protects the line and cable from rocks and allows them to be easily replaced without having to completely retrench the yard.

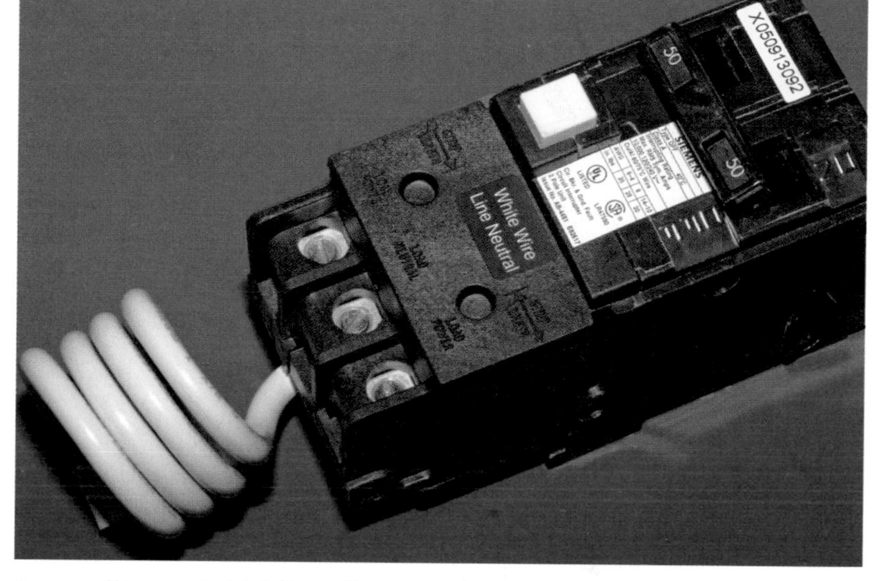

A 240-volt two-pole GFCI (typically 50 amps) is many times larger than a common GFCI—used in spas, hot tubs, and whirlpools with large pumps. Make sure your panel has room for it.

to lower the resistance in the circuit on long runs to a well. For instance, for a ⅓-hp pump (a 4-amp load), do not install a 30-amp breaker just because you ran 10-gauge wire to reach the pump; instead, use a 15-amp breaker. The smaller the breaker, the faster it will kick if something happens.

Whirlpool Tubs and Spas

Whirlpool tubs need to have their motors powered through GFCI-protected receptacles or breakers. Some whirlpool tubs come with motors that are connected with a cord and plug. Bring the cable, normally 12-2 w/g NMB for whirlpools and much larger for spas, to the motor and then install a GFCI circuit breaker at the main panel. The receptacle and the motor are maintainable items and need to remain accessible. Remember this, because it may affect the way the tub is oriented in the room. Small whirlpools come with a built-in skirt and access panel. Larger units must have a skirt and access panel added on-site.

Spas also need GFCI protection. Sometimes the protection is included with the spa; other times it is the responsibility of the installer. If it is the latter, you may have a problem. High-amperage GFCIs are expensive and can be hard to find. And don't forget that if you have a receptacle outlet within 6 ft. of a whirlpool tub lip, it must have GFCI protection. If it is referred to as an indoor spa or hot tub, then any 125-volt receptacle, 30 amps or less, must have GFCI protection and at least one 125-volt receptacle is required between 6 and 10 ft. from the water for maintenance.

Always install a dedicated circuit for whirlpools and spas. If either one is hardwired and a disconnect is not on the unit, you'll need to install it. For a spa, you'll also need to install an emergency cutoff switch and label it prominently.

Some whirlpool tubs and spas will come with heaters. Sometimes they are integral, other times you have to wire them in yourself. Run the heater on its own separate GFCI dedicated circuit.

Lightning and Surge Protection

When I was a kid I thought that whoever controlled lightning, controlled the world. Little did I know then that one cannot control the uncontrollable. Instead, we just try to understand it.

In pursing understanding, some people chase storms. I seem to chase the storm's effects. I get to a house after the damage is done—telephone systems gone up in smoke, pump control boxes blown across the room, fried computers, fried pumps, wood blown off the floors and walls, explosions, windows blown in—from various strikes of ball lightning, sheet lightning, and, my favorite, when lightning ran across the top of uncut hay in the field and all who were standing in the field went down like bowling pins.

Though nobody can stop a direct strike, I have never had an induced-voltage lightning problem I couldn't solve—and I have had some pretty bad ones. The worst is always a house at the end of a long tap down in the woods (sometimes over a mile). Lightning will induce voltage in the long line and the single house on the line catches the full surge. These are houses where people unplug everything during a lightning storm because they are tired of replacing it all. If I can solve those problems, I can solve yours.

Protecting yourself and your home from the effects of lightning is not a big deal. If you understand how the surges get into the house, it's an easy jump to adding equipment to keep them out. The trick here is choosing the right equipment (I've installed a lot that didn't work). You can learn from my mistakes. In this chapter, I'll tell you what has been field-tested and actually does what it claims.

Induced Voltage

We all know that lightning is dangerous up close, but I had no idea that its magnetic lines of force were powerful even at a long distance.

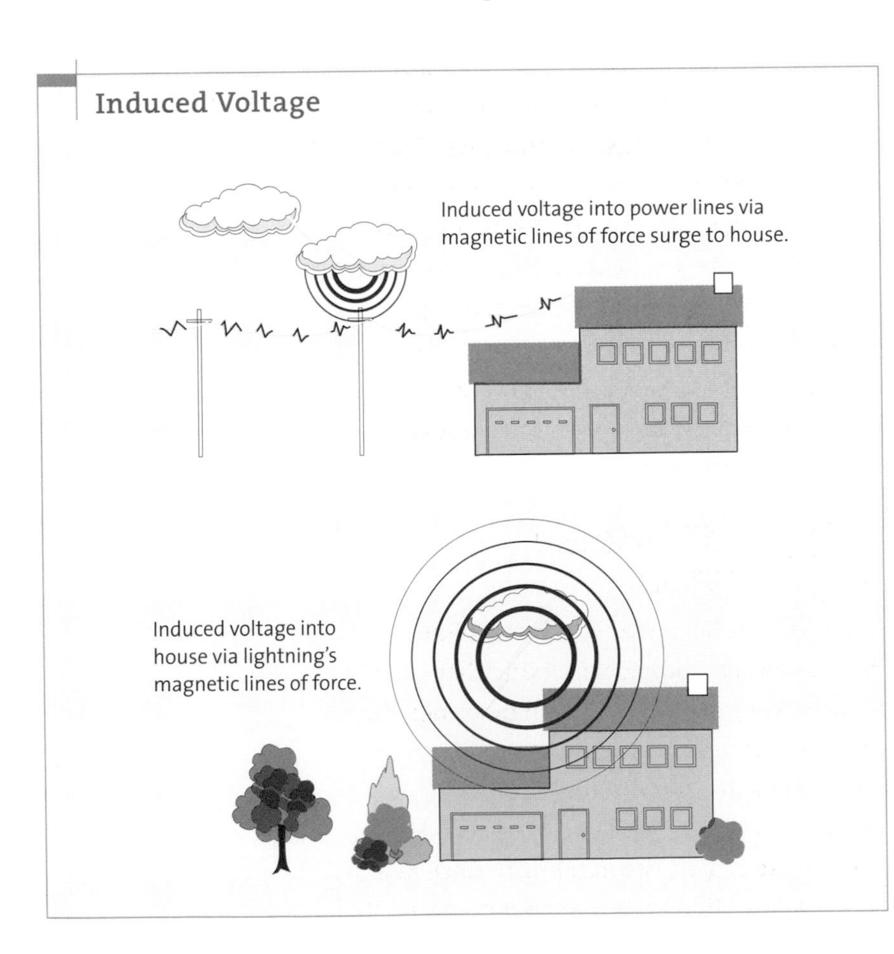

Take Note • A high-quality, low-resistance grounding system is an essential part of any lightning protection system. A million-dollar lightning suppressor won't work without a low-resistance ground.

I was first introduced to lightning's far-reaching effects in the mountains of Virginia. I was checking newly buried telephone cable to make sure the plastic jacket hadn't been ripped during installation. Thus, the cable ends were just sticking straight up out of the ground. Thunder rolled far off in the distance but I thought it was too far away to affect me—I thought wrong. As I touched the ungrounded cable shield, it zapped me hard. Lesson number one: The magnetic lines of force from distant lighting can build a charge on the cable's metal shield (about a mile of it in this one section) and it can discharge through a person to ground.

Because it had already discharged once, I thought it would be safe to touch it again—and it zapped me again. Lesson number two: Drain the shield with a jumper and give it a longer time to discharge the static voltage buildup.

I then ran a jumper from the shield to earth, waited a couple of minutes for the charge on the shield to completely drain away, and then removed the jumper. As I touched the shield a third time, it bit me again—hard. Third lesson: Lightning can instantly rebuild a charge on a long, continuous conductor from very far away. Lightning three, Rex zero. I went home.

Take Note • I know of no codes that require grounding a metal roof or metal siding, however, the NEC now has a fine print note on it.

The invisible lines of force that I encountered on that cable are the same ones that can wreak havoc on the conductors and appliances in your home. This is why it is imperative to install surge protection on the current-carrying conductors and ground all non-current-carrying conductors. A non-current-carrying conductor is any metal that isn't used as part of the wiring system but can conduct electricity, such as an appliance frame, steel beams, furnace duct work, metal water or drain line, and metal flues. Connect a #4 copper wire from what needs to be grounded to the grounding electrode (normally a ground rod), to the grounding electrode conductor (the big copper wire going from the service panel to the ground rod), or to the service panel grounding bus.

Direct strikes

No matter how "biting" the long-distance effects of lightning may feel, they cannot compare to close strikes. Massive amounts of induced voltage from lightning can build up on every wire, metal appliance frame, and metal water pipe in your house, just as it did on that telephone cable shield. That's one of many reasons, as we will find out later, that all such items in your house need to be connected to a low-resistance grounding system. Picture the grounding system as a drain—one that takes away current rather than water. If the drain isn't working properly, the electricity can arc over and cause electrical damage as it finds another path.

In addition to lightning-induced voltage saturating your house, which can create voltage spikes everywhere, surges can enter via the utility expressways: the power, coaxial cable, and telephone lines. Massive spikes can run into your house, look around, find all kinds of damage to do—and then zap! Fortunately, it is possible to limit the zaps that come in on the utility lines by clipping off their highest peaks and sending them to earth.

The neutral, as it is grounded at various poles and at the utility transformer, is thought by many to have no pulse coming into the house. In my opinion, this is not true. The theory that the neutral pulse voltage is zero stems from the assumption that the ground resistance provided by the utility to drain away all neutral surges is 0 ohms. But those who have ever measured ground resistance know this is not true (5 ohms is considered very low, less than 50 ohms is considered adequate by some, and 100 ohms or more is considered high). In truth, there is no such thing as a perfect ground. We drive one or more ground rods, tie into whatever else is considered grounded, and live with what we get.

Utility and In-House Surge Creators

Power utilities have to work on their lines. They are always adding this, fixing that, installing new poles, replacing broken ones, tapping into lines for new service, and so on. Each time they come online (and sometimes offline as well) a surge is placed on the utility conductors. For example, a capacitor bank (installed to reduce voltage fluctuations) produces a significant pulse and initially charges as it is brought online. Trees falling on the lines also create voltage and current surges that can be as damaging as lightning.

The equipment we try to protect also produces surges that can hurt other equipment. For example, a washing machine solenoid snapping in and out sends pulses on the in-house electrical wiring. These pulses may not be as massive as a lightning surge, but they are a lot more numerous. Sometimes the problem is so simple that we overlook it (such as a common snap switch). If you can help it, never put a snap switch on the same circuit as an electronic device.

Here is a simple test: Plug a radio into an outlet and turn the room lights on and off. The crack you hear coming from the radio is the pulse created by the switch. If you have a computer on that circuit, you are sending a pulse to it as well.

In the past, the only way to ensure that appliances were safe from damaging surges was to unplug them. Back then, most people had

Take Note • Sometimes the problem of incoming surges lies with the utility. Have the utility verify that it has properly installed a surge arrester on the pole and a multiple-ground-rod system. You may have to get an electrician to make the official complaints, following the age-old adage of "It's not what's said but who's saying it."

Noise Pulses

Surges and noise pulses can flow from the load to the service panel. Without point-of-entry protection, pulses and electrical noise generated in-house, such as those from a vacuum cleaner, washing machine solenoid, or shop motor, flow from the branch circuit on which they originate to the main service panel. From there, they flow onto every branch circuit in the house.

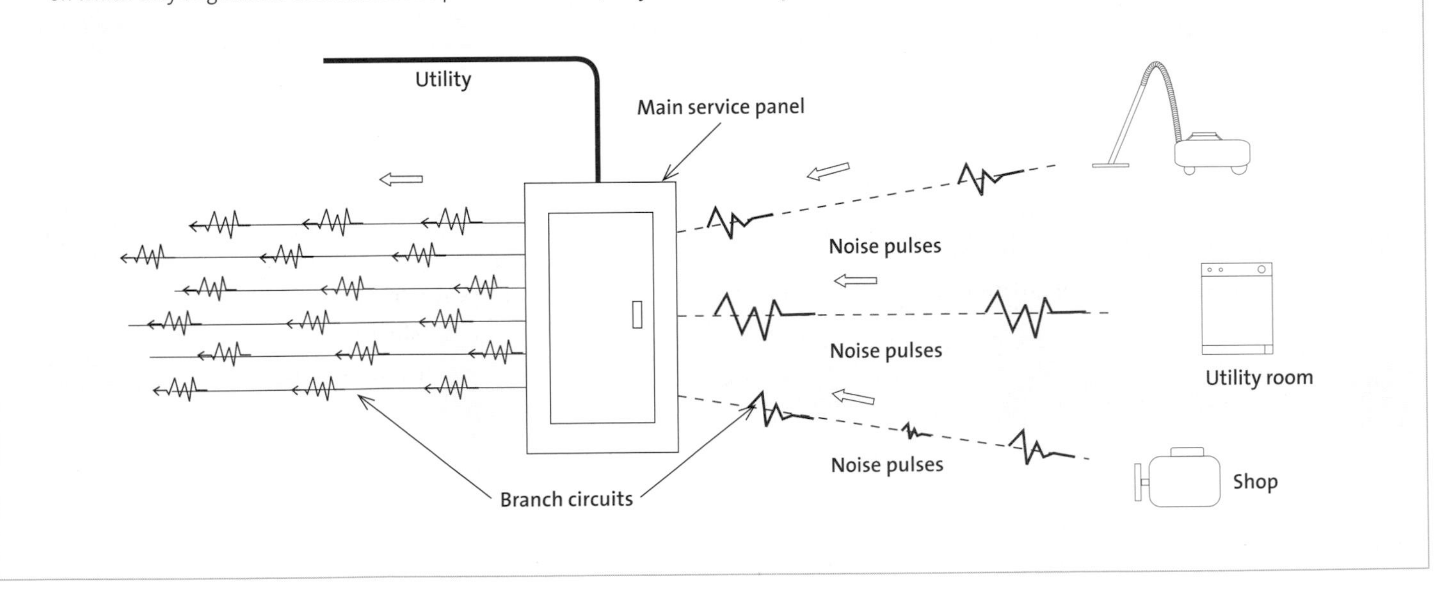

only two major appliances: a stove and a washing machine. Not too long after came electric dryers and television sets. Today, unplugging appliances is no longer practical. Our appliances are integral with the house, but they are also very fragile. Operating at only a few volts, they can't survive a battle with lightning or any other surge source. This cause and effect brings on the newest step in home evolution—surge protection.

Adding surge protection

A house needs surge protection in a minimum of two locations: at the point of entry (the main panel or subpanel) and at the point of use (where the equipment is used). Point-of-entry devices will have a longer life if you use a heavy-duty protector with a high clamping level that affects only large pulses. If a supersensitive system is installed at the point of entry (say, one that cuts and filters every surge down to just above a nominal 120 volts), the protective device will degrade and die an early death from the excessive number of times it must divert surges. Let the point-of-use devices deal with just the smaller pulses.

Point-of-entry protection

This is the first line of defense against surges. Without a point-of-entry protection system, one voltage spike can affect the entire house. This happens through the service panel. The service panel is the central hub, and every wire, receptacle, switch, and load is connected to this panel. Although this may be an efficient system for wire distribution, it also allows any incoming surge to branch out and run on every circuit throughout the house.

Surge Protection

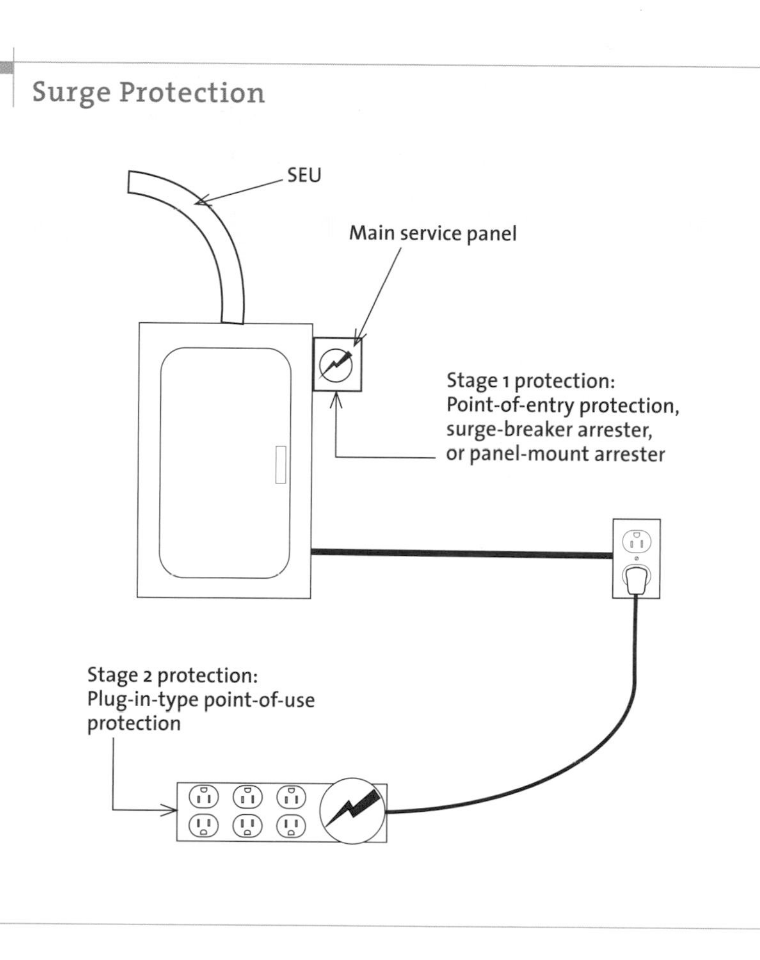

A surge can come into the panel from the utility (as in a lightning strike) or from in-house equipment, such as a washing machine solenoid or switching appliance. But no matter where it comes from, once a surge hits the main panel it will go through the house like a virus, infecting all the circuits. The only thing that can stop it is an effective point-of-entry protection device. There are two basic types of point-of-entry protection: plug-in and hard-wired protection.

Plug-in surge protection Plug-in surge protection is the easiest type of protection to install. A surge breaker has a tremendous advantage

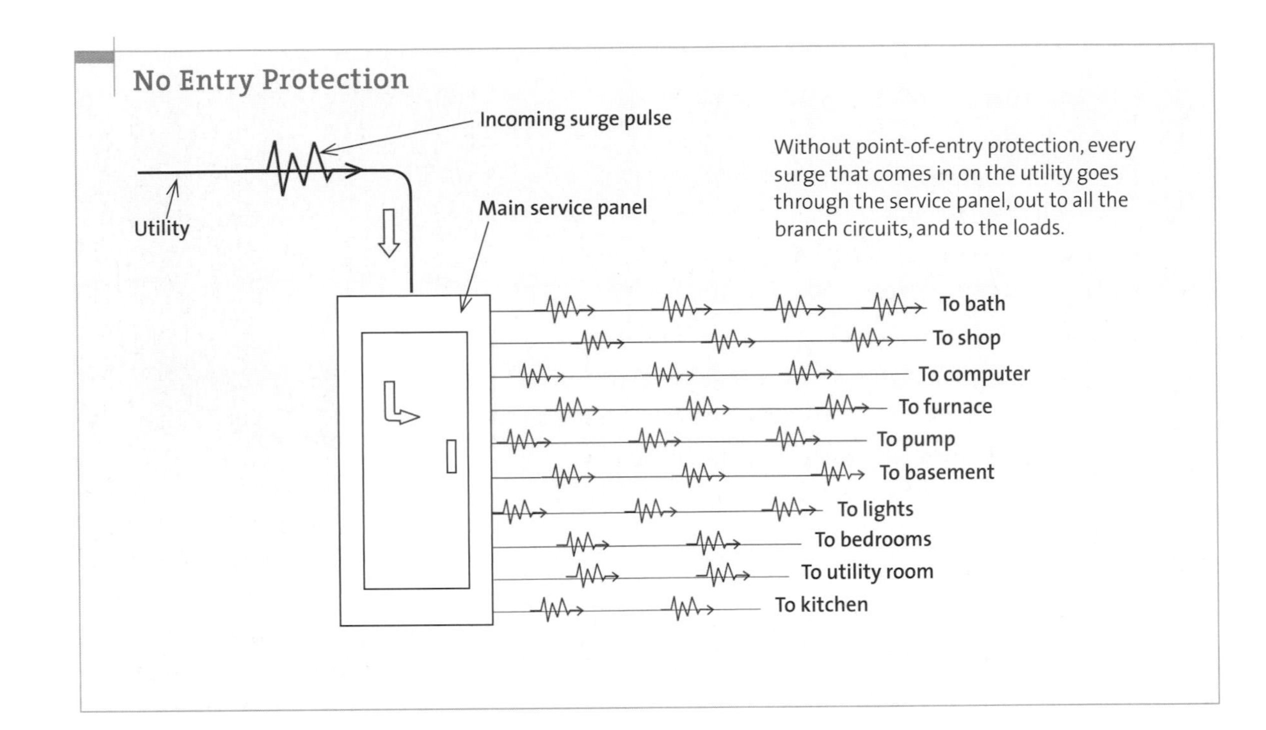

This two single-pole breaker module, made by Siemens, mounts in the main service panel like a standard breaker, but it also works as a surge arrester to protect the entire house. The lights indicate that the surge-protection system is working.

over a hardwired arrester because it snaps right onto the tabs of the service panel and has no long leads that may develop excessive voltages caused by massive surge currents taken off the line. It has just one wire connecting the breaker to the ground bus. This wire must be as short as possible.

Above Code • The surge breaker made by Siemens, officially called a Circuit Breaker and Secondary Surge Arrester, is what I use in my Above Code system. It consists of two single-pole breakers (to protect each phase) and installs just like a common circuit breaker. A red light indicates whether the protection system is working. The nice thing about this system is that it protects the entire panel.

Keep the Wires Short

The length of the pigtail wires on an arrester may dictate whether the unit will work or not. Massive surge currents flowing through these wires develop a voltage proportional to their length. A wire with resistance of only 1 ohm is still subject to Ohm's Law: Voltage = Current x Resistance. For example, 10,000 amps flowing through a 1-ohm-resistance wire develops a whopping 10,000 volts—and that's enough to blow anything apart.

The rule to remember here is the shorter the wire, the smaller the voltage. Always mount a hardwired device as close as you can to the main service panel. Once the leads enter the panel, keep the leads as short as possible, with no extra slack wound in the panel.

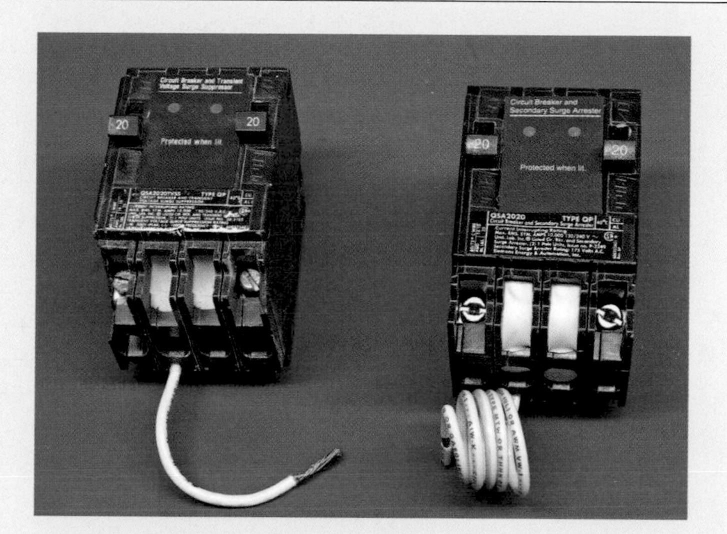

Keep the pigtail short. During a lightning strike, there will be tens of thousands of amps going through the white pigtail wire of the surge breaker as it strips the surge off the AC line. Keeping this wire short will send the surge to ground faster and create less developed voltage.

If there is more than one breaker box (say, two 200-amp boxes), you need a surge breaker in each box. If there is a subpanel downstream from the main panel (in a different room), you should put one in there as well.

Hardwired surge protection Hardwired protection is exactly what it sounds like: The wires are stripped and connected into some type of device. These are further broken down into two subcategories: those encased within metal boxes (either one of its own or the service panel itself) and those that are not. You want the former, not the latter. Any point-of-entry surge-protection device that takes massive surges through its body has the potential to explode. (I have seen the damage caused by the resulting shrapnel.) For that reason,

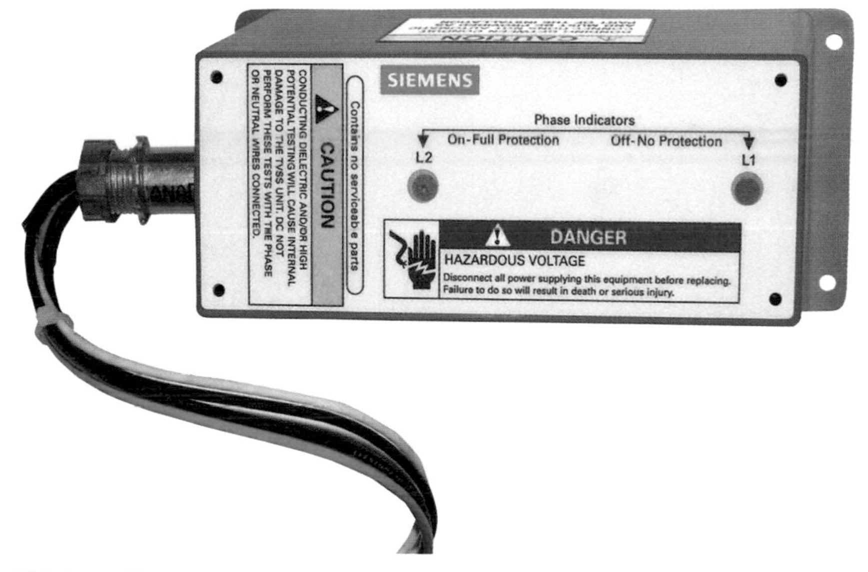

This type of suppressor mounts on the exterior of the service panel or box through a knockout. The wires then enter the panel through the knockout. The trouble with this type of suppressor mounting is that you have to cut the finished wall, if any, around it. Courtesy Siemens Corp.

Verifying a Low-Resistance Ground System to Panel

Suppressors (at point of use) take surges off the line (power, telephone, data, etc.) and send them into the grounding system. More specifically, they send the surge to the equipment grounding conductor (EGC). Thus, if the EGC is not a low-resistance conductor from receptacle to service panel, surge protection will not work, regardless of how good it is. So you can either hope for a low-resistance conductor or you can test it. If you opt for the latter you need the Ideal 61-165 (this is the old SureTest unit). With this tester, you can take resistance readings on all wiring, but especially the equipment grounding conductor, from outlet to panel. You want a reading of less than 1 ohm—the lower the better.

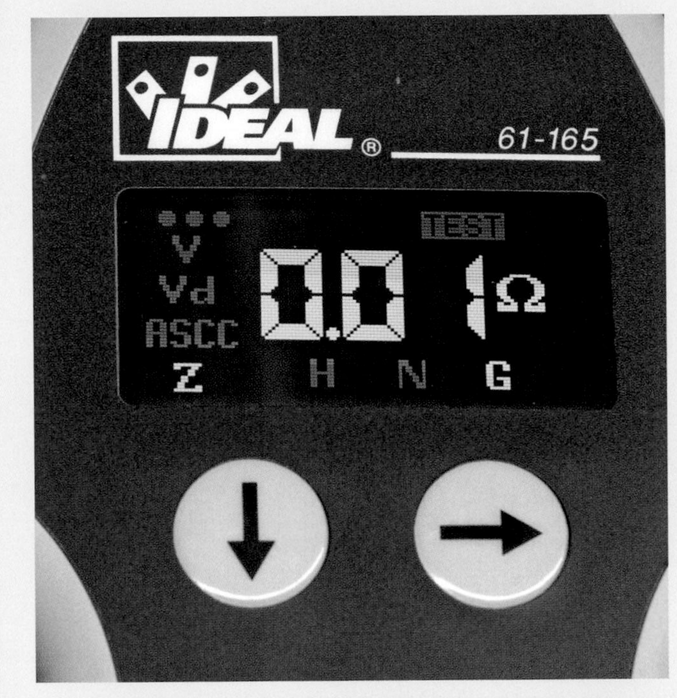

For the pros: by using a specialized tester one can test the ground resistance from receptacle back to the panel. A very low number (like the 0.01 ohm shown here) assures me that the equipment grounding-conductor splices are made properly and will conduct the surge currents away from the equipment under protection.

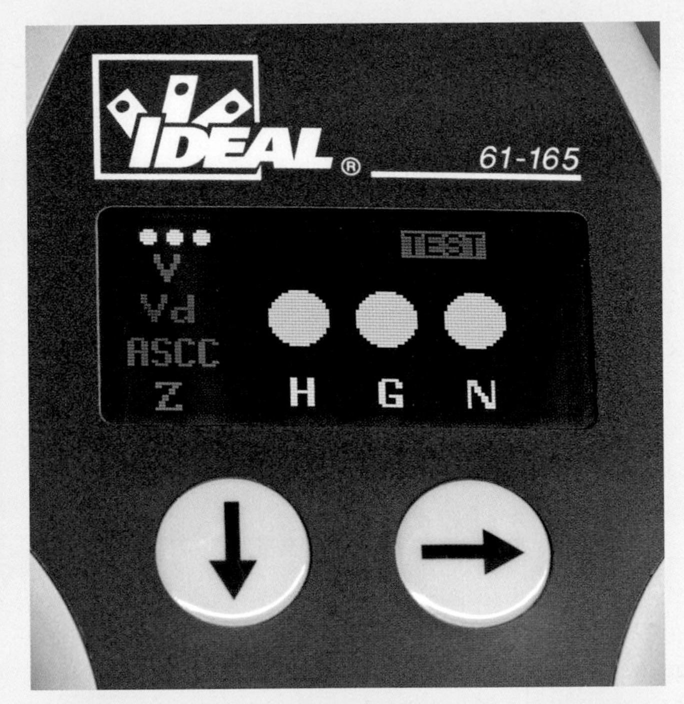

The outlet where your point-of-use protection is being used must be wired properly or the protection will not work. Here, three dots indicate proper receptacle wiring.

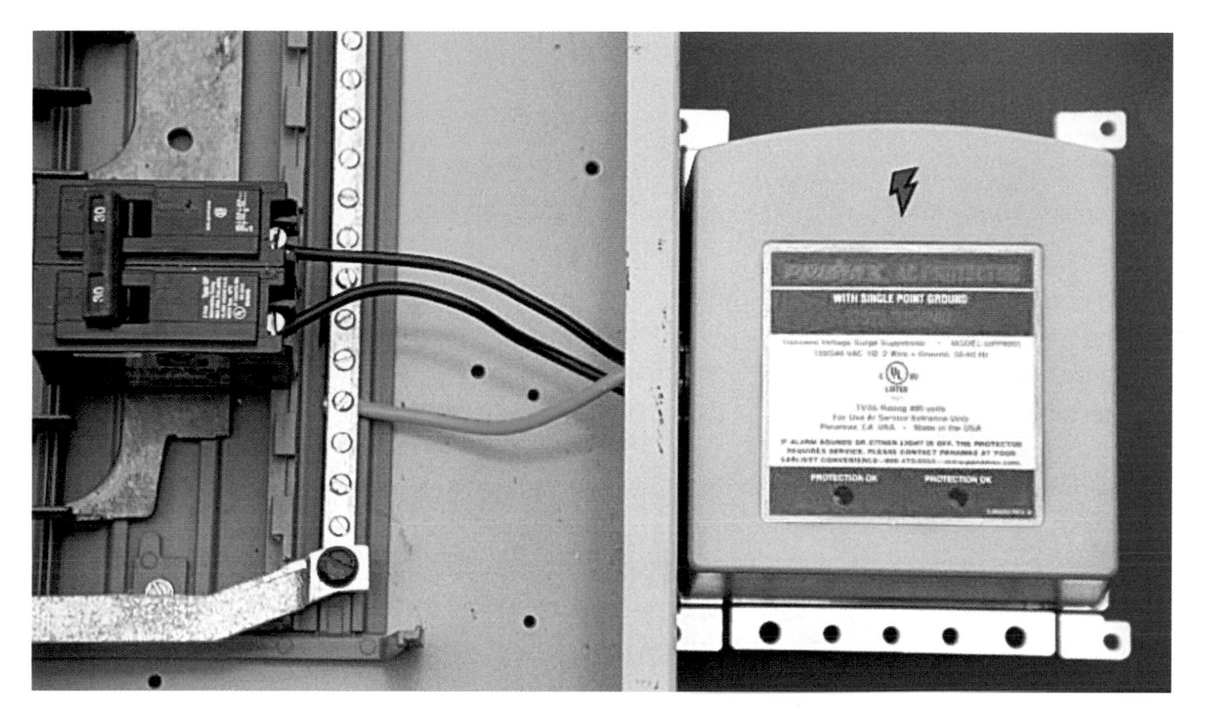

Although this Primax mounts outside the panel, it is encased in a metal box. The hardwired surge protector mounts on the side of the breaker panel and clips off surges coming in from the utility line. It receives power from a 30-amp double-pole breaker. Do not hide these within the walls.

hardwired point-of-entry protection must be mounted either within a metal service panel or within its own metal box adjoining the panel.

When mounted on the outside frame of the panel, make sure that the surge protector is not hidden within a wall, which makes it inaccessible for inspection. (I don't like the idea of having the device explode inside a wall where it could start a fire.) This is the most significant disadvantage to using hardwired surge-protection devices, because the finished wall has to be cut around the device, which creates an ugly installation. If the panel is mounted against or within an unfinished wall, the problem does not exist.

Point-of-use protection

Point-of-use protection is the second stage of surge protection. Once the massive surges have been taken off the line at the service panel, the more sensitive components of point-of-use protection can come into play.

Point-of-use protection normally comes on a receptacle strip with a long plug-in cord. The specific type of point-of-use protection you need depends on the equipment you want to protect (AC, telephone, cable, satellite, and so on). You need point-of-use protection for each appliance or group of equipment you want to protect. For example, I have two devices in my office behind the computer to protect it and all the equipment around it. Then I have another at the portable phone in the kitchen, one for each of the two TV sets, one for the copying machine, one for the stereo system, and so on.

Not installing point-of-use protection shortens the life of your electronic equipment and entertainment playthings. It is folly to go out and buy a $2,500 top-of-the-line TV set and not spend a few more dollars to protect it against surges that will either burn it up or hasten its demise.

All-in-one surge protectors give a variety of types of protection. On the left are outlets that are wide apart to allow transformers to fit. On the right are receptacle outlets.

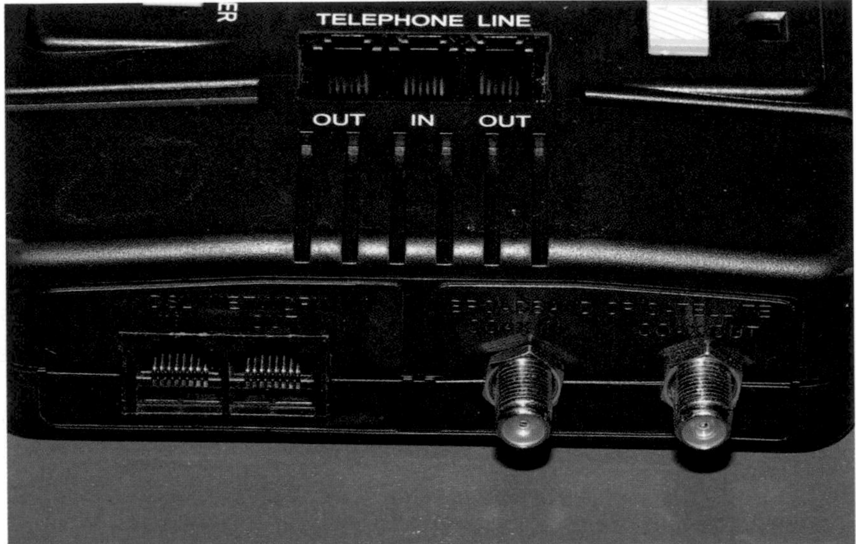

On the front are telephone, coax, and broadband outlets.

"Joules" refers to the amount of heat (from a surge) the protection unit can take before it malfunctions. The larger the number, the better. This one is quite large.

AC protection comes in various-size outlet strips, including two, four, six, and eight plug-ins. My experience is that you need at least six; eight is better. Make sure the strip is designed to allow room for transformers to be mounted side by side or back to back.

Picking the right protection The biggest challenge in selecting a point-of-use protection device is choosing from such a large assortment of products. Unfortunately, some of the devices on the marketplace are just junk and others are a fire hazard. Here is a list of prerequisites to help you separate the wheat from the chaff:

- UL approval.
- Automatic power disconnect. If the unit fails, which can happen when a massive surge or electrical fault occurs

on the line, the unit will disconnect all power to the load.

- A self-diagnostic system. Look for units that indicate whether the receptacle is properly grounded and the protection system is operating.
- An abundance of receptacle slots for all the AC, phone, and coaxial cable lines you want to protect. Make sure it can fit several transformers at once.
- A good warranty and guarantee. In addition to insuring itself, look for a unit that insures the objects it's protecting. Some come with a $100,000 insurance policy.

Problems with point-of-use protection Nothing is without problems, and Murphy's Law seems to work overtime with anything electrical. Here are the most common problems I see in installing point-of-use protection:

- Poor or nonexistent grounding. No low-resistance grounding means no protection.
- Wrong surge device used. Having an unprotected appliance, such as a telephone line or a coaxial cable to the TV or DVD, endangers the entire system.
- Ground skew. This situation is caused by using two or more different circuits for interconnected equipment. In the drawing, there are two ground references. This could cause a difference in ground potential and even a ground loop (current flowing on the ground wires between circuits).
- Used outside, around water, or in high heat. Keep any point-of-use device (or anything electrical, for that matter) away from water and high humidity. Also, don't put the unit next to a wood-burning stove, which you might do to protect the blower.
- Connected to a two-prong receptacle through a cheater plug or by cutting off the ground lug. Again, no ground means no protection.

Missing Protection

The AC is protected but the coaxial cable is not. You can have the same problem at a computer by not protecting the phone line.

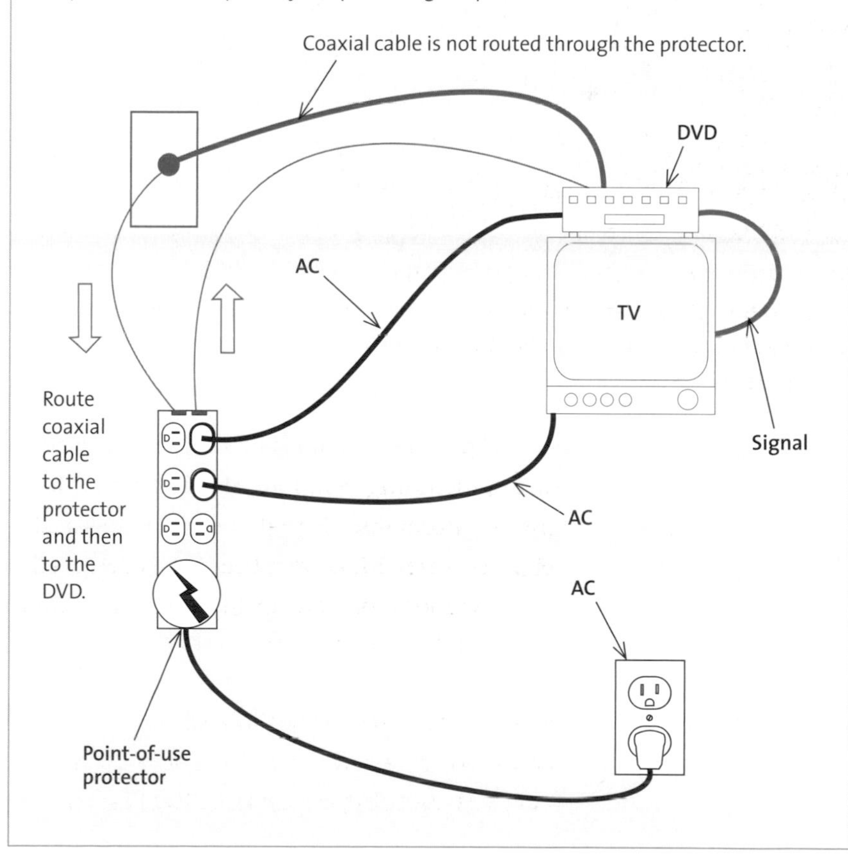

- Connected to an outlet that uses cheap push-in connections. This type of receptacle overheats easily and can produce intermittent connections. Rewire the receptacle by putting the wires under the screws, or even better, replace the receptacle with a good-quality one.

Ground Skew

Here, interconnected appliances—the computer and the printer—are plugged into two different circuits. The printer is not protected and it has a different ground reference from the computer. This is known as ground skew. As a result, the equipment may not work correctly at times.

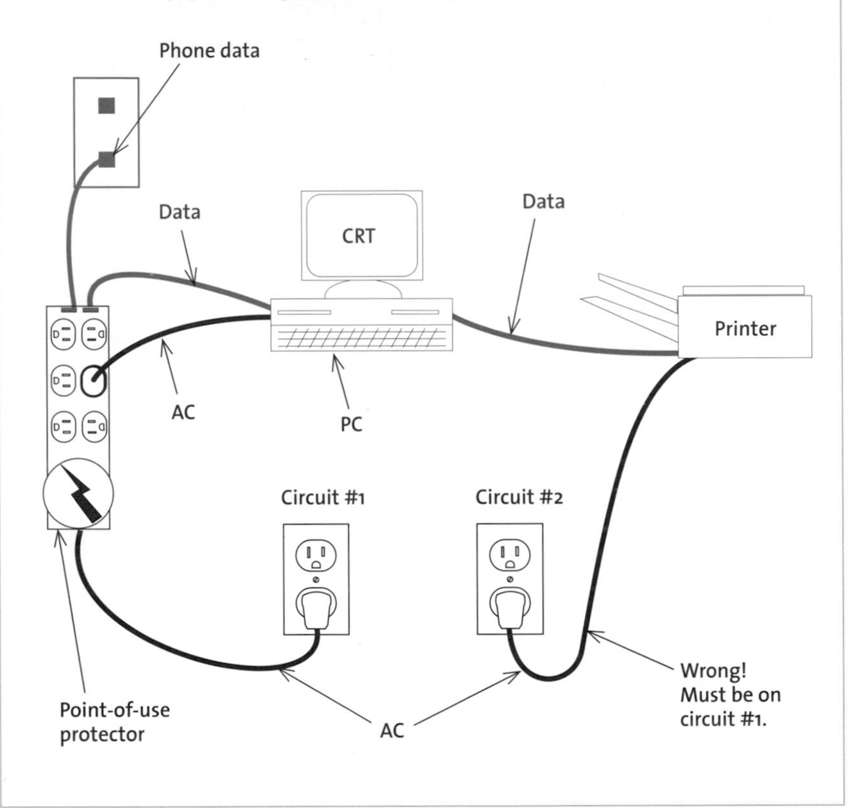

Phone and Coaxial Cable Protection

Surges coming into the house via the phone and cable lines are quite common and problematic. They destroy computers, TVs, DVDs, modems, answering systems, telephones, and just about anything to which these lines are connected. Although both aerial and buried cables are shielded to protect the wires inside them from lightning-induced surges, the system obviously is not without problems and can pick up major surges. One problem is that many utility systems do not have enough ground rods at their poles and pedestals.

For the phone system, another problem lies with the network interface device (also called the NID, or the network interface unit). The NID is mounted on the house siding and is where the telephone cable comes in and the house telephone wire goes out. It provides catastrophic surge protection and an interface point between the house telephone wiring and the utility cable. Back in the old days, an NID was just a tall metal can, and lightning protection was provided via carbon blocks or fuses. Except for providing protection from massive strikes, they did not work well. But then they didn't have to—our telephones were not as sensitive as they are now.

As the electronic era emerged, better methods of telephone surge control were needed. The carbon blocks were replaced with gas tubes, but the first models were a bit too sensitive and put a lot of noise on the line. Today's NIDs work best for long-duration massive surges (I've seen many a protector box literally blown off the house), but they aren't designed to protect your in-house equipment from quick, small surges. It's up to you to provide your own protection. And, as with AC protection, phone surge protectors start with low-resistance grounding.

Telephone protection

You can protect your own phone and cable system by simply adding point-of-use protection (always needed) and point-of-entry protection (optional; you'll have to use a commercial unit). Most residential and small- or home-business installations don't have thousands of feet of telephone lines running throughout the building and can normally get by with point-of-use protection for each piece of equipment or line. For extremely large homes and businesses, I recommend commercial point-of-entry protection in addition to point-of-use protection.

At your point of use, you have a choice of an inline unit or one that is attached to a plug-in receptacle. The inline unit will have to be grounded via the green lead that hangs off the side (it installs under a screw), whereas the plug-in unit is automatically grounded via the grounded receptacle.

Plug-in type telephone-cable surge protectors are incorporated with a receptacle so they can use the outlet's ground pin and EGC to take away surges.

Coaxial cable protection

Most houses have some type of coaxial cable, either from their own antenna or satellite or from the utility. Residential cable point-of-entry protection doesn't exist unless you use a commercial product. The blocks that the cable connects to on your house are grounding blocks, not surge-protection blocks. Their purpose is to send to ground any noise or surges that their shields pick up.

In some areas, installers have stopped putting in grounding blocks and simply use them as a coupling. There are two reasons for this: First, installers don't want to go to the trouble and expense of running the ground wire from the block to the other side of the house where the grounding system is located. Second, the runs from the dish to the house equipment are sometimes very short and don't have the potential to pick up surges that longer runs do. But this logic is thrown out the window as soon as a lightning strike hits a nearby tree and all your equipment goes up in smoke.

Don't assume that the equipment the coaxial cable connects to after it enters the house

Stand-alone coax (left) and telephone (right) surge suppressors. Note the ground wire: Each must be connected to grounded metal or a grounding pin in a receptacle outlet. Courtesy Siemens Corp.

will ground the cable's shield and thus take the surges to ground. Some electronic equipment is two-wire AC (no ground), so check to make sure. The bottom line is that without point-of-use protection to provide both protection and grounding, the coax shield will remain ungrounded, waiting for a surge to burn up the electronic gear.

Point-of-use protection for both phone and cable is quite easy to obtain. Many multioutlet point-of-use AC line protection systems safeguard these circuits as well. In addition, you can use stand-alone protection devices for both telephone and coax. These mount on the equipment being protected or at the wall outlet. Remember, though, that the devices have to be grounded. Thus, a computer with a metal case works just fine, but not so an answering machine with a plastic base. For the latter, and any other plastic-base electronic device you are trying to protect, look for a ground at the nearest grounded receptacle outlet.

Well Casings

If you are having a well drilled, the installer will give you a choice of casings: galvanized steel, nongalvanized steel, and plastic (PVC). Always choose galvanized steel. Nongalvanized steel can rust and contaminate the well. PVC is less expensive and doesn't rust, but it has two big disadvantages. First, you lose the ability to include the casing, which makes the absolute best ground possible, in the grounding system. Second, PVC doesn't seal well in bedrock. If you use steel casing, the driller can pound it hard into the rock for a good, watertight seal. You cannot do that with plastic, because it will shear off and block the casing hole.

Protecting Pumps

In high-lightning areas, submersible pumps can easily burn up, even if they are under water. Protecting pumps from lightning is a little different from protecting appliances within the house. In addition to surges coming down the utility lines, pumps are bombarded with heavy surges from a variety of directions.

After years of experimenting with brands and designs, I finally came up with a system that worked. (My apologies to those who received the systems that didn't.) The first part of my system is a Siemens surge breaker (the breaker for the pump) or a hardwired Tytewadd unit in the main service panel to stop massive surges from coming in on the utility line. In high-lightning areas, not having one of these means certain death for a submersible pump.

Pumps can get major hits from other sources as well, especially when the pump system is a long distance from the service panel. Typical problems include induced surges into the pump wire from overhead lightning, strikes hitting a metal vent line and jumping to the pump wire, strikes running along metal water lines, lightning running along fence lines and jumping to the pump wire, and lightning hitting a tree and running into the ground to the well. Thus, a surge comes into the house backward and blows apart the first thing it sees—namely, the pump control box and pressure switch. (Both of these units are located at the water pressure tank.)

The control box starts the motor in three-wire pumps (two-wire pumps have this circuitry in the motor) and the pressure switch turns the system on and off as water is needed. I've actually seen control boxes that have blown across a room and punched holes in the drywall. Thus if you have a cutoff switch you can add protection there.

For wells nowhere near the house, pumps need protection at the wellhead. Here I use a hardwired Tytewadd suppressor spliced right into the wires under the well cap.

My last piece of magic is an Above Code grounding system to make sure surges always have a low-resistance ground. Be sure to include the metal well casing in the grounding system. If the well is close to the house, run a #4 copper wire from the last ground rod to the casing.

A submersible pump deep in a well has its own protection in the motor, but it still can be damaged. When the surge comes in to the

How Lightning Finds a Pump

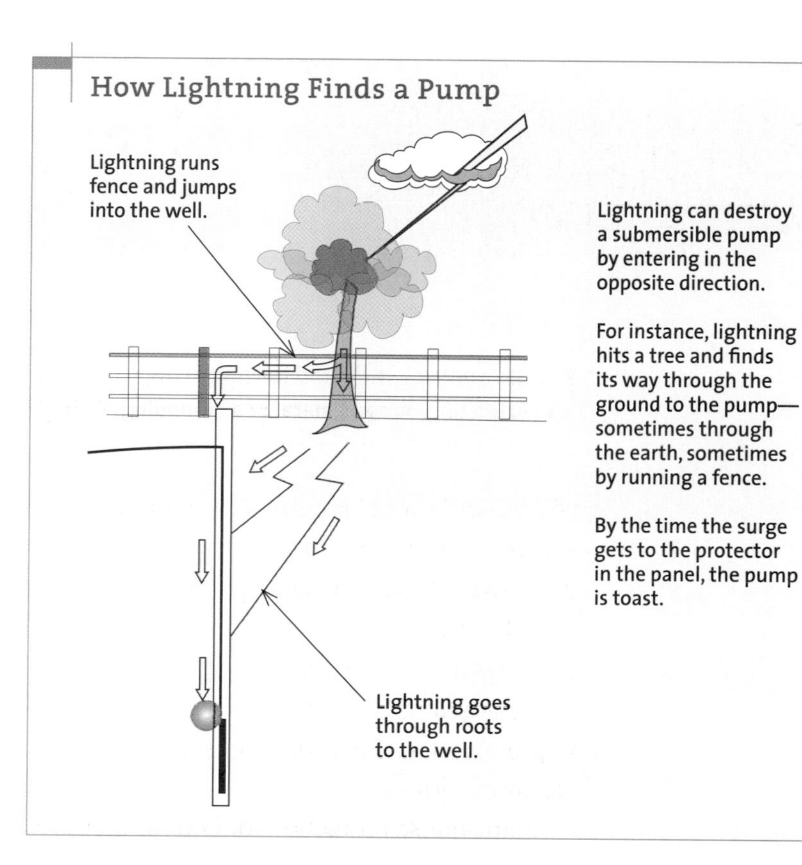

Lightning can destroy a submersible pump by entering in the opposite direction.

For instance, lightning hits a tree and finds its way through the ground to the pump—sometimes through the earth, sometimes by running a fence.

By the time the surge gets to the protector in the panel, the pump is toast.

This burned hulk is what's left after a surge came in backward from a well. The surge-protection system within the pump protected the motor but not the control box (the surge also burned out the pressure switch). I've seen the lids of these boxes explode across a room, denting the wall on the other side.

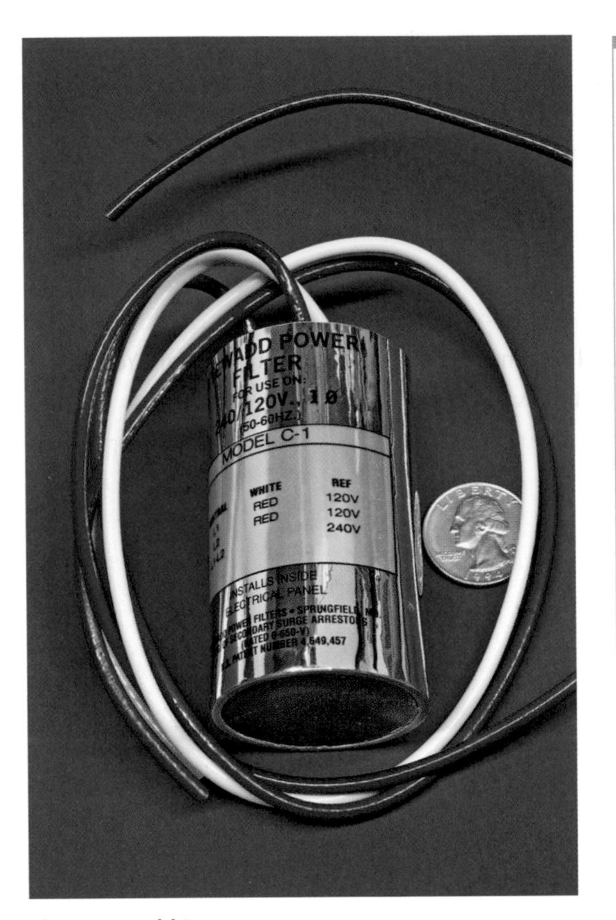

The Tytewadd Power Filter works well inside the panel, subpanels, and large switch boxes, and under the well cap.

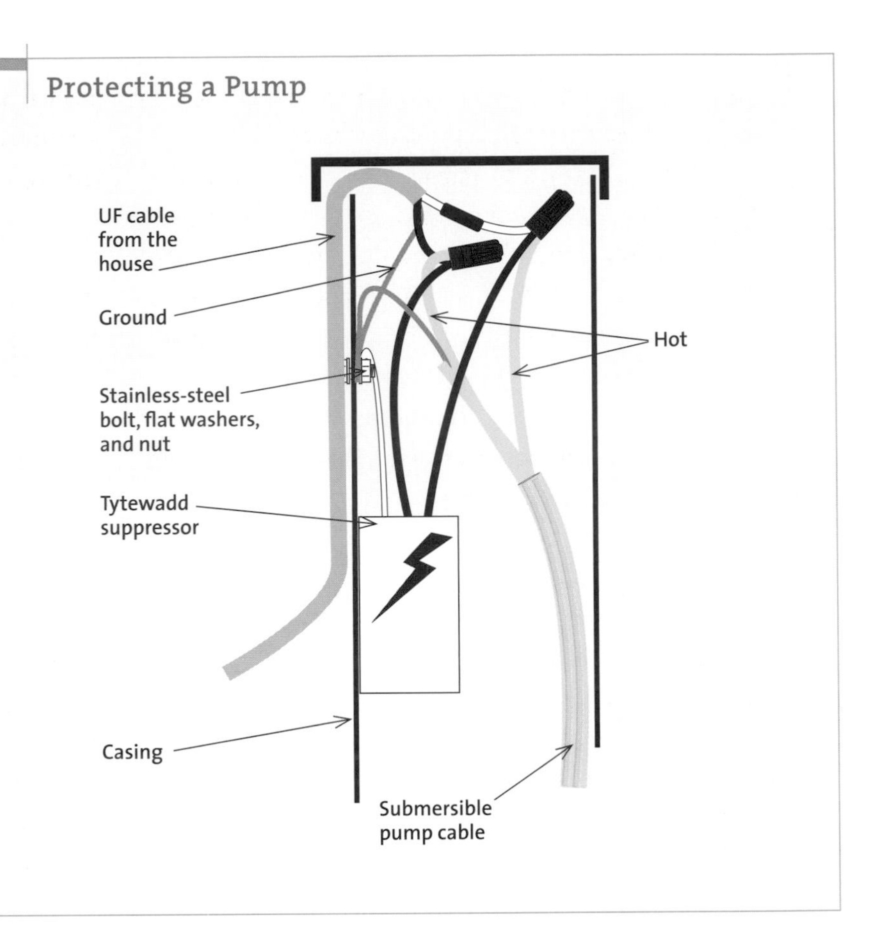

pump's motor, the pulse races to the end of the winding and doubles in value each time it bounces back to the opposite end of the winding—10,000 volts, 20,000 volts, 40,000 volts—until the windings fail.

Many pump motors don't fail because of just one thunderstorm. The damage is cumulative, with the windings becoming more susceptible to catastrophic damage with each surge until the unit just dies one day. Without some type of surge arrester mounted somewhere around the pump, tank, or wellhead to limit the surges before they reach the pump motor, lightning surges stream down into the motor time and time again and overstress the motor windings. The motor may continue to work, but the day will come when it takes one surge too many—or just gives up the ghost during Thanksgiving or Christmas.

Lightning Protection for Pumps

One of my customers went through three submersible pumps in three days—all in the same storm.

As I was replacing the original pump (between rain showers), I told the customer of my multistage system of lightning protection for pumps. He thought it sounded great, but he didn't want to buy it right then and there. The next day he called me and said he had lost the new pump I had just installed—again to lightning.

As I was replacing the pump for the second time, I again suggested my special system. This time he agreed, but said that he would install it himself in order to save on labor. I sold him the material, told him what to do, and went on my way.

The very next day, just like clockwork, he called again (by now we were becoming close friends). He had lost the second pump to lightning. I asked him if he had installed my special system because I had never lost anything with it installed properly. He said he hadn't gotten around to it yet, but he would now. After I installed the third pump and he installed my multistage pump-protection system, I never heard from him again (so much for good friends). Bottom line: Put the system in before you lose your pump, not after.

Pump-Protection System

A pump lightning-protection system works in layers, just like all other protection systems. The first layer is at the service panel and the second is at the pump controls. You are required to have a cutoff at the pressure switch if the switch is not within eyesight of the panel. Make this switch a Siemens surge breaker if the cutoff is a breaker/switch. If the cutoff is just a knife-blade throw switch, hardwire in a Tytewadd filter. The third (optional) layer is at the wellhead—but should be your option of choice if you are in a high lightning area or if the wellhead is a long distance from the house. Here you will have to cut a Tytewadd directly into the wiring under the well cap.

Install an Above Code grounding system by including a ground to the metal well casing.

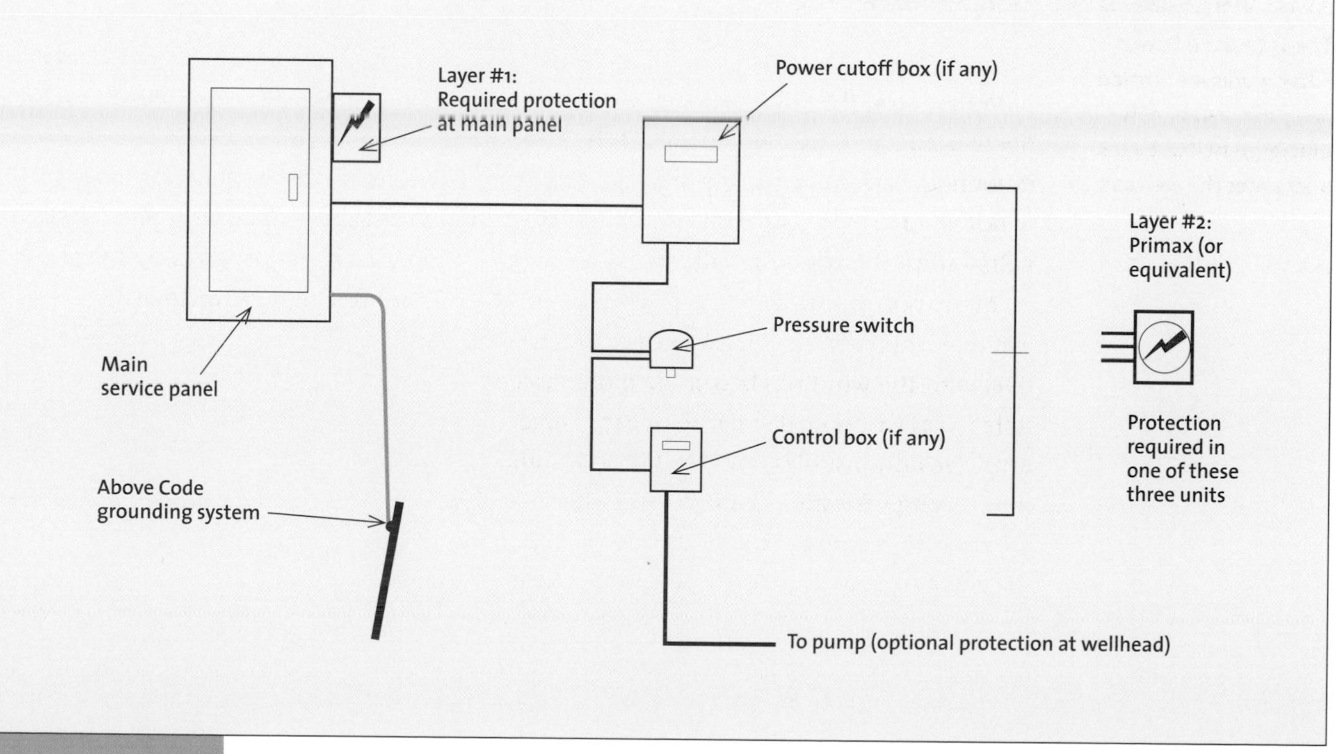

CHAPTER

15

Standby Generators

Just as there are no shortages of people wanting standby power, there are no shortages of those who do not know which generator to buy or how to install one properly. Buying the wrong size generator can be a waste of money (one of my customers bought three before he acquired the right one.) After selection, the next hurdle is installation. An improper installation can be dangerous to both the utility worker and the homeowner and can even cause the generator to go up in flames.

History of Incorrect Connections

The first affordable household generators were small. To use them, people could only run extension cords from the generator to the appliances that needed power. With the advent of newer, more powerful generators, people began experimenting with ways to plug the generator directly into the house wiring via a household outlet. This method, called backfeeding, is both illegal and dangerous; unfortunately, it became an all-too-common occurrence. This direct-to-receptacle method, though low cost, also allows power to run backward out of the house and through the utility transformer, and it can electrocute workers repairing the line. When the utility brings the power back online, the generator will go up in flames.

This on/off switch is not legal. Two-position switches can put voltage on the utility line that can disrupt the utility power and electrocute power-company workers.

Backfeeding evolved into using switches to turn the power on and off. To keep the generated power in the house, the homeowner would pull the main fuses or turn off the main breaker. A large on/off switch then supplied power to the house. This system is almost as bad as the extension-cord-plugged-into-an-outlet method. If someone accidentally turns on the switch when the utility voltage is coming into the house, sparks will fly, brownouts will occur down the utility line, and the generator will go up in smoke. It has happened. We cannot have the lives of utility workers dependent on the whim of every homeowner who might accidentally throw the wrong switch. Thus came the need for a special switch that automatically disconnects the main utility circuit from the house the moment the generator switch is turned on—a three-position switch with the off position in the middle, such as "on/off/on" (it could also say "line/off/gen"). The middle off position is crucial. It is the part that prevents you from cross-connecting the generator with the utility. This has now evolved into full-size breaker subpanels that contain specially interconnected disconnect breakers. This means that as one disconnect breaker is turned on, such as the utility disconnect, it forces the generator disconnect breaker to be turned off, and vice versa.

In the past, the standard legal way to connect a generator to the house was to install a transfer switch between the meter base and the service panel. However, this system no longer meets code, unless the panel is rated as a service entrance and is at the full current flow of the house service panel. This means that if you have a 400-amp service entrance the switch has to be service rated at 400 amps. The labor and material costs would be astronomical. That leaves two options: A subpanel with permanently installed pop-out-type breakers that feeds from the service panel, or the newer-style subpanel with full-size removable breakers that does the same.

Transfer-Switch Style

The first prewired transfer switches I installed had six to eight fixed pop-out breaker circuits. I put one in my house many years ago and it still works great. But now something new is on the market—a replaceable breaker-transfer switch. Though the fixed system is good and installs fast and easy, it is fixed by design. That is, you are stuck with whatever amperage is on the pop-out breaker.

The Old and the New

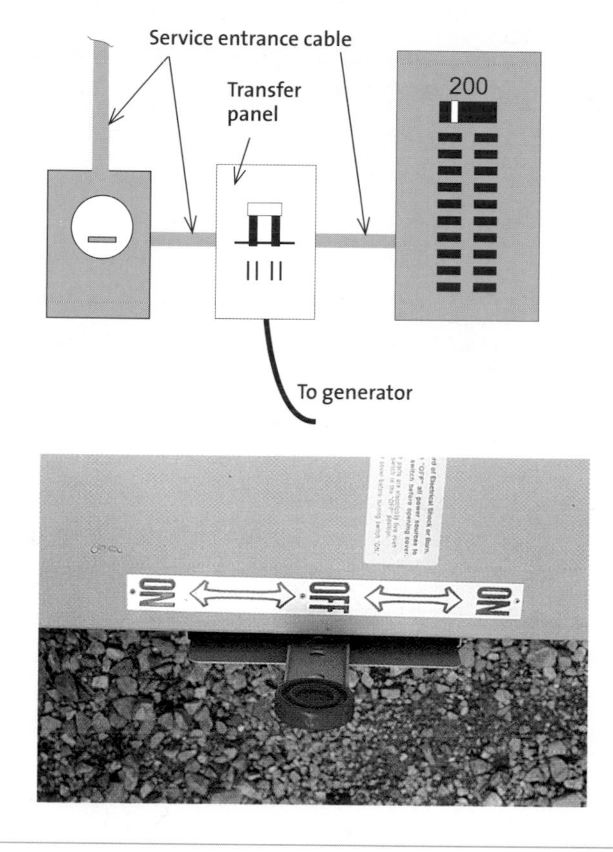

The New Method

Utility meter

Service panel

200

The new method places transfer panel adjacent to service panel and connects via a pigtail. Breakers are wired individually.

Transfer panel

To generator

In the past, we used to cut the service-entrance cable and insert the outdoor-rated transfer switch (a giant double-pole, double-throw switch) between the utility meter and the service panel. The generator was fed into the transfer panel and you turned off the breakers in the service panel that you did not want to feed. With code changes, we no longer do this.

With full-size removable breakers, if you don't like the ones that came in the transfer panel, you can change them. This gives you a lot more flexibility. You can now design a system you want rather than living with what you get. Typically these breakers are available in six to twelve circuits of various amperages.

Replaceable breaker-style transfer switches are available prewired (a short pigtail that connects the transfer-panel breakers to the service panel is already wired in) or unwired. For the latter, you create your own pigtail to the service panel and hardwire into the generator location. Though this takes longer, it allows

This replaceable breaker-style transfer switch comes out of the box with a prewired pigtail, load meters, and a twist-lock receptacle. This style, which is quite popular, is available in many wattages and circuits.

This style of transfer panel I call "fixed pop-out" simply because the little round breakers are fixed (cannot be changed or moved) and the center white pin will pop out if the breaker kicks.

you to put the transfer switch in locations other than adjacent to the service panel.

Another advantage of the replaceable breaker is that some, like the Gen/Tran®, can be field modified to allow more circuits by simply changing out full-size breakers for half-size breakers. During normal operation power will leave the service panel via a 60-amp breaker in the service panel (you buy and install), go through an integral 60-amp breaker in the transfer panel, go through individual breakers in the transfer panel, and finally go to the blue wires that splice into the load wires back at the service panel.

If utility is lost, the operator throws the combo switch in the transfer panel to the left. Utility is now disconnected and the generator is now connected to the transfer-panel breakers. Power will flow as described.

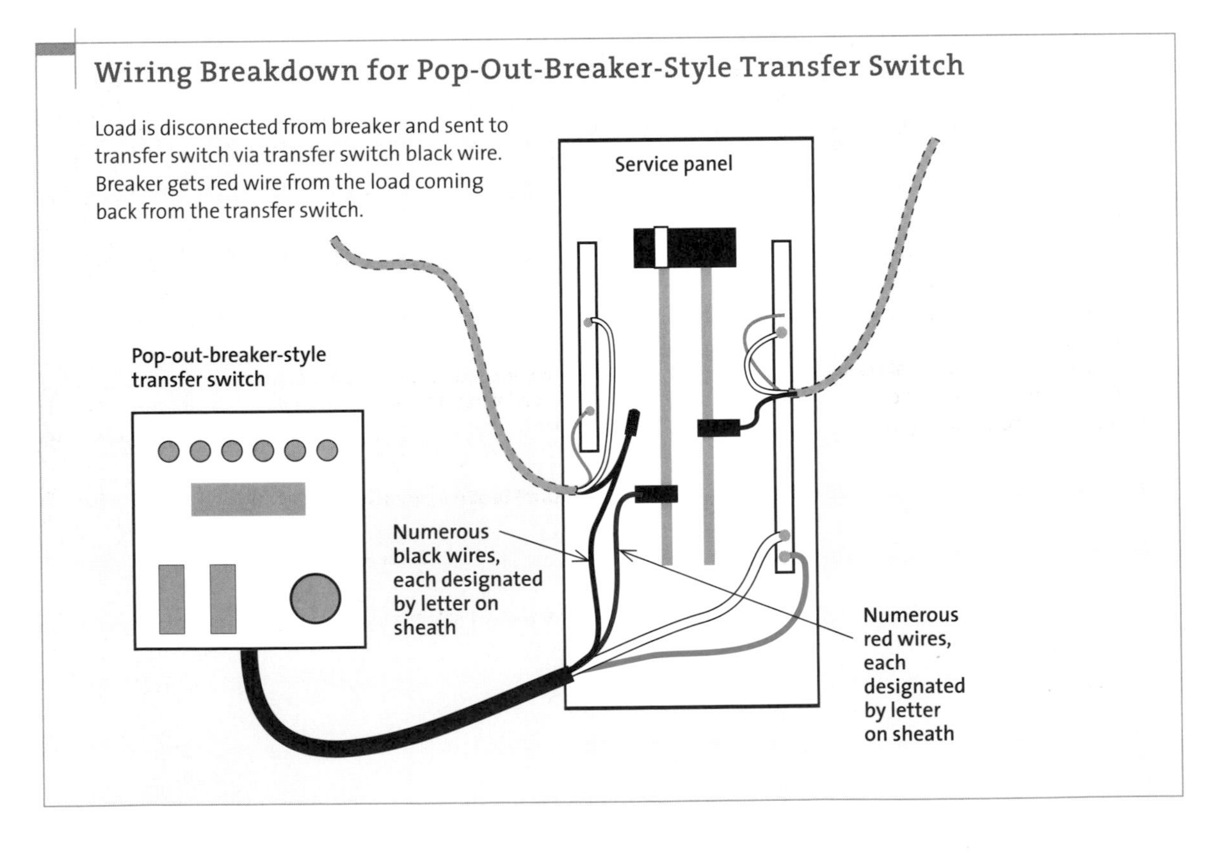

Wiring Breakdown for Replaceable-Breaker-Style Transfer Switch

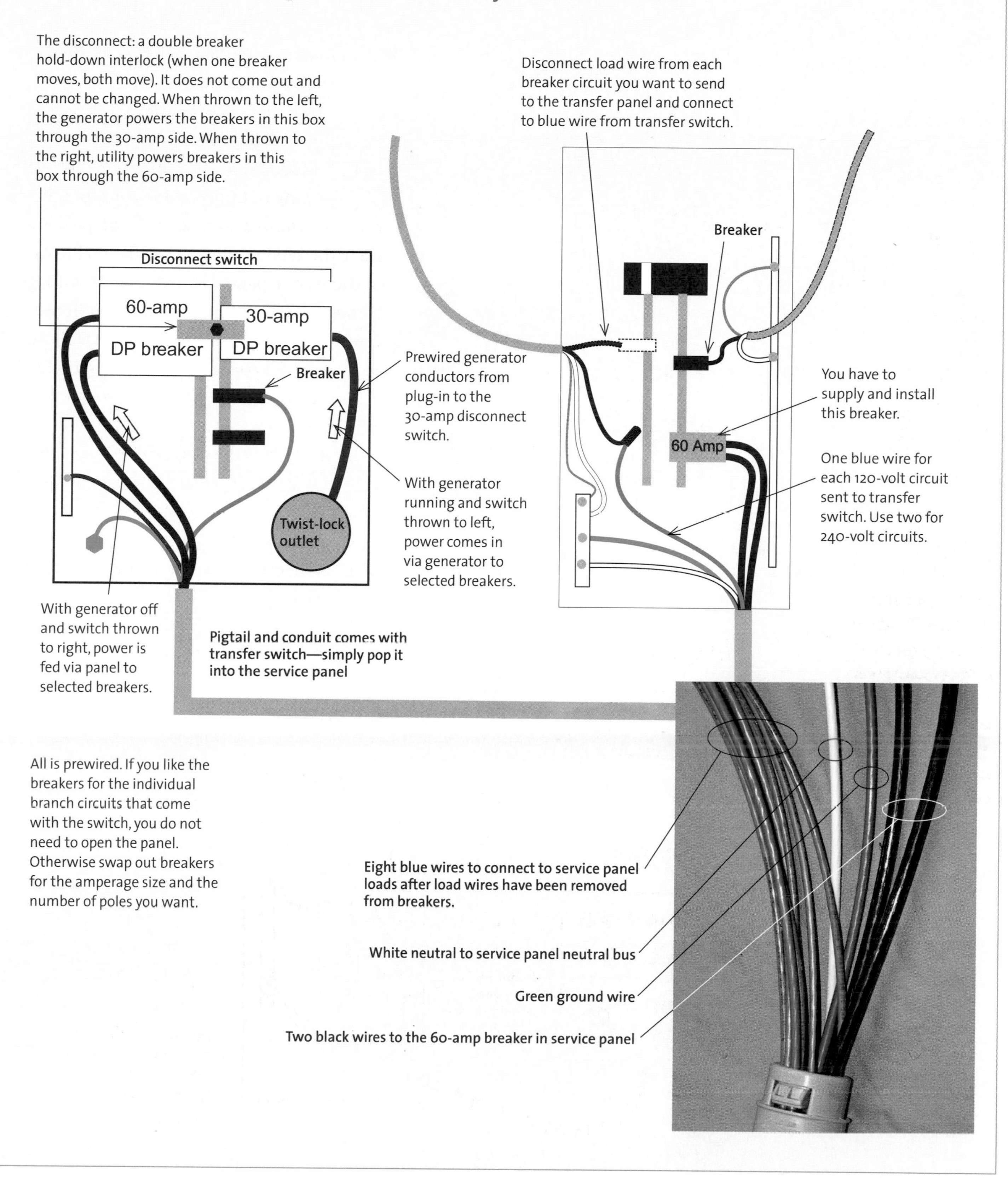

Breaker-Style Transfer Switch Allows Flexibility in Picking Breakers to Fit Loads

Throw both breakers the same direction and tighten down nut.

Note: Since designs and methods are ever changing, always follow the manufacturer's directions on installations.

Disconnects. Once the horizontal metal bar is tightened down, throwing one on throws the other off.

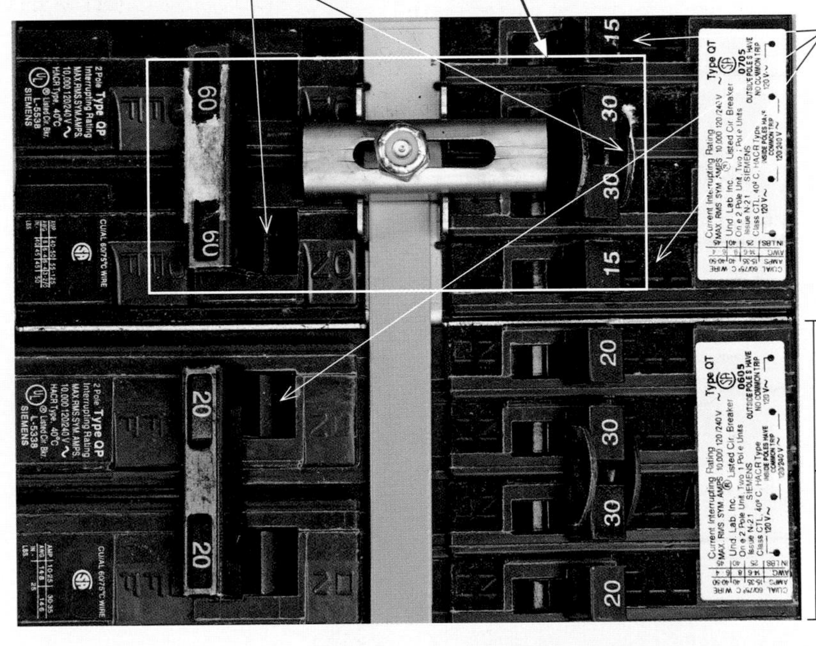

Transfer switch comes with several removable breakers for the emergency circuits. Loads that go to these breakers will be the loads that will be cut over to the generator as soon as the disconnect switch is thrown from 60 to 30. Until that time they are fed via the utility through the 60-amp disconnect.

15-amp breaker (half size). Can be replaced with a 20-amp breaker if needed.

30-amp DP breaker (half size) serving as generator disconnect. When on, current flows through this breaker from generator to loads in box. Utility is disconnected. Why 30? Because that is all a typical portable generator can produce. That and the generator cord to this box is rated at 30 amps max.

60-amp double-pole breaker serving as utility disconnect. When on, current flows thru this breaker from utility to loads in box. Generator is disconnected.

20-amp DP breaker for any house circuit that needs it. Can be replaced with two SP breakers if needed.

Quad breaker. This one is two SP 20-amp breakers (top and bottom) and a 30-amp DP in the center. This breaker can be replaced with other breakers if different amperages are needed.

From Generator to Switch

Typically you will connect the generator to the transfer switch via a twist-lock cord, though it's assumed that the generator is reasonably close-by. Sometimes, however, it is not. You can create your own extra long cord using outdoor-approved 10-gauge cord and twist-lock ends—and there is nothing wrong with this.

You can also run conduit from the transfer switch to the part of the house that's closest to the generator and run THHN conductor within (typically 10 gauge). If you run conduit, you will have to wire in an outlet box at the generator to plug in the generator cord (you can buy premade outlet boxes for this). If the outlet box is outside, you will have to use an approved box. An inside box will have the cord

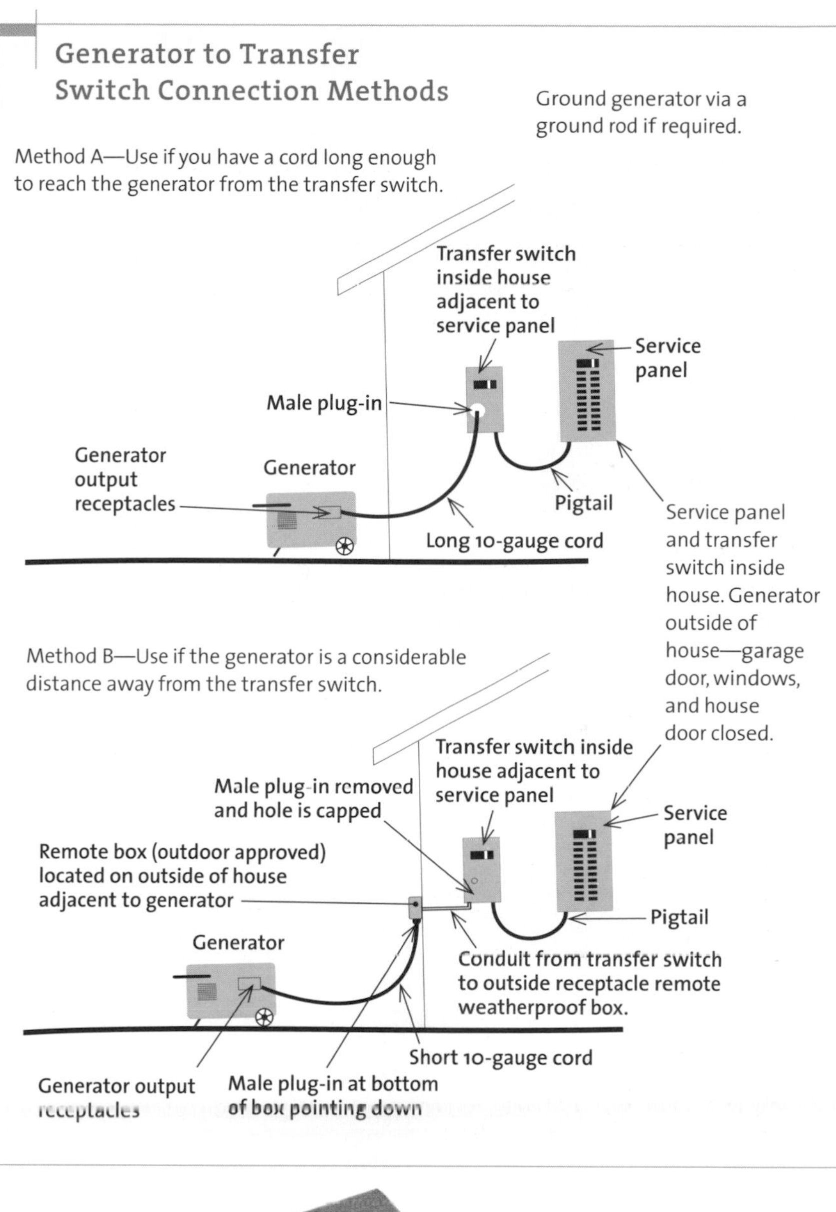

This type of generator plug outlet box must be used for inside locations. Note that its lid, when open for the plug, would allow water and snow to enter the cover if installed outside.

This type of generator plug outlet box can be used for both inside and outside locations.

Two House Circuits for the Price of One

It's possible to have two individual house circuits powered by a single transfer-panel circuit by rearranging and combining circuits in the service panel. For example, if you have two light-load circuits like a hallway and office, you can combine them as one circuit in the service panel for the transfer switch. Simply splice the two together with a wire connector, then add the wire to the transfer switch. You'll now power two emergency circuits for the price of one. Remember to spread the compressor-load appliances on different transfer circuits—don't wire them together or they will kick the breaker when both start at the same time.

Here we have two light-duty loads coming off their respective breakers and both need to be in my transfer panel as emergency circuits. If I carry them over as is, I will use up two slots in the transfer panel. Instead, I remove the two load wires from the breakers in the service panel.

I twist both load wires together along with one of the blue wires from the transfer panel and cap it. I now have the transfer panel seeing both loads as a singular load and I am taking up only one slot in the transfer panel.

plugging straight into it, whereas an outside box will have the cord plugging in from the bottom. For transfer boxes with no cord plug-in, it is assumed all will be hardwired.

Installing the transfer switch

In most situations, the transfer switch is installed next to the main service panel and wired directly to it via a prewired pigtail connecting the generator transfer switch to the main service panel house wiring. The pigtail's flex conduit can be metal, which has a habit of opening at the spirals, or plastic. Newer designs are leaning toward plastic to alleviate this problem. I've installed many such switches (including in my own house), and the average installation time is about 1 hour for systems with prewired pigtails. You'll spend more time trying to decide exactly which circuits to cut over.

Picking a Generator

Wattage, wattage, wattage. That's all most people think about when they pick out a generator. Wattage is important. Without the proper wattage you can't power the house or an appliance. However, wattage is not all-important. There are many other variables that must be considered before you spend your hard-earned cash.

Good. If you don't have much to power for emergency circuits then this size generator will do. It will power both a fridge and freezer, though it might not be able to start both at the same time. This size wattage can start a $^3/_4$-HP pump—but just barely. I would not buy any generator smaller than this one if you live in the country because it takes at least 3,500 watts to start a $^3/_4$-HP submersible pump. One significant advantage of a smaller wattage is the run time per gallon of gas. This particular generator has a 4-gal. tank with a 12-hour run time. Courtesy Briggs & Stratton

Better. Compared to 3,500 watts, this 5,500 watt will give you a lot more lights and receptacles to power up. Both fridge and freezer can start at the same time. I use a 5,500 watt to power my entire house, including $^1/_2$-HP pump, lights, fridge, freezer, TV, and computer. I would call this generator the average wattage needed for emergency circuits. This particular generator has a 7-gal. tank and a 13-hour run time. Courtesy Briggs & Stratton

Best. For houses with a lot of circuits to run, this is the generator you want. At 8,000 watts run and 13,500 start, you can run some significant loads. But remember, the larger the wattage, the larger the motor, and the more gas you use. This particular generator has a 7-gal. tank and a 7-hour run time. Compare that to the 3,500 watt unit and you can see that smaller-wattage generators do have an advantage in running cost. Courtesy Briggs & Stratton

Wattage and circuits

To figure out the wattage you need, think about all the appliances you want to power. While you're doing it, remember that each appliance may be an individual circuit unto itself and you are limited in the number of circuits you can power. Thus, your two limitations: wattage and power. Realize that you will not be able to power everything. Make an emergency list of appliances that simply must have power. The freezer and refrigerator are the two most common. After that come a few lights and water (if you have a pump). Hot water (for electric water heaters) is hard to obtain unless you buy a very-high-wattage generator.

Rather than adding up large lists of wattages, you can buy a basic 3,500-watt generator. This size generator is large enough for almost any short-duration emergency situation. If you want more wattage, by all means go for it, but I don't recommend anything smaller than a 3,500-watt unit.

Some appliances need a "kick" to start and thus require a little extra wattage. Most compressors and motors need two to four times their run wattage in order to start. For example, a ¾-hp submersible pump needs a 3,500-watt generator to start pumping water into the house. Knowing this, many generator manufacturers have surge-current capabilities built into their generators. Some unscrupulous salesmen will try to hype the surge wattage, which is usually 200 watts to 500 watts greater than the run wattage, as the run wattage. Don't be fooled. Surge wattage is available from the generator for only a few seconds.

For every appliance you "had to have," you also added one circuit for each 120-volt appliance or load and two circuits for each 240-volt appliance or load. These circuits add up fast. In most cases, go with six or ten circuits for the Gen/Tran. An average house with ordinary loads can get by with six emergency circuits, which is all that a typical generator

This generator proudly advertises 6,300 watts, but to find the real wattage read the nameplate on the generator. It's 5,900 watts—a big difference.

Power Tester

To know when the utility power comes back on, place a lightbulb and fixture (such as a 20-watt refrigerator bulb) next to the main service-panel box on a circuit that is always powered by the utility. When you see the light on, you'll know power has been restored and your generator can be turned off.

Work Safe • Unless you buy a very large generator, it just isn't practical to run a typical electric water heater. Replacing the standard 4,500-watt elements with 3,500-watt elements and investing in a larger (6,000-watt to 7,500-watt) generator may make it manageable. The most economical way to get hot water is to shut off all other loads and turn the water heater on an hour before you want hot water. After an hour, turn off the water heater and turn on everything else.

(for example, the 4,500-watt unit mentioned earlier) can power. Typical emergency loads and the number of circuits for a house may include the following:

- Water pump; two circuits
- Water heater, if any; two circuits
- Lights and receptacles; one area per circuit (combine light-duty load circuits, if needed)
- Refrigerator; one circuit
- Freezer; one circuit
- Computer or living room; one circuit

For more circuits, you can jump up to a 7,500-watt or greater generator and get a 10- or 12-circuit panel. You can use a 10-circuit panel with a smaller 4,500-watt generator, but you'll have less wattage to spread across the number of circuits. If you are building a new home, design the house wiring so that you can install a maximum number of circuits into a 10- or 12-circuit system.

Tank size

You want a large fuel tank—one that provides the generator with enough fuel so you don't have to keep shutting the system down every hour or two to fill it with gas. But just getting a large tank isn't enough. The actual time between fills depends on a variety of factors, including the motor's horsepower and efficiency. Check the generator's spec sheet for the "run time," which takes all of these factors into consideration.

You will also want a way to tell how much gas is left in the tank without having to stop the engine, open the lid, and insert a stick to see if it comes up wet. Fuel gauges for generators are commonly built right into the lid. If your generator doesn't have one, get one. They are usually sold adjacent to the generator display. (Or even better, have the salesperson throw it in with the package.)

Generators are noisy

When you fire up your generator, the entire neighborhood knows. Do not be tempted to put the generator in the garage and close the door to minimize the noise—this is deadly. If noise is an issue, look for a generator that is advertised as being quiet, or at least somewhat more quiet than the standard noisy ones. When shopping, check out the muffler on the generator, which in some cases will be called a "Super Silencer" muffler.

Work Safe • Never fill the tank with the generator running—the gas can ignite if some sloshes over the side or if the fumes swirl over to the exhaust.

Work Safe • Portable standby generators need gasoline. If a disaster is eminent, stock up. But don't stockpile gasoline inside the house. It needs to be outside the house so fumes don't build up.

Look for this outlet on a generator if you are shopping around for one to power your house (via the round 30-amp 120/240-volt twist-lock receptacle) and give duplex power to individual appliances if needed.

Exhaust

People have died from generator exhaust. You probably already know not to run the generator in the house, entry, or garage. But when placing the unit outside, make sure you take the wind into account. You don't want the wind blowing exhaust gases back into an open garage door, house door, or window. So keep all doors and windows closed.

You can run the generator cord to the transfer panel under the closed garage door. When working around the generator, remember that the exhaust pipe will be super hot, so don't bump into it. When filling it with gas, don't spill any onto the hot exhaust. And no, you cannot put the generator in the garage and pipe the exhaust outside—unless you have a death wish.

Receptacles

Generators come with a great number and type of receptacles, including 120-volt, 15-amp and 20-amp duplex (with and without GFCIs); 240-volt, 15 amp and 20 amp; 120/240 volt, 20-amp and 30-amp twist-lock receptacles; and 50-amp 120/240 plug-in. Don't let the wide variety worry you. If you are only going to use the generator to run an appliance or two off an extension cord, then all you need is a 120-volt, 15-amp or 20-amp receptacle on the generator. Almost all generators have that.

For safety reasons GFCIs are better than a common duplex. However, many people do not want to use a GFCI on a refrigerator, freezer, or sump pump because of false tripping. If your portable generator does not come with GFCIs, you can always use an inline GFCI. Do not run any power tool off a generator without GFCI protection. If you are going to run an entire house, you will need a 30-amp, four-prong, twist-lock receptacle outlet rated at 120/240 volts—meaning 120 volts and 240 volts (which requires two insulated hot wires, an insulated neutral, and a ground) all in one outlet.

Automatic Generators

With the popularity of portable standby generators increasing, so has the popularity of automatic generators. It is said that 40 percent of those who buy automatic generators have had a portable generator at one time. Personally, I think the percentage is much higher because I believe people are buying an automatic generator because they know firsthand the disadvantages of portables.

Exhaust Fumes Can Kill

A few years back, when a heavy snow took down power lines to his house, a Vermont timber framer started a standby generator that he stored in his attached garage. Even with the garage door open a few feet for ventilation, a draft carried exhaust gases into the house. Wakened by the smell of fumes, the man went to investigate and was overcome by carbon monoxide and died. He left behind a wife and two daughters.

To keep the generator (and its fumes) a safe distance from your house, use a long extension cord to reach from the switch panel to the generator or hardwire a cable from inside the switch panel to the distant generator location. If the cord approaches 50 ft., consider the latter.

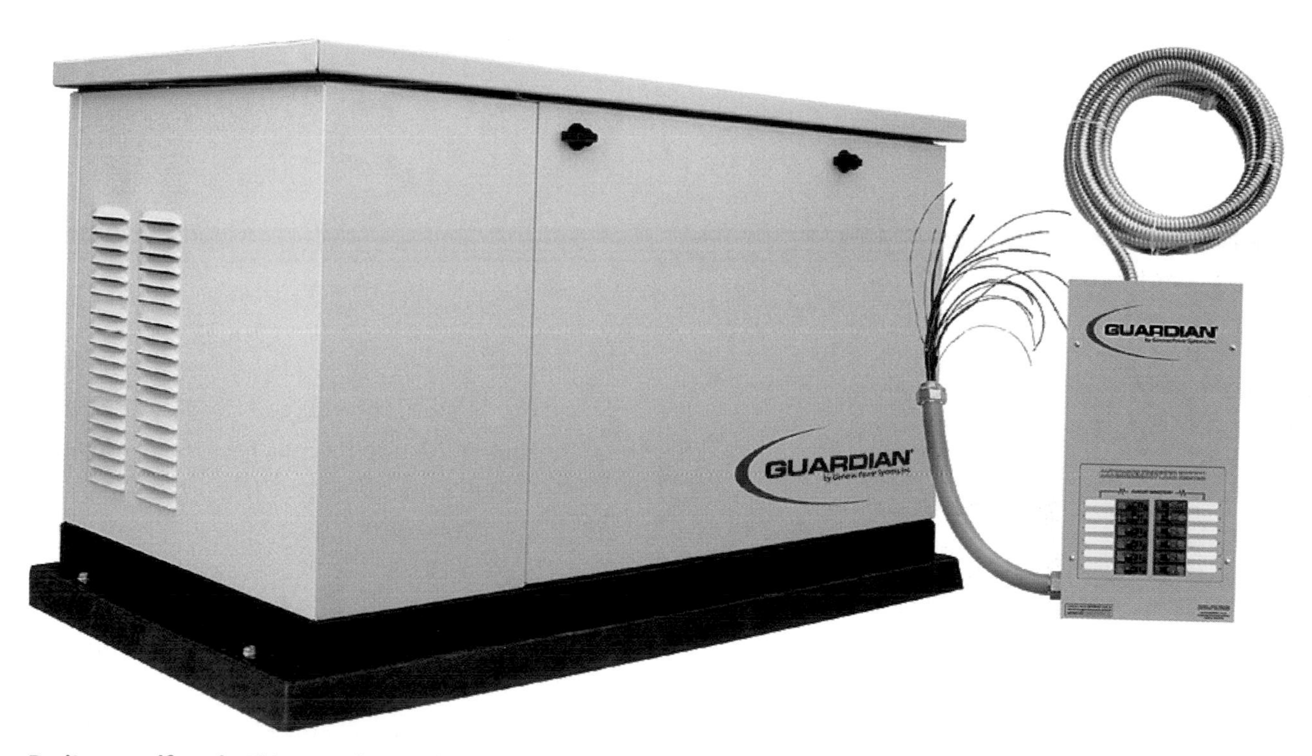

Do-it-yourself ready. This manufacturer's smaller automatic generators (approximately 7kw to 15kw) come with the transfer panel already wired and ready to go. Just connect to the main service panel and connect the gas. Courtesy GenTran

Index